GENERAL TOPOLOGY

JOHN L. KELLEY

DOVER PUBLICATIONS, INC.
MINEOLA, NEW YORK

Bibliographical Note

This Dover edition, first published in 2017, is an unabridged republication of the work originally published in 1955 by the Van Nostrand Reinhold Company, New York.

Library of Congress Cataloging-in-Publication Data

Names: Kelley, John L.
Title: General topology / John L. Kelley.
Description: Dover edition. | Mineola, New York : Dover Publications, Inc., 2017. | Originally published: New York : Van Nostrand Reinhold Company, 1955. | Includes bibliographical references and index.
Identifiers: LCCN 2016040931| ISBN 9780486815442 | ISBN 0486815447
Subjects: LCSH: Topology.
Classification: LCC QA611 .K4 2017 | DDC 514--dc23 LC record available at https://lccn.loc.gov/2016040931

Manufactured in the United States by LSC Communications
81544701 2017
www.doverpublications.com

PREFACE

This book is a systematic exposition of the part of general topology which has proven useful in several branches of mathematics. It is especially intended as background for modern analysis, and I have, with difficulty, been prevented by my friends from labeling it: What Every Young Analyst Should Know.

The book, which is based on various lectures given at the University of Chicago in 1946–47, the University of California in 1948–49, and at Tulane University in 1950–51, is intended to be both a reference and a text. These objectives are somewhat inconsistent. In particular, as a reference work it offers a reasonably complete coverage of the area, and this has resulted in a more extended treatment than would normally be given in a course. There are many details which are arranged primarily for reference work; for example, I have taken some pains to include all of the most commonly used terminology, and these terms are listed in the index. On the other hand, because it is a text the exposition in the earlier chapters proceeds at a rather pedestrian pace. For the same reason there is a preliminary chapter, not a part of the systematic exposition, which covers those topics requisite to the main body of work that I have found to be new to many students. The more serious results of this chapter are theorems on set theory, of which a systematic exposition is given in the appendix. This appendix is entirely independent of the remainder of the book, but with this exception each part of the book presupposes all earlier developments.

There are a few novelties in the presentation. Occasionally the title of a section is preceded by an asterisk; this indicates that the section constitutes a digression. Other topics, many of equal or greater interest, have been treated in the problems. These problems are supposed to be an integral part of the discussion. A few of them are exercises which are intended simply to aid in understanding the concepts employed. Others are counter examples, marking out the boundaries of possible theorems. Some are small theories which are of interest in themselves, and still others are introductions to applications of general topology in various fields. These last always include references so that the interested reader (that elusive creature) may continue his reading. The bibliography includes most of the recent contributions which are pertinent, a few outstanding earlier contributions, and a few "cross-field" references.

I employ two special conventions. In some cases where mathematical content requires "if and only if" and euphony demands something less I use Halmos' "iff." The end of each proof is signalized by ▌. This notation is also due to Halmos.

J. L. K.

Berkeley, California
February 1, 1955

ACKNOWLEDGMENTS

It is a pleasure to acknowledge my indebtedness to several colleagues.

The theorems surrounding the concept of even continuity in chapter 7 are the joint work of A. P. Morse and myself and are published here with his permission. Many of the pleasanter features of the appended development of set theory are taken from the unpublished system of Morse, and I am grateful for his permission to use these; he is not responsible for inaccuracies in my writing. I am also indebted to Alfred Tarski for several conversations on set theory and logic.

I owe thanks to several colleagues who have read part or all of the manuscript and made valuable criticisms. I am particularly obliged to Isaku Namioka, who has corrected a grievous number of errors and obscurities in the text and has suggested many improvements. Hugo Ribeiro and Paul R. Halmos have also helped a great deal with their advice.

Finally, I tender my very warm thanks to Tulane University and to the Office of Naval Research for support during the preparation of this manuscript. This book was written at Tulane University during the years 1950–52; it was revised in 1953, during tenure of a National Science Foundation Fellowship and a sabbatical leave from the University of California.

J. L. K.

April 21, 1961

A number of corrections have been made in this printing of the text. I am indebted to many colleagues, and especially to Krehe Ritter, for bringing errors to my attention.

J. L. K.

CONTENTS

CHAPTER 0: PRELIMINARIES

CHAPTER 1: TOPOLOGICAL SPACES

CHAPTER 2: MOORE-SMITH CONVERGENCE

CHAPTER 5: COMPACT SPACES

Chapter 0

PRELIMINARIES

The only prerequisites for understanding this book are a knowledge of a few of the properties of the real numbers and a reasonable endowment of that invaluable quality, mathematical maturity. All of the definitions and basic theorems which are assumed later are collected in this first chapter. The treatment is reasonably self-contained, but, especially in the discussion of the number system, a good many details are omitted. The most profound results of the chapter are theorems of set theory, of which a systematic treatment is given in the appendix. Because the chapter is intended primarily for reference it is suggested that the reader review the first two sections and then turn to chapter one, using the remainder of the chapter if need arises. Many of the definitions are repeated when they first occur in the course of the work.

SETS

We shall be concerned with sets and with members of sets. "Set," "class," "family," "collection," and "aggregate" are synonymous,* and the symbol ε denotes membership. Thus $x \varepsilon A$ if and only if x is a member (an element, a point) of A. Two sets are identical iff they have the same members, and equality is

* This statement is not strictly accurate. There are technical reasons, expounded in the appendix, for distinguishing between two different sorts of aggregates. The term "set" will be reserved for classes which are themselves members of classes. This distinction is of no great importance here; with a single non-trivial exception, each class which occurs in the discussion (prior to the appendix) is also a set.

always used to mean identity. Consequently, $A = B$ if and only if, for each x, $x \in A$ when and only when $x \in B$.

Sets will be formed by means of braces, so that $\{x: \cdots$ *(proposition about x)* $\cdots\}$ is the set of all points x such that the proposition about x is correct. Schematically, $y \in \{x: \cdots$ *(proposition about x)* $\cdots\}$ if and only if the corresponding proposition about y is correct. For example, if A is a set, then $y \in \{x: x \in A\}$ iff $y \in A$. Because sets having the same members are identical, $A = \{x: x \in A\}$, a pleasant if not astonishing fact. It is to be understood that in this scheme for constructing sets "x" is a dummy variable, in the sense that we may replace it by any other variable that does not occur in the proposition. Thus $\{x: x \in A\} = \{y: y \in A\}$, but $\{x: x \in A\} \neq \{A: A \in A\}$.

There is a very useful rule about the construction of sets in this fashion. If sets are constructed from two different propositions by the use of the convention above, and if the two propositions are logically equivalent, then the constructed sets are identical. The rule may be justified by showing that the constructed sets have the same members. For example, if A and B are sets, then $\{x: x \in A \text{ or } x \in B\} = \{x: x \in B \text{ or } x \in A\}$, because y belongs to the first iff $y \in A$ or $y \in B$, and this is the case iff $y \in B$ or $y \in A$, which is correct iff y is a member of the second set. All of the theorems of the next section are proved in precisely this way.

SUBSETS AND COMPLEMENTS; UNION AND INTERSECTION

If A and B are sets (or families, or collections), then A is a **subset (subfamily, subcollection)** of B if and only if each member of A is a member of B. In this case we also say that A is **contained** in B and that B **contains** A, and we write the following: $A \subset B$ and $B \supset A$. Thus $A \subset B$ iff for each x it is true that $x \in B$ whenever $x \in A$. The set A is a **proper subset** of B (A is properly contained in B and B properly contains A) iff $A \subset B$ and $A \neq B$. If A is a subset of B and B is a subset of C, then clearly A is a subset of C. If $A \subset B$ and $B \subset A$, then $A = B$, for in this case each member of A is a member of B and conversely.

The **union (sum, logical sum, join)** of the sets A and B, written $A \cup B$, is the set of all points which belong either to A or to B; that is, $A \cup B = \{x: x \in A \text{ or } x \in B\}$. It is understood that "or" is used here (and always) in the non-exclusive sense, and that points which belong to both A and B also belong to $A \cup B$. The **intersection (product, meet)** of sets A and B, written $A \cap B$, is the set of all points which belong to both A and B; that is, $A \cap B = \{x: x \in A \text{ and } x \in B\}$. The **void set (empty set)** is denoted 0 and is defined to be $\{x: x \neq x\}$. (Any proposition which is always false could be used here instead of $x \neq x$.) The void set is a subset of every set A because each member of 0 (there are none) belongs to A. The inclusions, $0 \subset A \cap B \subset A \subset A \cup B$, are valid for every pair of sets A and B. Two sets A and B are **disjoint**, or **non-intersecting,** iff $A \cap B = 0$; that is, no member of A is also a member of B. The sets A and B **intersect** iff there is a point which belongs to both, so that $A \cap B \neq 0$. If $\mathcal{A}$ is a family of sets (the members of $\mathcal{A}$ are sets), then $\mathcal{A}$ is a **disjoint family** iff no two members of $\mathcal{A}$ intersect.

The **absolute complement** of a set A, written $\sim A$, is $\{x: x \notin A\}$. The **relative complement** of A with respect to a set X is $X \cap \sim A$, or simply $X \sim A$. This set is also called the **difference** of X and A. For each set A it is true that $\sim\sim A = A$; the corresponding statement for relative complements is slightly more complicated and is given as part of 0.2.

One must distinguish very carefully between "member" and "subset." The set whose only member is x is called **singleton** x and is denoted $\{x\}$. Observe that $\{0\}$ is not void, since $0 \in \{0\}$, and hence $0 \neq \{0\}$. In general, $x \in A$ if and only if $\{x\} \subset A$.

The two following theorems, of which we prove only a part, state some of the most commonly used relationships between the various definitions given above. These are basic facts and will frequently be used without explicit reference.

1 THEOREM *Let A and B be subsets of a set X. Then $A \subset B$ if and only if any one of the following conditions holds:*

$$A \cap B = A, \quad B = A \cup B, \quad X \sim B \subset X \sim A,$$

$$A \cap X \sim B = 0, \quad \textit{or} \quad (X \sim A) \cup B = X.$$

2 **Theorem** *Let A, B, C, and X be sets. Then:*

(a) $X \sim (X \sim A) = A \cap X$.
(b) *(Commutative laws)* $A \cup B = B \cup A$ *and* $A \cap B = B \cap A$.
(c) *(Associative laws)* $A \cup (B \cup C) = (A \cup B) \cup C$ *and* $A \cap (B \cap C) = (A \cap B) \cap C$.
(d) *(Distributive laws)* $A \cap (B \cup C) = (A \cap B) \cup (A \cap C)$ *and* $A \cup (B \cap C) = (A \cup B) \cap (A \cup C)$.
(e) *(De Morgan formulae)* $X \sim (A \cup B) = (X \sim A) \cap (X \sim B)$ *and* $X \sim (A \cap B) = (X \sim A) \cup (X \sim B)$.

PROOF Proof of (a): A point x is a member of $X \sim (X \sim A)$ iff $x \in X$ and $x \notin X \sim A$. Since $x \notin X \sim A$ iff $x \notin X$ or $x \in A$, it follows that $x \in X \sim (X \sim A)$ iff $x \in X$ and either $x \notin X$ or $x \in A$. The first of these alternatives is impossible, so that $x \in X \sim (X \sim A)$ iff $x \in X$ and $x \in A$; that is, iff $x \in X \cap A$. Hence $X \sim (X \sim A) = A \cap X$. Proof of first part of (d): A point x is a member of $A \cap (B \cup C)$ iff $x \in A$ and either $x \in B$ or $x \in C$. This is the case iff either x belongs to both A and B or x belongs to both A and C. Hence $x \in A \cap (B \cup C)$ iff $x \in (A \cap B) \cup (A \cap C)$, and equality is proved. ▮

If $A_1, A_2, \cdots, A_n$ are sets, then $A_1 \cup A_2 \cup \cdots \cup A_n$ is the union of the sets and $A_1 \cap A_2 \cap \cdots \cap A_n$ is their intersection. It does not matter how the terms are grouped in computing the union or intersection because of the associative laws. We shall also have to consider the union of the members of non-finite families of sets and it is extremely convenient to have a notation for this union. Consider the following situation: for each member a of a set A, which we call an index set, we suppose that a set X_a is given. Then the union of all the X_a, denoted $\bigcup \{X_a: a \in A\}$, is defined to be the set of all points x such that $x \in X_a$ for some a in A. In a similar way the intersection of all X_a for a in A, denoted $\bigcap \{X_a: a \in A\}$, is defined to be $\{x: x \in X_a \textit{ for each } a \textit{ in } A\}$. A very important special case arises when the index set is itself a family $\mathcal{A}$ of sets and X_A is the set A for each A in $\mathcal{A}$. Then the foregoing definitions become: $\bigcup \{A: A \in \mathcal{A}\} = \{x: x \in A \textit{ for some } A \textit{ in } \mathcal{A}\}$ and $\bigcap \{A: A \in \mathcal{A}\} = \{x: x \in A \textit{ for each } A \textit{ in } \mathcal{A}\}$.

There are many theorems of an algebraic character on the union and intersection of the members of families of sets, but we shall need only the following, the proof of which is omitted.

3 THEOREM *Let A be an index set, and for each a in A let X_a be a subset of a fixed set Y. Then:*

(a) *If B is a subset of A, then $\bigcup\{X_b: b \in B\} \subset \bigcup\{X_a: a \in A\}$ and $\bigcap\{X_b: b \in B\} \supset \bigcap\{X_a: a \in A\}$.*

(b) *(De Morgan formulae) $Y \sim \bigcup\{X_a: a \in A\} = \bigcap\{Y \sim X_a: a \in A\}$ and $Y \sim \bigcap\{X_a: a \in A\} = \bigcup\{Y \sim X_a: a \in A\}$.*

The De Morgan formulae are usually stated in the abbreviated form: the complement of the union is the intersection of the complements, and the complement of an intersection is the union of the complements.

It should be emphasized that a reasonable facility with this sort of set theoretic computation is essential. The appendix contains a long list of theorems which are recommended as exercises for the beginning student. (See the section on elementary algebra of classes.)

4 *Notes* In most of the early work on set theory the union of two sets A and B was denoted by $A + B$ and the intersection by AB, in analogy with the usual operations on the real numbers. Some of the same algebraic laws do hold; however, there is compelling reason for not following this usage. Frequently set theoretic calculations are made in a group, a field, or a linear space. If A and B are subsets of an (additively written) group, then $\{c: c = a + b \text{ for some } a \text{ in } A \text{ and some } b \text{ in } B\}$ is a natural candidate for the label "$A + B$," and it is natural to denote $\{x: -x \in A\}$ by $-A$. Since the sets just described are used systematically in calculations where union, intersection, and complement also appear, the choice of notation made here seems the most reasonable.

The notation used here for construction of sets is the one most widely used today, but "$\underset{x}{E}$" for "the set of all x such that" is also used. The critical feature of a notation of this sort is the following: one must be sure just which is the dummy variable. An example will clarify this contention. The set of all squares of positive numbers might be denoted quite naturally by $\{x^2: x > 0\}$, and, proceeding, $\{x^2 + a^2: x < 1 + 2a\}$ also has a natural meaning. Unfortunately, the latter has three possible natural meanings, namely: $\{z: \textit{for some } x \textit{ and some } a,\ z = x^2 + a^2 \textit{ and } x < 1 + 2a\}$, $\{z: \textit{for some } x,\ z = x^2 + a^2 \textit{ and } x < 1 + 2a\}$, and $\{z: \textit{for some } a,\ z = x^2 + a^2 \textit{ and } x < 1 + 2a\}$. These sets are quite different, for the first depends on neither x nor a, the second is dependent on a, and the third depends on x. In slightly more technical terms one says that "x" and "a" are both dummies in the first, "x" is a dummy in the second, and "a" in the third. To avoid ambiguity, in each use of the brace notation the first space after the brace and preceding the colon is always occupied by the dummy variable.

Finally, it is interesting to consider one other notational feature. In reading such expressions as "$A \cap (B \cup C)$" the parentheses are essential. However, this could have been avoided by a slightly different choice of notation. If we had used "$\cup AB$" instead of "$A \cup B$," and similarly for intersection, then all parentheses could be omitted. (This general method of avoiding parentheses is well known in mathematical logic.) In the modified notation the first distributive law and the associative law for unions would then be stated: $\cap A \cup BC = \cup \cap AB \cap AC$ and $\cup A \cup BC = \cup \cup ABC$. The shorthand notation also reads well; for example, $\cup AB$ is the union of A and B.

RELATIONS

The notion of set has been taken as basic in this treatment, and we are therefore faced with the task of defining other necessary concepts in terms of sets. In particular, the notions of ordering and function must be defined. It turns out that these may be treated as relations, and that relations can be defined rather naturally as sets having a certain special structure. This section is therefore devoted to a brief statement of the definitions and elementary theorems of the algebra of relations.

Suppose that we are given a relation (in the intuitive sense) between certain pairs of objects. The basic idea is that the relation may be represented as the set of all pairs of mutually related objects. For example, the set of all pairs consisting of a number and its cube might be called the cubing relation. Of course, in order to use this method of realization it is necessary that we have available the notion of ordered pair. This notion can be defined in terms of sets.* The basic facts which we need here are: each ordered pair has a first coordinate and a second coordinate, and two ordered pairs are equal (identical) if and only if they have the same first coordinate and the same second coordinate. The ordered pair with first coordinate x and second coordinate y is denoted (x,y). Thus $(x,y) = (u,v)$ if and only if $x = u$ and $y = v$.

It is convenient to extend the device for the formation of sets so that $\{(x,y): \cdots\}$ is the set of all pairs (x,y) such that $\cdots$. This convention is not strictly necessary, for the same set is obtained by the specification: $\{z: \textit{for some } x \textit{ and some } y,\ z = (x,y) \textit{ and } \cdots\}$.

A **relation** is a set of ordered pairs; that is, a relation is a set, each member of which is an ordered pair. If R is a relation we write $\boldsymbol{xRy}$ and $(x,y) \varepsilon R$ interchangeably, and we say that x is **R-related** to y if and only if xRy. The **domain** of a relation R is the set of all first coordinates of members of R, and its **range** is the set of all second coordinates. Formally, $\textit{domain } R = \{x: \textit{for some } y, (x,y) \varepsilon R\}$ and $\textit{range } R = \{y: \textit{for some } x, (x,y) \varepsilon R\}$. One of the simplest relations is the set of all pairs (x,y) such that x is a member of some fixed set A and y is a member of some fixed set B. This relation is the **cartesian product** of A and B and is denoted by $A \times B$. Thus $A \times B = \{(x,y): x \varepsilon A \textit{ and } y \varepsilon B\}$. If B is non-void the domain of $A \times B$ is A. It is evident that every relation is a subset of the cartesian product of its domain and range.

The **inverse** of a relation R, denoted by R^{-1}, is obtained by reversing each of the pairs belonging to R. Thus $R^{-1} = \{(x,y): (y,x) \varepsilon R\}$ and xRy if and only if $yR^{-1}x$. For example, $(A \times B)^{-1} = B \times A$ for all sets A and B. The domain of the inverse of a relation R is always the range of R, and the range of R^{-1} is the domain of R. If R and S are relations their **composition,** $R \circ S$ (sometimes written RS), is defined to be the set of all pairs (x,z)

* An honest treatment of the problem is given in the appendix, where N. Wiener's definition of ordered pair is used. The ingenious notion of representing relations in this fashion is due to C. S. Peirce. A very readable account of the elementary relation algebra will be found in A. Tarski [1].

such that for some y it is true that $(x,y) \in S$ and $(y,z) \in R$. Composition is generally not commutative. For example, if $R = \{(1,2)\}$ and $S = \{(0,1)\}$, then $R \circ S = \{(0,2)\}$ and $S \circ R$ is void. The **identity** relation on a set X (the **identity on X**), denoted Δ or $\Delta(X)$, is the set of all pairs of the form (x,x) for x in X. The name is derived from the fact that $\Delta \circ R = R \circ \Delta = R$ whenever R is a relation whose range and domain are subsets of X. The identity relation is also called the **diagonal,** a name suggestive of its geometric position in $X \times X$.

If R is a relation and A is a set, then $R[A]$, the set of all R-relatives of points of A, is defined to be $\{y: xRy \text{ for some } x \text{ in } A\}$. If A is the domain of R, then $R[A]$ is the range of R, and for arbitrary A the set $R[A]$ is contained in the range of R. If R and S are relations and $R \subset S$, then clearly $R[A] \subset S[A]$ for every A.

There is an extensive calculus of relations, of which the following theorem is a fragment.

5 THEOREM *Let R, S, and T be relations and let A and B be sets. Then:*

(a) $(R^{-1})^{-1} = R$ *and* $(R \circ S)^{-1} = S^{-1} \circ R^{-1}$.
(b) $R \circ (S \circ T) = (R \circ S) \circ T$ *and* $(R \circ S)[A] = R[S[A]]$.
(c) $R[A \cup B] = R[A] \cup R[B]$ *and* $R[A \cap B] \subset R[A] \cap R[B]$.

More generally, if there is given a set X_a for each member a of a non-void index set A then:

(d) $R[\bigcup \{X_a: a \in A\}] = \bigcup \{R[X_a]: a \in A\}$ *and* $R[\bigcap \{X_a: a \in A\}] \subset \bigcap \{R[X_a]: a \in A\}$.

PROOF As an example we prove the equality: $(R \circ S)^{-1} = S^{-1} \circ R^{-1}$. A pair (z,x) is a member of $(R \circ S)^{-1}$ iff $(x,z) \in R \circ S$, and this is the case iff for some y it is true that $(x,y) \in S$ and $(y,z) \in R$. Consequently $(z,x) \in (R \circ S)^{-1}$ iff $(z,y) \in R^{-1}$ and $(y,x) \in S^{-1}$ for some y. This is precisely the condition that (z,x) belong to $S^{-1} \circ R^{-1}$. ∎

There are several special sorts of relations which occur so frequently in mathematics that they have acquired names. Aside from orderings and functions, which will be considered in detail in the following sections, the types listed below are probably the most useful. Throughout the following it will be convenient to suppose that R is a relation and that X is the set of all points which belong to either the domain or the range of R; that is, $X =$ (domain R) $\cup$ (range R). The relation R is **reflexive** if and only if each point of X is R-related to itself. This is entirely equivalent to requiring that the identity Δ (or $\Delta(X)$) be a subset of R. The relation R is **symmetric,** provided that xRy whenever yRx. Algebraically, this requirement may be phrased: $R = R^{-1}$. At the other extreme, the relation R is **anti-symmetric** iff it is never the case that both xRy and yRx. In other words, R is anti-symmetric iff $R \cap R^{-1}$ is void. The relation R is **transitive** iff whenever xRy and yRz then xRz. In terms of the composition of relations, the relation R is transitive if and only if $R \circ R \subset R$. It follows that, if R is transitive, then $R^{-1} \circ R^{-1} = (R \circ R)^{-1} \subset R^{-1}$, and hence the inverse of a transitive relation is transitive. If R is both transitive and reflexive, then $R \circ R \supset R \circ \Delta$ and hence $R \circ R = R$; in the usual terminology, such a relation is idempotent under composition.

An **equivalence** relation is a reflexive, symmetric, and transitive relation. Equivalence relations have a very simple structure, which we now proceed to describe. Suppose that R is an equivalence relation and that X is the domain of R. A subset A of X is an **equivalence class** (an R-equivalence class) if and only if there is a member x of A such that A is identical with the set of all y such that xRy. In other words, A is an equivalence class iff there is x in A such that $A = R[\{x\}]$. The fundamental result on equivalence relations states that the family $\mathcal{A}$ of all equivalence classes is disjoint, and that a point x is R-related to a point y if and only if both x and y belong to the same equivalence class. The set of all pairs (x,y) with x and y in a class A is simply $A \times A$, which leads to the following concise formulation of the theorem.

6 THEOREM *A relation R is an equivalence relation if and only if there is a disjoint family $\mathcal{A}$ such that $R = \bigcup\{A \times A: A \in \mathcal{A}\}$.*

PROOF If R is an equivalence relation, then R is transitive: if yRx and zRy, then zRx. In other words, if xRy, then $R[\{y\}] \subset R[\{x\}]$. But R is symmetric (xRy whenever yRx), from which it follows that, if xRy, then $R[\{x\}] = R[\{y\}]$. If z belongs to both $R[\{x\}]$ and $R[\{y\}]$, then $R[\{x\}] = R[\{z\}] = R[\{y\}]$, and consequently two equivalence classes either coincide or are disjoint. If y and z belong to the equivalence class $R[\{x\}]$, then, since $R[\{y\}] = R[\{x\}]$, it follows that yRz or, in other words, $R[\{x\}] \times R[\{x\}] \subset R$. Hence the union of $A \times A$ for all equivalence classes A is a subset of R, and since R is reflexive, if xRy, then $(x,y) \in R[\{x\}] \times R[\{x\}]$. Hence $R = \bigcup\{A \times A: A \in \mathfrak{A}\}$. The straightforward proof of the converse is omitted. ∎

We are frequently interested in the behavior of a relation for points belonging to a subset of its domain, and frequently the relation possesses properties for these points which it fails to have for all points. Given a set X and a relation R one may construct a new relation $R \cap (X \times X)$ whose domain is a subset of X. For convenience we will say that a relation R has a **property on** X, or that R **restricted to** X has the property iff $R \cap (X \times X)$ has the property. For example, R is transitive on X iff $R \cap (X \times X)$ is a transitive relation. This amounts to asserting that the defining property holds for points of X; in this case, whenever x, y, and z are points of X such that xRy and yRz, then xRz.

FUNCTIONS

The notion of function must now be defined in terms of the concepts already introduced. This offers no difficulty if we consider the following facts. Whatever a function is, its graph has an obvious definition as a set of ordered pairs. Moreover, there is no information about the function which cannot be derived from its graph. In brief, there is no reason why we should attempt to distinguish between a function and its graph.

A **function** is a relation such that no two distinct members have the same first coordinate. Thus f is a function iff the members of f are ordered pairs, and whenever (x,y) and (x,z) are members of f, then $y = z$. We do not distinguish between a function and its graph. The terms **correspondence, transformation, map, operator,** and **function** are synonymous. If f is a function and x is a point of its domain (the set of all first coordinates of members of f), then **$f(x)$**, or **f_x** is the second coordinate of the unique member of f whose first coordinate is x. The point $f(x)$

is the **value** of f at x, or the **image** of x under f, and we say that f **assigns** the value $f(x)$ to x, or **carries** x into $f(x)$. A function f is **on** X iff X is its domain and it is **onto** Y iff Y is its range (the set of second coordinates of members of f, sometimes called the set of values). If the range of f is a subset of Y, then f is **to** Y, or **into** Y. In general a function is many to one, in the sense that there are many pairs with the same second coordinate or, equivalently, many points at which the function has the same value. A function f is **one to one** iff distinct points have distinct images; that is, if the inverse relation, f^{-1}, is also a function.

A function is a set, and consequently two functions, f and g, are identical iff they have the same members. It is clear that this is the case iff the domain of f is identical with the domain of g and $f(x) = g(x)$ for each x in this domain. Consequently, we may define a function by specifying its domain and the value of the function at each member of the domain. If f is a function on X to Y and A is a subset of X, then $f \cap (A \times Y)$ is also a function. It is called the **restriction** of f to A, denoted $f|A$, its domain is A, and $(f|A)(x) = f(x)$ for x in A. A function g is the restriction of f to some subset iff the domain of g is a subset of the domain of f, and $g(x) = f(x)$ for x in the domain of g; that is, iff $g \subset f$. The function f is called an **extension** of g iff $g \subset f$. Thus f is an extension of g iff g is the restriction of f to some subset of the domain of f.

If A is a set and f is a function, then, following the definition given for arbitrary relations, $f[A] = \{y: \textit{for some } x \textit{ in } A, (x,y) \in f\}$; equivalently, $f[A]$ is $\{y: \textit{for some } x \textit{ in } A, y = f(x)\}$. The set $f[A]$ is called the image of A under f. If A and B are sets, then, by theorem 0.5, $f[A \cup B] = f[A] \cup f[B]$ and $f[A \cap B] \subset f[A] \cap f[B]$, and similar formulae hold for arbitrary unions and intersections. It is not true in general that $f[A \cap B] = f[A] \cap f[B]$, for disjoint sets may have intersecting images. If f is a function, then the set $f^{-1}[A]$ is called the **inverse (inverse image, counter image)** of A under f. The inverse satisfies the following algebraic rules.

7 THEOREM *If f is a function and A and B are sets then*

(a) $f^{-1}[A \sim B] = f^{-1}[A] \sim f^{-1}[B]$,

(b) $f^{-1}[A \cup B] = f^{-1}[A] \cup f^{-1}[B]$, *and*
(c) $f^{-1}[A \cap B] = f^{-1}[A] \cap f^{-1}[B]$.

More generally, if there is given a set X_c for each member c of a non-void index set C then

(d) $f^{-1}[\bigcup\{X_c: c \in C\}] = \bigcup\{f^{-1}[X_c]: c \in C\}$, *and*
(e) $f^{-1}[\bigcap\{X_c: c \in C\}] = \bigcap\{f^{-1}[X_c]: c \in C\}$.

PROOF Only part (e) will be proved. A point x is a member of $f^{-1}[\bigcap\{X_c: c \in C\}]$ if and only if $f(x)$ belongs to this intersection, which is the case iff $f(x) \in X_c$ for each c in C. But the latter condition is equivalent to $x \in f^{-1}[X_c]$ for each c in C; that is, $x \in \bigcap\{f^{-1}[X_c]: c \in C\}$. ∎

The foregoing theorem is often summarized as: the inverse of a function preserves relative complements, unions, and intersections. It should be noted that the validity of these formulae does not depend upon the sets A and B being subsets of the range of the function. Of course, $f^{-1}[A]$ is identical with the inverse image of the intersection of A with the range of f. However, it is convenient not to restrict the notation here (and the corresponding notation for images under f) to subsets of the range (respectively, the domain).

The composition of two functions is again a function by a straightforward argument. If f is a function, then $f^{-1} \circ f$ is an equivalence relation, for $(x,y) \in f^{-1} \circ f$ if and only if $f(x) = f(y)$. The composition $f \circ f^{-1}$ is a function; it is the identity on the range of f.

8 *Notes* There are other notations for the value of a function f at a point x. Besides $f(x)$ and f_x, all of the following are in use: (f,x), (x,f), fx, xf, and $\cdot fx$. The first two of these are extremely convenient in dealing with certain dualities, where one is considering a family F of functions, each on a fixed domain X, and it is desirable to treat F and X in a symmetric fashion. The notations "fx" and "xf" are obvious abbreviations of the notation we have adopted; whether the "f" is written to the left or to the right of "x" is clearly a matter of taste. These two share a disadvantage which is possessed by the "$f(x)$" notation. In certain rather complicated situations the notation is ambiguous, unless parentheses are interlarded liberally. The last notation (used by A. P. Morse) is free from this difficulty. It is unambiguous and does not require parentheses. (See the comments on union and intersection in 0.4.)

There is a need for a bound variable notation for a function. For example, the function whose domain is the set of all real numbers and which has the value x^2 at the point x should have a briefer description. A possible way out of this particular situation is to agree that x is the identity function on the set of real numbers, in which case x^2 might reasonably be the squaring function. The classical device is to use x^2 both for the function and for its value at the number x. A less confusing approach is to designate the squaring function by $x \to x^2$. This sort of notation is suggestive and is now coming into common use. It is not universal and, for example, the statement $(x \to x^2)(t) = t^2$ would require explanation. Finally it should be remarked that, although the arrow notation will undoubtedly be adopted as standard, the λ-convention of A. Church has technical advantages. (The square function might be written as λx: x^2.) No parentheses are necessary to prevent ambiguity.

ORDERINGS

An ordering **(partial ordering, quasi-ordering)** is a transitive relation. A relation $<$ **orders (partially orders)** a set X iff it is transitive on X. If $<$ is an ordering and $x < y$, then it is customary to say that x **precedes** y or x is **less than** y (relative to the order $<$) and that y **follows** x and y is **greater than** x. If A is contained in a set X which is ordered by $<$, then an element x of X is an **upper bound** of A iff for each y in A either $y < x$ or $y = x$. Similarly an element x is a **lower bound** of A if x is less than or equal to each member of A. Of course, a set may have many different upper bounds. An element x is a **least upper bound** or **supremum** of A if and only if it is an upper bound and is less than or equal to every other upper bound. (In other words, a supremum is an upper bound which is a lower bound for the set of all upper bounds.) In the same way, a **greatest lower bound** or **infimum** is an element which is a lower bound and is greater than or equal to every other lower bound. A set X is **order-complete** (relative to the ordering $<$) if and only if each non-void subset of X which has an upper bound has a supremum. It is a little surprising that this condition on upper bounds is entirely equivalent to the corresponding statement for lower bounds. That is:

9 THEOREM *A set X is order-complete relative to an ordering if and only if each non-void subset which has a lower bound has an infimum.*

PROOF Suppose that X is order-complete and that A is a non-void subset which has a lower bound. Let B be the set of all lower bounds for A. Then B is non-void and surely every member of the non-void set A is an upper bound for B. Hence B has a least upper bound, say, b. Then b is less than or equal to each upper bound of B, and in particular b is less than or equal to each member of A, and hence b is a lower bound of A. On the other hand, b is itself an upper bound of B; that is, b is greater than or equal to each lower bound of A. Hence b is a greatest lower bound of A. The converse proposition may be proved by the same sort of argument, or, directly, one may apply the result just proved to the relation inverse to $<$. ∎

It should be remarked that the definition of ordering is not very restrictive. For example, $X \times X$ is an ordering of X, but a rather uninteresting one. Relative to this ordering each member of X is an upper bound, and in fact a supremum, of every subset. The more interesting orderings satisfy the further condition: if x is less than or equal to y and y is also less than or equal to x, then $y = x$. In this case there is at most one supremum for a set, and at most one infimum.

A **linear ordering (total, complete,** or **simple** ordering) is an ordering such that:

(a) *If $x < y$ and $y < x$, then $x = y$, and*
(b) *$x < y$ or $y < x$ whenever x and y are distinct members of the union of the domain and the range of $<$.*

It should be noticed that a linear ordering is not necessarily reflexive. However, agreeing that $x \leq y$ iff $x < y$ or $x = y$, the relation $\leq$ is always a reflexive linear ordering if $<$ is a linear ordering. Following the usual convention, a relation is said **to linearly order** a set X iff the relation restricted to X is a linear ordering. A set with a relation which linearly orders it is called a **chain.** Clearly suprema and infima are unique in chains. The remaining theorems in this section will concern chains, although it will be evident that many of the considerations apply to less restricted orderings.

A function f on a set X to a set Y is **order preserving (monotone, isotone)** relative to an order $<$ for X and an order $\prec$ for Y iff $f(u) \prec f(v)$ or $f(u) = f(v)$ whenever u and v are points of X such that $u \leqq v$. If the ordering $\prec$ of Y is simply $Y \times Y$, or if the ordering $<$ of X is the void relation, then f is necessarily order preserving. Consequently one cannot expect that the inverse of a one-to-one order preserving function will always be order preserving. However, if X and Y are chains and f is one to one and isotone, then necessarily f^{-1} is isotone, for if $f(u) \prec f(v)$ and $f(u) \neq f(v)$, then it is impossible that $v < u$ because of the order-preserving property.

Order-complete chains have a very special property. Suppose that X and Y are chains, that X_0 is a subset of X, and that f is an order-preserving function on X_0 to Y. The problem is: Does there exist an isotone extension of f whose domain is X? Unless some restriction is made on f the answer is "no," for, if X is the set of all positive real numbers, X_0 is the subset consisting of all numbers which are less than one, $Y = X_0$ and f is the identity map, then it is easy to see that there is no isotone extension. (Assuming an extension f^-, what is $f^-(1)$?) But this example also indicates the nature of the difficulty, for X_0 is a subset of X which has an upper bound and $f[X_0]$ has no upper bound. If an isotone extension f^- exists, then the image under f^- of an upper bound for a set A is surely an upper bound for $f[A]$. A similar statement holds for lower bounds, and it follows that, if a subset A of X_0 is **order-bounded** in X (that is, it has both an upper and lower bound in X), then the image $f[A]$ is order-bounded in Y. The following theorem asserts that this condition is also sufficient for the existence of an isotone extension.

10 THEOREM *Let f be an isotone function on a subset X_0 of a chain X to an order-complete chain Y. Then f has an isotone ex-*

tension whose domain is X if and only if f carries order-bounded sets into order-bounded sets. (More precisely stated, the condition is that, if A is a subset of X_0 which is order-bounded in X, then $f[A]$ is order-bounded in Y.)

PROOF It has already been observed that the condition is necessary for the existence of an isotone extension, and it remains to prove the sufficiency. We must construct an isotone extension of a given function f. First we note that if A is a subset of X_0 which has a lower bound in X, then $f[A]$ has a lower bound, for, choosing a point x in A, the set $\{y: y \in A \text{ and } y \leqq x\}$ is order-bounded, hence its image under f is order-bounded, and a lower bound for this image is also a lower bound for $f[A]$. A similar statement applies to upper bounds. For each x in X let L_x be the set of all members of X_0 which are less than or equal to x; that is, $L_x = \{y: y \leqq x \text{ and } y \in X_0\}$. If L_x is void, then x is a lower bound for X_0, hence $f[X_0]$ has an infimum v, and we define $f^-(x)$ to be v. If L_x is not void, then, since x is an upper bound for L_x, the set $f[L_x]$ has an upper bound and hence a supremum, and we define $f^-(x) = \sup f[L_x]$. The straightforward proof that f^- is an isotone extension of f is omitted. ∎

In certain cases the isotone extension of a function is unique. One such case will occur in treating the decimal expansion of a real number. Without attempting to get the best result of the sort, we give a simple sufficient condition for uniqueness which will apply.

11 THEOREM *Let f and g be isotone functions on a chain X to a chain Y, let X_0 be a subset of X on which f and g agree, and let Y_0 be $f[X_0]$. A sufficient condition that $f = g$ is that Y_0 intersect every set of the form $\{y: u < y < v, u \neq y \text{ and } y \neq v\}$, where u and v are members of Y such that $u < v$.*

PROOF If $f \neq g$, then $f(x) \neq g(x)$ for some x in X, and we may suppose that $f(x) < g(x)$. Each point of X_0 which is less than or equal to x maps under f into a point less than or equal to $f(x)$, because f is isotone, and each point which is greater than or equal to x maps under g into a point greater than or equal to $g(x)$, because g is isotone. It follows that no point of X_0 maps into the set $\{y: f(x) < y < g(x), f(x) \neq y \text{ and } y \neq g(x)\}$, and the theorem is proved. ▌

12 *Notes* There is a natural way to embed a chain in an order-complete chain which is an abstraction of Dedekind's construction of the real numbers from the set of rational numbers. The process can also be applied to less restricted orderings, as shown by H. M. MacNeille (see Birkhoff [1; 58]). The pattern is very suggestive of the compactification procedure for topological spaces (chapter 5).

ALGEBRAIC CONCEPTS

In this section a few definitions from elementary algebra are given. For the most part these notions are used in the problems. The terminology is standard, and it seems worth while to summarize the few notions which are required.

A **group** is a pair, $(G,\cdot)$ such that G is a non-void set and $\cdot$, called the group operation, is a function on $G \times G$ to G such that: (a) the operation is associative, that is, $x\cdot(y\cdot z) = (x\cdot y)\cdot z$ for all elements x, y and z of G; (b) there is a neutral element, or identity, e, such that $e\cdot x = x\cdot e = x$ for each x in G; and (c) for each x in G there is an inverse element x^{-1} such that $x\cdot x^{-1} = x^{-1}\cdot x = e$. If the group operation is denoted $+$, then the element inverse to x is usually written $-x$. Following the usual custom, the value of the function $\cdot$ at (x,y) is written $x\cdot y$ instead of the usual functional notation $\cdot(x,y)$, and if no confusion seems likely, the symbol $\cdot$ may be omitted entirely and the group operation indicated by juxtaposition. We shall sometimes say (imprecisely) that G is a group. If A and B are subsets of G, then $A\cdot B$, or simply AB, is the set of all elements of the form $x\cdot y$ for some x in A and some y in B. The set $\{x\}\cdot A$ is also denoted by $x\cdot A$ or simply xA, and similarly for operation on the right. The group is **abelian,** or **commutative,** iff $x\cdot y = y\cdot x$ for all members x and y of G. A group H is a subgroup of G iff $H \subset G$ and the group operation of H is that of G, restricted to $H \times H$. A subgroup H is **normal (distinguished, invariant)** iff $x\cdot H = H\cdot x$ for each x in G. If H is a subgroup of G a **left coset** of H is a subset which is of the form $x\cdot H$ for some x in G. The family of all left cosets is denoted by G/H. If H is normal and A and B belong to G/H, then $A \cdot B$ is also a member, and, with this definition of group operation, G/H is a group, called the **quotient** or **factor** group. A function f on a group G to a group H is a **homomorphism,** or **representation,** iff $f(x \cdot y) = f(x) \cdot f(y)$ for all members x and y of G. The **kernel** of f is the set $f^{-1}[e]$; it is always an invariant subgroup. If H is an invariant subgroup of G, then the function whose value at x is $x \cdot H$ is a homomorphism, usually called the **projection,** or **quotient map,** of G onto G/H.

A **ring** is a triple $(R,+,\cdot)$ such that $(R,+)$ is an abelian group and $\cdot$ is a function on $R \times R$ to R such that: the operation is associative, and the distributive laws $u \cdot (x + y) = u \cdot x + u \cdot y$ and $(u + v) \cdot x = u \cdot x + v \cdot x$ hold for all members x, y, u, and v of R. A **subring** is a subset which, under the ring operations restricted, is a ring, and a ring **homomorphism** or **representation** is a function f on a ring to another ring such that $f(x + y) = f(x) + f(y)$ and $f(x \cdot y) = f(x) \cdot f(y)$ for all members x and y of the domain. An additive subgroup I of a ring R is a **left ideal** iff $xI \subset I$ for each x in R, and is a **two-sided ideal** iff $xI \subset I$ and $Ix \subset I$ for each x in R. If I is a two-sided ideal, R/I is, with the proper addition and multiplication, a ring, and the projection of R onto R/I is a ring homomorphism. A **field** is a ring $(F,+,\cdot)$ such that F has at least two members, and $(F \sim \{0\},\cdot)$, where 0 is the element neutral with respect to $+$, is a commutative group. The operation $+$ is the **addition** operation, $\cdot$ is the **multiplication,** and the element neutral with respect to multiplication is the **unit,** 1. It is customary, when no confusion results, to replace $\cdot$ by juxtaposition, and, ignoring the operations, to say that "F is a field." A **linear space,** or **vector space,** over a field F (the **scalar** field of the space) is a quadruple $(X,\oplus,\cdot,F)$, such that $(X,\oplus)$ is an abelian group and $\cdot$ is a function on $F \times X$ to X such that for all members x and y of X, and all members a and b of F, $a \cdot (b \cdot x) = (a \cdot b) \cdot x$, $(a + b) \cdot x = a \cdot x \oplus b \cdot x$, $a \cdot (x \oplus y) = a \cdot x \oplus a \cdot y$, and $1 \cdot x = x$. A **real linear space** is a linear space over the field of real numbers. The notion of linear space can also be formulated in a slightly different fashion. The family of all homomorphisms of an abelian group into itself becomes, with addition defined pointwise and with composition of functions as multiplication, a ring, called the **endomorphism ring** of the group. A linear space over a field F is a quadruple $(X, \oplus, \cdot, F)$ such that $(X, \oplus)$ is an abelian group and $\cdot$ is a ring homomorphism of F into the endomorphism ring of $(X, \oplus)$ which carries the unit, 1, into the identity homomorphism.

A linear space $(Y, \oplus, \odot, F)$ is a **subspace** of a linear space $(X, +, \cdot, F)$ iff $Y \subset X$ and the operations $+$ and $\cdot$ agree with $\oplus$ and $\odot$ where the latter are defined. The family X/Y of cosets of X modulo a subspace Y may be made into a linear space if addition and scalar multiplication are defined in the obvious way. The projection f of X onto X/Y then has the property that $f(a \cdot x + b \cdot y) = a \cdot f(x) + b \cdot f(y)$ for all members a and b of F and all x and y in X. Such a function is called a **linear function.** If f is a linear function the set $f^{-1}[0]$ is called the **null space** of f; the null space of a linear function is a linear subspace of the domain (provided the operations of addition and scalar multiplication are properly defined).

Suppose f is a linear function on X to Y and g is a linear map of X onto Z such that the null space of f contains the null space of g. Then there is a unique linear function h on Z to Y such that $f = h \circ g$ (explicitly, $h(z)$ is the unique member of $f \circ g^{-1}[z]$). (The function h is said to be **induced** by f and g.) A particular consequence of this fact is that each linear function may be written as a projection into a quotient space followed by a one-to-one linear function.

THE REAL NUMBERS

This section is devoted to the proof of a few of the most important results concerning the real numbers.

An **ordered field** is a field F and a subset P, called the set of **positive elements,** such that

(a) *if x and y are members of P, then $x + y$ and xy are also members; and*

(b) *if x is a member of F, then precisely one of the following statements is true:* $x \in P$, $-x \in P$, or $x = 0$.

One easily verifies that $<$ is a linear ordering of F, where, by definition, $x < y$ iff $y - x \in P$. The usual simple propositions about adding and multiplying inequalities hold. The members x of F such that $-x \varepsilon P$ are **negative.**

It will be assumed that the real numbers are an ordered field which is order-complete, in the sense that every non-void subset which has an upper bound has a least upper bound, or supremum. By 0.9 this last requirement is entirely equivalent to the statement that each non-void subset which has a lower bound has a greatest lower bound, or infimum.

We first prove a few propositions about integers. An **inductive set** is a set A of real numbers such that $0 \varepsilon A$, and whenever $x \varepsilon A$, then $x + 1 \varepsilon A$. A real number x is a **non-negative integer** iff x belongs to every inductive set. In other words, the set ω of non-negative integers is defined to be the intersection of the members of the family of all inductive sets. Each member of ω is actually non-negative because the set of all non-negative numbers is inductive. It is evident that ω is itself an inductive set and is a subset of every other inductive set. It follows that **(principle of mathematical induction)** each inductive subset of ω is identical with ω. A proof which relies on this principle is a **proof by induction.** We prove the following little theorem as an example: if p and q are non-negative integers and $p < q$, then $q - p \varepsilon \omega$. First observe that the set consisting of 0 and all numbers of the form $p + 1$ with p in ω is inductive, and hence each non-zero member of ω is of the form $p + 1$. Next, let A be the set of all non-negative integers p such that $q - p \varepsilon \omega$ for each larger member q of ω. Surely $0 \varepsilon A$, and let us suppose that p is a member of A and that q is an arbitrary member of ω which is larger than $p + 1$. Then $p < q - 1$ and therefore $q - 1 - p \varepsilon \omega$, because $p \varepsilon A$ and $q - 1 \varepsilon \omega$. Consequently $p + 1 \varepsilon A$, hence A is an inductive set, and therefore $A = \omega$. It is equally simple to show that the sum of two members of ω is a member of ω, and it follows that the set $\{x: x \varepsilon \omega \text{ or } -x \varepsilon \omega\}$ is a group. It is the **group of integers.**

There is another form of the principle of mathematical induction which is frequently convenient, namely: *each non-void subset A of ω has a smallest member.* To prove this proposition consider the set B of all members of ω which are lower bounds for A; that is, $B = \{p: p \varepsilon \omega \text{ and } p \leqq q \text{ for all } q \text{ in } A\}$. The set B

is not inductive, for, if $q \in A$, then $q + 1 \notin B$. Since $0 \in B$ it follows that there is a member p of B such that $p + 1 \notin B$. If $p \in A$, then clearly p is the smallest member of A; otherwise there is a member q of A such that $p < q < p + 1$. But then $q - p$ is a non-zero member of ω and hence $q - p - 1$ is a negative member of ω, which is impossible.

It is possible to **define a function by induction** in the following sense. For each non-negative integer p let $\omega_p = \{q: q \in \omega \text{ and } q \leqq p\}$. Suppose that we seek a function on ω, that the functional value a at 0 is given, and for each function g on a set ω_p there is given $F(g)$, the value of the desired function at $p + 1$. Thus the value desired at $p + 1$ may depend on all of the values for smaller integers. In these circumstances it is true that there is a unique function f on ω such that $f(0) = a$ and $f(p + 1) = F(f \mid \omega_p)$ for each p in ω. (The function $f \mid \omega_p$ is the function f restricted to the set ω_p.) This proposition is frequently considered to be obvious, but the proof is not entirely trivial.

13 THEOREM *Suppose a is given and $F(g)$ is given whenever g is a function whose domain is of the form ω_p for some p in ω. Then there is a unique function f such that $f(0) = a$ and $f(p + 1) = F(f \mid \omega_p)$ for each p in ω.*

PROOF Let $\mathfrak{F}$ be the family of all functions g such that the domain of g is a set ω_p for some p in ω, $g(0) = a$, and for each member q of ω such that $q \leqq p - 1$, $g(q + 1) = F(g \mid \omega_q)$. (Intuitively, the members of $\mathfrak{F}$ are initial segments of the desired function.) The family $\mathfrak{F}$ has the very important property: if g and h are members of $\mathfrak{F}$, then either $g \subset h$ or $h \subset g$. To prove this it is necessary to show that $g(q) = h(q)$ for each q belonging to the domain of both. Suppose this is false, and let q be the smallest integer such that $g(q) \neq h(q)$. Then $q \neq 0$, because $g(0) = h(0) = a$, and hence $g(q) = F(g \mid \omega_{q-1})$ which, since g and h agree for values smaller than q, is $F(h \mid \omega_{q-1}) = h(q)$, and this is a contradiction. Now let $f = \bigcup \{g: g \in \mathfrak{F}\}$. Then the members of f are surely ordered pairs, and if $(x,y) \in g \in \mathfrak{F}$ and $(x,z) \in h \in \mathfrak{F}$, then (x,y) and (x,z) both belong to g or both to h, and hence $y = z$. Consequently f is a function, and it must be shown that it is the required function. First, because $\{(0,a)\} \in \mathfrak{F}$, $f(0) = a$.

Next, if $q + 1$ belongs to the domain of f, then for some g in $\mathfrak{F}$, $q + 1$ is a member of the domain of g, and hence $f(q + 1) = g(q + 1) = F(g \mid \omega_q) = F(f \mid \omega_q)$. Finally, to show that the domain of f is ω, suppose that q is the first member of ω which is not in the domain of f. Then $q - 1$ is the last member of the domain of f, and $f \cup \{(q,F(f))\}$ is a member of $\mathfrak{F}$. Hence q belongs to the domain of f, which is a contradiction. ▮

The foregoing theorem can be used systematically in showing the elementary properties of the real numbers. For example, if b is a positive number and p an integer, b^p is defined as follows. In the foregoing theorem, let $a = 1$ and for each function g with domain ω_p let $F(g) = bg(p)$. Then $f(0) = 1$ and $f(p + 1) = bf(p)$ for each p in ω, if f is the function whose existence is guaranteed by the theorem. Letting $b^p = f(p)$, it follows that $b^0 = 1$, and $b^{p+1} = bb^p$, from which one can show by induction that $b^{p+q} = b^pb^q$ for all members p and q of ω. If b^{-p} is defined to be $1/b^p$ for each non-negative integer p, then the usual elementary proof shows that $b^{p+q} = b^pb^q$ for all integers p and q.

So far in this discussion of the real numbers we have not used the fact that the field of real numbers is order-complete. We now prove a simple, but noteworthy, consequence of order completeness. First, the set ω of non-negative integers does not have an upper bound, for, if x were a least upper bound of ω, then $x - 1$ would not be an upper bound, and hence $x - 1 < p$ for some p in ω. But then $x < p + 1$ and this contradicts the fact that x was supposed to be an upper bound. Consequently, if x is a positive real number and y is a real number, then $px > y$ for some positive integer p because there is a member p of ω which is larger than y/x. An ordered field for which this proposition is true is said to have an **Archimedean order.**

We will need the fact that each non-negative real number has a b-adic expansion, where b is an arbitrary integer greater than one. Roughly speaking, we want to write a number x as the sum of multiples of powers of b, the multiples (digits) being non-negative integers less than b. Of course, the b-adic expansion of a number may fail to be unique—in the decimal expansion, .9999··· (all nines) and 1.000··· (all zeros) are to be expansions of the same real number. The expansion itself is a function which assigns to each integer an integer between 0 and $b - 1$, such that (since we want only a finite number of non-zero integers before the decimal point) there is a first non-zero digit. Formally, a is a b-**adic expansion** iff a is a function on the integers to ω_{b-1} ($= \{q: q \in \omega \text{ and } q \leq b - 1\}$), such that there is a smallest integer p for which a_p ($= a(p)$) is not zero. A b-adic expansion a is **rational** iff there is a last non-zero digit (that is, for some integer p, $a_q = 0$ whenever $q > p$). For each rational b-adic expansion a there is a simple way of assigning a corresponding real number $r(a)$. Except for a finite number of integers p the number $a_p b^{-p}$ is zero, and the sum of $a_p b^{-p}$ for p in this finite set is the real number $r(a)$. We write $r(a) = \Sigma\{a_p b^{-p}: p \text{ an integer}\}$. A real number which is of this form is a b-**adic rational.** These numbers are precisely those of the form, qb^{-p}, for integers p and q. Let E be the set of all b-adic expansions. Then E is linearly ordered by dictionary order; in detail, a b-adic expansion a precedes a b-adic expansion c in **dictionary order (lexicographic order)** iff for the smallest integer p such that $a_p \neq c_p$ it is true that $a_p < c_p$. It is very easy to see that, like a dictionary, E is actually linearly ordered by $<$. The correspondence r is order preserving, and this is the key to the following proposition.

14 THEOREM *Let E be the set of b-adic expansions, let R be the set of rational expansions, and for a in R let $r(a) = \Sigma\{a_p b^{-p}$: p an integer$\}$. Then there is a unique isotone extension $\bar{r}$ of r whose domain is E, and $\bar{r}$ maps $E \sim R$ onto the positive real numbers in a one-to-one fashion.*

PROOF According to theorem 0.10 there will be an isotone extension $\bar{r}$ of r iff r carries each subset of R which is order-bounded in E into an order-bounded subset of the real numbers. But for each a in E there is evidently b in R such that $b > a$, and it follows that, if a subset A of R has a for an upper bound, then $r(b)$ is an upper bound for $f[A]$. A similar argument applies to lower bounds, and we conclude that r carries order-bounded sets into order-bounded sets and consequently has an isotone extension $\bar{r}$ whose domain is E.

To show the extension is unique it is sufficient, by 0.11, to prove that, for non-negative real numbers x and y, if $x < y$, then there is a in R such that $x < r(a) < y$. Because $b^p > p$ for each non-negative integer p (a fact which is easily proved by induction), and because the set of non-negative integers is not bounded, there is an integer p such that $b^p > 1/(y - x)$. Then $b^{-p} < (y - x)$. There is an integer q such that $qb^{-p} \geqq y$ because the ordering is Archimedean, and since there is a smallest such integer q, it may be supposed that $(q - 1)b^{-p} < y$. It follows that $(q - 1)b^{-p} > x$ because b^{-p} is less than $(y - x)$ and this proves that there is a b-adic rational, $(q - 1)b^{-p}$, which is the image of a member of R and lies between x and y. Consequently the correspondence $\bar{r}$ is unique.

Next, we show that the correspondence $\bar{r}$ is one to one on $E \sim R$. It is straightforward to see that $\bar{r}$ is one to one on R, and this fact is assumed in the following. Suppose that $a \in E$, $c \in E \sim R$, and $a < c$. Then for the first value of p such that a_p and c_p are different, $a_p < c_p$. The expansion d, such that for $q < p$, $d_q = a_q$, for $q > p$, $d_q = 0$, and $d_p = a_p + 1$, is a member of R which is greater than a, and since c does not have a last non-zero digit, $a < d < c$. Repeating, there is a member e of R such that $a < d < e < c$. Then, since on R the function $\bar{r}$ is one to one, $\bar{r}(a) \leqq \bar{r}(d) < \bar{r}(e) \leqq \bar{r}(c)$, and $\bar{r}$ is therefore one to one on $E \sim R$.

Finally, it must be shown that the image of $E \sim R$ under $\bar{r}$ is the set of all positive numbers. First notice that for every pair of members c and d of R for which $c < d$ there is a in $E \sim R$ such that $c < a < d$, and consequently for positive real numbers x and y with $x < y$ there is a in $E \sim R$ such that $x < \bar{r}(a) < y$. If now x is a positive real number which is not the image of a member of $E \sim R$, let $F = \{a\colon a \in E \sim R \text{ and } \bar{r}(a) < x\}$. If the set F has a supremum c then, if $\bar{r}(c) < x$ no point of $E \sim R$ maps into the interval $(\bar{r}(c),x)$, and if $\bar{r}(c) > x$, then ($\bar{r}$ is order preserving) no point of $E \sim R$ maps into the interval $(x,\bar{r}(c))$. In either event a contradiction results, and the theorem will follow if it is shown that each non-empty subset of $E \sim R$ which has an upper bound has a supremum: that is, $E \sim R$ is order-complete. Suppose then that F is a non-void subset of $E \sim R$ which has an upper bound. Then there is a smallest integer p

such that $a_p \neq 0$ for some a in F. Define c_q to be zero for $q < p$, let F_p be the set of all members a of F with non-zero p-th digit a_p, and let $c_p = \max \{a_p: a \in F_p\}$. Continue inductively, letting F_{p+1} be the set of all members a of F_p such that $a_q = c_q$ for $q = p$, and let $c_{p+1} = \max \{a_{p+1}: a \in F_{p+1}\}$. No one of the sets F_p can be void and without difficulty one sees that the expansion c obtained by this construction is an upper bound of F, and in fact a supremum, and that $c \in E \sim R$. ∎

The foregoing theorem will be used for b equal to two, three, and ten. The b-adic expansions are then called **dyadic, triadic,** and **decimal,** respectively.

COUNTABLE SETS

A set is finite iff it can be put into one-to-one correspondence with a set of the form $\{p: p \in \omega \text{ and } p < q\}$, for some q in ω. A set A is **countably infinite** iff it can be put into one-to-one correspondence with the set ω of non-negative integers; that is, iff A is the range of some one-to-one function on ω. A set is **countable** iff it is either finite or countably infinite.

15 THEOREM *A subset of a countable set is countable.*

PROOF Suppose A is countable, f is one to one on ω with range A, and that $B \subset A$. Then f, restricted to $f^{-1}[B]$, is a one-to-one function on a subset of ω with range B, and if it can be shown that $f^{-1}[B]$ is countable, then a one-to-one function onto B can be constructed by composition. The proof therefore reduces to showing that an arbitrary subset C of ω is countable. Let $g(0)$ be the first member of C, and proceeding inductively, for p in ω, let $g(p)$ be the first member of C different from $g(0), g(1), \cdots, g(p-1)$. If this choice is impossible for some p then g is a function on $\{q: q \in \omega \textit{ and } q < p\}$ with range C, and C is finite. Otherwise (using 0.13 on the construction of functions by induction) there is a function g on ω such that, for each p in ω, $g(p)$ is the first member of C different from $g(0), g(1), \cdots, g(p-1)$. Clearly g is one to one. It is easily verified by induction that $g(p) \geqq p$ for all p, and hence it follows from the choice of $g(p+1)$ that each member p of C is one of the numbers $g(q)$ for $q \leqq p$. Therefore the range of g is C. ∎

16 Theorem *If the domain of a function is countable, then the range is also countable.*

PROOF It is sufficient to show that, if A is a subset of ω and f is a function on A onto B, then B is countable. Let C be the set of all members x of A such that, if $y \in A$ and $y < x$, then $f(x) \neq f(y)$; that is, C consists of the smallest member of each of the sets $f^{-1}[y]$ for y in B. Then $f \mid C$ maps C onto B in a one-to-one fashion, and since C is countable by 0.15, so is B. ▮

17 Theorem *If $\mathcal{A}$ is a countable family of countable sets, then $\bigcup\{A: A \in \mathcal{A}\}$ is countable.*

PROOF Because $\mathcal{A}$ is countable there is a function F whose domain is a subset of ω and whose range is $\mathcal{A}$. Since $F(p)$ is countable for each p in ω, it is possible to find a function G_p on a subset of $\{p\} \times \omega$ whose range is $F(p)$. Consequently there is a function (the union of the functions G_p) on a subset of $\omega \times \omega$ whose range is $\bigcup\{A: A \in \mathcal{A}\}$, and the problem reduces to showing that $\omega \times \omega$ is countable. The key to this proof is the observation that, if we think of $\omega \times \omega$ as lying in the upper right-hand part of the plane, the diagonals which cross from upper left to lower right contain only a finite number of members of $\omega \times \omega$. Explicitly, for n in ω, let $B_n = \{(p,q): (p,q) \in \omega \times \omega \text{ and } p + q = n\}$. Then B_n contains precisely $n + 1$ points, and the union $\bigcup\{B_n: n \in \omega\}$ is $\omega \times \omega$. A function on ω with range $\omega \times \omega$ may be constructed by choosing first the members of B_0, next those of B_1, and so on. The explicit definition of such a function is left to the reader. ▮

The **characteristic function** of a subset A of a set X is the function f such that $f(x) = 0$ for x in $X \sim A$ and $f(x) = 1$ for x in A. A function f on a set X which assumes no value other than zero and one is called a characteristic function; it is clearly *the* characteristic function of $f^{-1}[1]$. The function which is zero everywhere is the characteristic function of the void set, and the function which is identically one on X is the characteristic function of X. Two sets have the same characteristic functions iff they are identical, and hence there is a one-to-one correspondence between the family of all characteristic functions on a set X and the family of all subsets of X.

If ω is the set of non-negative integers, the family of all characteristic functions on ω may be put into one-to-one correspondence with the set F of all dyadic expansions a such that $a_p = 0$ for $p < 0$. The family of all finite subsets of ω corresponds in a one-to-one way to the subfamily G of F consisting of rational dyadic expansions. We now use the classical Cantor process to prove that F is uncountable.

18 THEOREM *The family of all finite subsets of a countably infinite set is countable, but the family of all subsets is not.*

PROOF In view of the remarks preceding the statement of the theorem it is sufficient to show that the set F of all dyadic expansions a with $a_p = 0$ for p negative is uncountable, and that the subset G of F consisting of rational expansions is countable. Suppose that f is a one-to-one function on ω with range F. Let a be the member of F such that $a_p = 1 - f(p)_p$ for each non-negative integer p. That is, the p-th digit of a is one minus the p-th digit of $f(p)$. Then $a \in F$ and clearly, for each p in ω, $a \neq f(p)$ because a and $f(p)$ differ in the p-th digit. It follows that a does not belong to the range of f, and this is a contradiction. Hence F is uncountable.

It remains to be proved that G is countable. For p in ω let $G_p = \{a\colon a \in G \text{ and } a_q = 0 \text{ for } q > p\}$. Then G_0 contains just two elements, and since there are precisely twice as many members in G_{p+1} as in G_p, it follows that G_p is always finite. Hence $G = \bigcup\{G_p\colon p \in \omega\}$ is countable. ▌

The natural correspondence between F and a subset of the real numbers is, according to 0.14, one to one on $F \sim G$. Since G is countable, $F \sim G$ must be uncountable. Hence

19 COROLLARY *The set of all real numbers is uncountable.*

CARDINAL NUMBERS

Many of the theorems on countability are special cases of more general theorems on cardinal numbers. The set ω of non-negative integers played a special role in the above and, in a more general way, this role may be occupied by sets (of which ω is one) called cardinal numbers. Let us agree that two sets,

A and B, are **equipollent** iff there is a one-to-one function on A with range B. It turns out that for every set A there is a unique cardinal number C such that A and C are equipollent. If C and D are distinct cardinal numbers, then C and D are not equipollent but one of the cardinal numbers, say C, and a proper subset of the other are equipollent. In this case C is said to be the **smaller cardinal number** and we write $C < D$. With this definition of order the family of all cardinal numbers is linearly ordered, and even more, every non-void subfamily has a least member. (These facts are proved in the appendix.)

Accepting the facts in the previous paragraph for the moment it follows that, if A and B are sets, then there is a one-to-one function on A to a subset of B, or the reverse, because there are cardinal numbers C and D such that A and C, and B and D, respectively, are equipollent. Suppose now that there is a one-to-one function on A to a subset of B and also a one-to-one function on B to a subset of A. Then C and a subset of D are equipollent, and D and a subset of C are equipollent, from which it follows, since the ordering of the cardinal numbers is linear, that $C = D$. Hence A and B are equipollent. This is the classical Schroeder-Bernstein theorem. We give a direct proof of this theorem which is independent of the general theory of cardinal numbers because the proof gives non-trivial additional information.

20 THEOREM *If there is a one-to-one function on a set A to a subset of a set B and there is also a one-to-one function on B to a subset of A, then A and B are equipollent.*

PROOF Suppose that f is a one-to-one map of A into B and g is one to one on B to A. It may be supposed that A and B are disjoint. The proof of the theorem is accomplished by decomposing A and B into classes which are most easily described in terms of parthenogenesis. A point x (of either A or B) is an ancestor of a point y iff y can be obtained from x by successive application of f and g (or g and f). Now decompose A into three sets: let A_E consist of all points of A which have an even number of ancestors, let A_O consist of points which have an odd number of ancestors, and let A_I consist of points with infinitely many ancestors. Decompose B similarly and observe: f maps A_E onto B_O and A_I onto B_I, and g^{-1} maps A_O onto B_E. Hence the function which agrees with f on $A_E \cup A_I$ and agrees with g^{-1} on A_O is a one-to-one map of A onto B. ▮

21 *Notes* The foregoing proof does not use the axiom of choice, which is interesting but not very important. It is important to notice that the function desired was constructed from the two given functions by a countable process. Explicitly, if f is a one-to-one function on A to B and g is one to one on B to A, if $E_0 = A \sim g[B]$, $E_{n+1} = g \circ f[E_n]$ for each n, and if $E = \bigcup\{E_n : n \in \omega\}$, then the function h which is equal to f on E and equal to g^{-1} on $A \sim E$ is a one-to-one map of A onto B. (More precisely, $h = (f | E) \cup (g^{-1} | A \sim E)$.) The importance of this result lies in the fact that, if f and g have certain pleasant properties (such as being Borel functions), then h retains these properties.

The intuitively elegant form of the proof of theorem 0.20 is due to G. Birkhoff and S. MacLane.

ORDINAL NUMBERS

Except for examples, the ordinal numbers will not be needed in the course of this work. However, several of the most interesting counter examples are based on extremely elementary properties of the ordinals and it seems proper to state here the few facts which are necessary for these. (The ordinal numbers are constructed and these and other properties proved in the appendix.)

22 SUMMARY *There is an uncountable set Ω', which is linearly ordered by a relation $<$ in such a way that:*

(a) *Every non-void subset of Ω' has a smallest element.*
(b) *There is a greatest element Ω of Ω'.*
(c) *If $x \in \Omega'$ and $x \neq \Omega$, then the set of all members of Ω' which precede x is countable.*

The set Ω' is the set of all ordinals which are less than or equal to Ω, **the first uncountable ordinal.** A linearly ordered set such that every non-void subset has a least element is **well ordered.**

In particular, each non-void subset of a well-ordered set has an infimum. Since every subset of Ω' has an upper bound, namely, Ω, it follows by 0.9 that every non-void subset of Ω' has a supremum. One of the curious facts about Ω' is the following.

23 Theorem *If A is a countable subset of Ω' and $\Omega \notin A$, then the supremum of A is less than Ω.*

PROOF Assume that A is a countable subset of Ω' and that $\Omega \notin A$. For each member a of A the set $\{x: x \leqq a\}$ is countable and hence the union of all such sets is countable. This union is $\{x: x \leqq a \text{ for some } a \text{ in } A\}$ and the supremum b of the union is therefore an upper bound for A. The point b has only a countable number of predecessors relative to the ordering, and hence $b \neq \Omega$. It follows that the supremum of A is less than Ω. ▮

One member of Ω' deserves special notice. The first member of Ω' which does not have a finite number of predecessors is the **first non-finite ordinal** and is denoted ω. The symbol ω has already been used to denote the set of non-negative integers. In the construction of the ordinal numbers it turns out that the first non-finite ordinal is, in fact, the set ω of non-negative integers!

CARTESIAN PRODUCTS

If A and B are sets the cartesian product $A \times B$ has been defined as the set of all ordered pairs (x,y) such that $x \in A$ and $y \in B$. It is useful to extend the definition of cartesian product to families of sets, just as the notion of union and intersection was extended to arbitrary families of sets. Suppose that for each member a of an index set A there is given a set X_a. The **Cartesian product of the sets** X_a, written $\times\{X_a: a \in A\}$, is defined to be the set of all functions x on A such that $x(a) \in X_a$ for each a in A. It is customary to use subscript notation rather than the usual functional notation, so that $\times\{X_a: a \in A\} = \{x: x \text{ is a function on } A \text{ and } x_a \in X_a \text{ for } a \text{ in } A\}$. The definition is initially a little surprising but it is actually a precise statement of the intuitive concept: a point x of the product consists of a point (namely, x_a) selected from each of the sets X_a. The set X_a is the **a-th coordinate set,** and the point x_a is the **a-th coordi-**

nate of the point x of the product. The function P_a which carries each point x of the product onto its a-th coordinate x_a is the **projection** into the a-th coordinate set. That is, $P_a(x) = x_a$. The map P_a is also called the **evaluation at** $\boldsymbol{a}$.

There is an important special case of a cartesian product. Suppose that the coordinate set X_a is a fixed set Y for each a in the index set A. Then the cartesian product $\times\{X_a: a \in A\} = \times\{Y: a \in A\} = \{x: x \text{ is a function on } A \text{ to } Y\}$. Thus $\times\{Y: a \in A\}$ is precisely the set of all functions on A to Y, sometimes written Y^A. A familiar instance is real **Euclidean** $\boldsymbol{n}$**-space.** This is the set of all real-valued functions on a set consisting of the integers $0, 1, \cdots, n - 1$, and the i-th coordinate of a member x is x_i.

There is another interesting special case. Suppose the index set is itself a family $\mathfrak{A}$ of sets, and that for each A in $\mathfrak{A}$ the A-th coordinate set is A. In this case the cartesian product $\times\{A: A \in \mathfrak{A}\}$ is the family of all functions x on $\mathfrak{A}$ such that $x_A \in A$ for each A in $\mathfrak{A}$. These functions, members of the cartesian product, are sometimes called **choice functions** for $\mathfrak{A}$, since intuitively the function x "chooses" a member x_A from each set A. If the empty set is a member of $\mathfrak{A}$, then there is clearly no choice function for $\mathfrak{A}$; that is, the cartesian product is void. If the members of $\mathfrak{A}$ are not empty it is still not entirely obvious that the cartesian product is non-void, and, in fact, the question of the existence of a choice function for such a family turns out to be quite delicate. The next section is devoted to several propositions, each equivalent to a positive answer to the question. We shall assume as an axiom the most convenient one of these propositions. (A different choice is made in the appendix; together with the next section, this shows the equivalence of the various statements.) With unusual self-restraint we refrain from discussing the philosophical implications.

HAUSDORFF MAXIMAL PRINCIPLE

If $\mathfrak{A}$ is a family of sets (or a collection of families of sets) a member A is the **largest member** of $\mathfrak{A}$ if it contains every other member; that is, if A is larger than every other member of $\mathfrak{A}$.

Similarly, A is the **smallest member** of the family iff A is contained in each member. It is frequently of importance to know that a family has a largest member or a smallest member. Clearly the largest and smallest members are unique when they exist. However, even in cases where the family $\mathcal{A}$ has no largest member, there may be a member such that no other member properly contains A, although there are members which neither contain nor are contained in A. Such a member is called a maximal member of the family. Formally, A is a **maximal** member of $\mathcal{A}$ iff no member of $\mathcal{A}$ properly contains A. Similarly A is a **minimal** member of $\mathcal{A}$ iff no member of $\mathcal{A}$ is properly contained in A. It is very easy to make examples of families which have no maximal member, or families in which each member is both maximal and minimal (for example a disjoint family). In general, some special hypothesis must be added to ensure the existence of maximal members.

A family $\mathcal{N}$ of sets is a **nest** (sometimes called a **tower** or a **chain**) iff, whenever A and B are members of the family, then either $A \subset B$ or $B \subset A$. This is precisely the same thing as saying that the family $\mathcal{N}$ is linearly ordered by inclusion, or, in our terminology, that $\mathcal{N}$ with the inclusion relation is a chain. If $\mathcal{N} \subset \mathcal{A}$ and $\mathcal{N}$ is a nest, then $\mathcal{N}$ is a **nest in** $\mathcal{A}$. We know that a family of sets may fail to have a maximal element. Let us consider the collection of all nests in a fixed family $\mathcal{A}$ and ask if among these there is a maximal nest. That is, for each family $\mathcal{A}$, is there a nest $\mathcal{M}$ in $\mathcal{A}$ which is properly contained in no nest in $\mathcal{A}$? We assume the following statement as an axiom.

24 HAUSDORFF MAXIMAL PRINCIPLE *If $\mathcal{A}$ is a family of sets and $\mathcal{N}$ is a nest in $\mathcal{A}$, then there is a maximal nest $\mathcal{M}$ in $\mathcal{A}$ which contains $\mathcal{N}$.*

The next theorem lists a number of important consequences of the Hausdorff maximal principle. Before stating the results we review some of the terminology which is commonly used in this connection. A family $\mathcal{A}$ of sets is of **finite character** iff each finite subset of a member of $\mathcal{A}$ is a member of $\mathcal{A}$, and each set A, every finite subset of which belongs to $\mathcal{A}$, itself belongs to $\mathcal{A}$. If $<$ is an ordering of a set A, then a subset B which is linearly ordered by $<$ is called a chain in A. A **maximal element** of the ordered set A is an element x such that x follows each comparable element of A; that is, if $y \in A$, then either y precedes x or x does not precede y. A relation $<$ is a well ordering of a set A iff $<$ is a linear ordering of A such that each non-void subset has a first member (a member which is less than or equal to every other member). If there exists a well ordering of A, then we say that A can be well ordered.

25 THEOREM

(a) MAXIMAL PRINCIPLE *There is a maximal member of a family* $\mathcal{A}$ *of sets, provided that for each nest in* $\mathcal{A}$ *there is a member of* $\mathcal{A}$ *which contains every member of the nest.*

(b) MINIMAL PRINCIPLE *There is a minimal member of a family* $\mathcal{A}$, *provided that for each nest in* $\mathcal{A}$ *there is a member of* $\mathcal{A}$ *which is contained in every member of the nest.*

(c) TUKEY LEMMA *There is a maximal member of each non-void family of finite character.*

(d) KURATOWSKI LEMMA *Each chain in a (partially) ordered set is contained in a maximal chain.*

(e) ZORN LEMMA *If each chain in a partially ordered set has an upper bound, then there is a maximal element of the set.*

(f) AXIOM OF CHOICE *If* X_a *is a non-void set for each member* a *of an index set* A, *then there is a function* c *on* A *such that* $c(a) \in X_a$ *for each* a *in* A.

(g) ZERMELO POSTULATE *If* $\mathcal{A}$ *is a disjoint family of non-void sets, then there is a set* C *such that* $A \cap C$ *consists of a single point for every* A *in* $\mathcal{A}$.

(h) WELL-ORDERING PRINCIPLE *Each set can be well ordered.*

PROOF We sketch the proof of each of these propositions, leaving a good many of the details to the reader.

Proof of (a): Choose a maximal nest $\mathfrak{M}$ in $\mathcal{A}$ and let A be a member of $\mathcal{A}$ containing $\bigcup\{M: M \in \mathfrak{M}\}$. Then A is a maximal member of $\mathcal{A}$, for if A is properly contained in a member B of $\mathcal{A}$, then $\mathfrak{M} \cup \{B\}$ is a nest in $\mathcal{A}$ which properly contains $\mathfrak{M}$, which is a contradiction.

Proof of (b): A proof very much like the one above is clearly possible. However, one may use (a) instead, by letting $X =$

$\bigcup\{A: A \in \mathcal{A}\}$, letting $\mathcal{C}$ be the family of complements relative to X of members of $\mathcal{A}$, observing that because of the De Morgan formulae $\mathcal{C}$ satisfies the hypothesis of (a), hence has a maximal member M, and that $X \sim M$ is surely a minimal member of $\mathcal{A}$.

Proof of (c): The proof is based on the maximal principle (a). Let $\mathcal{A}$ be a family which is of finite character, let $\mathfrak{N}$ be a nest in $\mathcal{A}$, and let $A = \bigcup\{N: N \in \mathfrak{N}\}$. Each finite subset F of A is necessarily a subset of some member of $\mathfrak{N}$, for we may choose a finite subfamily of the nest $\mathfrak{N}$ whose union contains F, and this finite subfamily has a largest member which then contains F. Consequently $A \in \mathcal{A}$. Then $\mathcal{A}$ satisfies the hypothesis of (a) and therefore has a maximal member.

Proof of (d): Suppose B is a chain in the partially ordered set A. Let $\mathcal{A}$ be the family of all chains in A which contain B. If $\mathfrak{N}$ is a nest in $\mathcal{A}$, then it can be directly verified that $\bigcup\{N: N \in \mathfrak{N}\}$ is again a member, so that $\mathcal{A}$ satisfies the hypothesis of (a) and consequently has a maximal member.

Proof of (e): Choose an upper bound for a maximal chain.

Proof of (f): Recall that a function is a set of ordered pairs such that no two members have the same first coordinate. Let $\mathfrak{F}$ be the family of all functions f such that the domain of f is a subset of A and $f(a) \in X_a$ for each a in the domain of f. (The members of $\mathfrak{F}$ are "fragments" of the function we seek.) The following argument shows that $\mathfrak{F}$ is a family of finite character. If f is a member of $\mathfrak{F}$, then every subset of f, and in particular every finite subset, is also a member of $\mathfrak{F}$. On the other hand, if f is a set, each finite subset of which belongs to $\mathfrak{F}$, then the members of f are ordered pairs, no two different pairs have the same first coordinate, and consequently f is a function. Moreover, if a is a member of the domain of f, then $\{(a,f(a))\} \in \mathfrak{F}$ and hence $f(a) \in X_a$, and it follows that $f \in \mathfrak{F}$. Because $\mathfrak{F}$ is a family of finite character there is a maximal member c of $\mathfrak{F}$, and it is only necessary to show that the domain of c is A. If a is a member of A which is not a member of the domain of c, then, since X_a is non-void, there is a member y of X_a and $c \cup \{(a,y)\}$ is itself a function and is a member of $\mathfrak{F}$, which contradicts the fact that c is maximal.

Proof of (g): Apply the axiom of choice to the index set $\mathcal{A}$ with $X_A = A$ for each A in $\mathcal{A}$.

Proof of (h): Suppose that X is the (non-void) set which is to be well ordered. Let $\mathcal{A}$ be the family of all non-void subsets of X, and let c be a choice function for $\mathcal{A}$; that is, c is a function on $\mathcal{A}$ such that $c(A) \in A$ for each A in $\mathcal{A}$. The idea of the proof is to construct an ordering $\leqq$ such that for each "initial segment" A the first point which follows A in the ordering is $c(X \sim A)$. Explicitly, define a set A to be a *segment* relative to an order $<$ iff each point which precedes a member of A is itself a member of A. In particular the void set is a segment. Let $\mathcal{C}$ be the class of all reflexive linear orderings $\leqq$ which satisfy the conditions: the domain D of $\leqq$ is a subset of X and for each segment A other than D the first point of $D \sim A$ is $c(X \sim A)$. It is almost evident that each member of $\mathcal{C}$ is a well ordering, for if B is a non-void subset of the domain of a member $\leqq$ and $A = \{y: y \leqq x \text{ and } y \neq x \text{ for each } x \text{ in } B\}$, then $c(X \sim A)$ is the first member of B. Suppose that $\leqq$ and $\leq$ are members of $\mathcal{C}$, that D is the domain of $\leqq$, and that E is the domain of $\leq$. Let A be the set of all points x such that the sets $\{y: y \leqq x\}$ and $\{y: y \leq x\}$ are identical and such that on these sets the two orderings agree. Then A is a segment relative to both $\leqq$ and $\leq$. If A is not identical with either D or E, then $c(X \sim A)$ is the first point of each of these sets which does not belong to A; but then $c(X \sim A) \in A$ in view of the definition of A. It follows that $A = D$ or $A = E$. Thus any two members of $\mathcal{C}$ are related as follows: the domain of one member is a segment relative to the other, and the two orderings agree on this segment. Using this fact it is not hard to see that the union $\prec$ of the members of $\mathcal{C}$ is itself a member of $\mathcal{C}$; it is the largest member of $\mathcal{C}$. If F is the domain of $\prec$, then $F = X$, for otherwise the point $c(X \sim F)$ may be adjoined at the end of the ordering (more precisely, $\prec \cup (F \times \{c(X) \sim F)\}$ is a member of $\mathcal{C}$ which properly contains $\prec$). The theorem follows. ▮

26 *Notes* Each of the propositions listed above is actually equivalent to the Hausdorff maximal principle, and any one of these might reasonably be assumed as an axiom. In the appendix the maximal principle is derived from the axiom of choice.

The derivation of the well ordering principle from the choice axiom which is given above is essentially that of Zermelo [1]. A proof which uses 0.25(e) is also quite feasible. It may be noted that the union of a nest of well orderings is generally not a well ordering, so that a direct application of the maximal principle to the family of well orderings is impossible.

It should be remarked that the labelling of the various propositions in 0.25 is somewhat arbitrary. The Hausdorff maximal principle was used independently by C. Kuratowski, R. L. Moore, and M. Zorn in forms approximating those above.

Finally it may be noted that, although the formulation of Tukey's lemma which is given is more or less standard, it does not imply (directly) the most commonly cited applications (for example, each group contains a maximal abelian subgroup). There is a more general form which states (very roughly): if a family $\mathcal{A}$ of sets is defined by a (possibly infinite) number of conditions such that each condition involves only finitely many points, then $\mathcal{A}$ has a maximal member.

Chapter 1

TOPOLOGICAL SPACES

TOPOLOGIES AND NEIGHBORHOODS

A **topology** is a family $\mathfrak{I}$ of sets which satisfies the two conditions: the intersection of any two members of $\mathfrak{I}$ is a member of $\mathfrak{I}$, and the union of the members of each subfamily of $\mathfrak{I}$ is a member of $\mathfrak{I}$. The set $X = \bigcup \{U: U \in \mathfrak{I}\}$ is necessarily a member of $\mathfrak{I}$ because $\mathfrak{I}$ is a subfamily of itself, and every member of $\mathfrak{I}$ is a subset of X. The set X is called the **space** of the topology $\mathfrak{I}$ and $\mathfrak{I}$ is a **topology for *X***. The pair $(X,\mathfrak{I})$ is a **topological space.** When no confusion seems possible we may forget to mention the topology and write "X is a topological space." We shall be explicit in cases where precision is necessary (for example if we are considering two different topologies for the same set X).

The members of the topology $\mathfrak{I}$ are called **open** relative to $\mathfrak{I}$, or $\mathfrak{I}$-open, or if only one topology is under consideration, simply open sets. The space X of the topology is always open, and the void set is always open because it is the union of the members of the void family. These may be the only open sets, for the family whose only members are X and the void set is a topology for X. This is not a very interesting topology, but it occurs frequently enough to deserve a name; it is called the **indiscrete** (or **trivial**) topology for X, and $(X,\mathfrak{I})$ is then an **indiscrete topological space.** At the other extreme is the family of all subsets of X, which is the **discrete** topology for X (then $(X,\mathfrak{I})$ is a **discrete topological space**). If $\mathfrak{I}$ is the discrete topology, then every subset of the space is open.

The discrete and the indiscrete topology for a set X are respectively the largest and the smallest topology for X. That is, every topology for X is contained in the discrete topology and contains the indiscrete topology. If $\mathfrak{I}$ and $\mathfrak{U}$ are topologies for X, then, following the convention for arbitrary families of sets, $\mathfrak{I}$ is smaller than $\mathfrak{U}$ and $\mathfrak{U}$ is larger than $\mathfrak{I}$ iff $\mathfrak{I} \subset \mathfrak{U}$. In other words, $\mathfrak{I}$ is smaller than $\mathfrak{U}$ iff each $\mathfrak{I}$-open set is $\mathfrak{U}$-open. In this case it is also said that $\mathfrak{I}$ is **coarser** than $\mathfrak{U}$ and $\mathfrak{U}$ is **finer** than $\mathfrak{I}$. (Unfortunately, this situation is described in the literature by both of the statements: $\mathfrak{I}$ is **stronger** than $\mathfrak{U}$ and $\mathfrak{I}$ is **weaker** than $\mathfrak{U}$.) If $\mathfrak{I}$ and $\mathfrak{U}$ are arbitrary topologies for X it may happen that $\mathfrak{I}$ is neither larger nor smaller than $\mathfrak{U}$; in this case, following the usage for partial orderings, it is said that $\mathfrak{I}$ and $\mathfrak{U}$ are not **comparable.**

The set of real numbers, with an appropriate topology, is a very interesting topological space. This is scarcely surprising since the notion of a topological space is an abstraction of some interesting properties of the real numbers. The **usual topology** for the real numbers is the family of all those sets which contain an open interval about each of their points. That is, a subset A of the set of real numbers is open iff for each member x of A there are numbers a and b such that $a < x < b$ and the **open interval** $\{y: a < y < b\}$ is a subset of A. Of course, we must verify that this family of sets is indeed a topology, but this offers no difficulty. It is worth noticing that, conveniently, an open interval is an open set.

A set U in a topological space $(X,\mathfrak{I})$ is a **neighborhood** ($\mathfrak{I}$-neighborhood) of a point x iff U contains an open set to which x belongs. A neighborhood of a point need not be an open set, but every open set is a neighborhood of each of its points. Each neighborhood of a point contains an open neighborhood of the point. If $\mathfrak{I}$ is the indiscrete topology the only neighborhood of a point x is the space X itself. If $\mathfrak{I}$ is the discrete topology, then every set to which a point belongs is a neighborhood of it. If X is the set of real numbers and $\mathfrak{I}$ is the usual topology, then a neighborhood of a point is a set containing an open interval to which the point belongs.

1 THEOREM *A set is open if and only if it contains a neighborhood of each of its points.*

PROOF The union U of all open subsets of a set A is surely an open subset of A. If A contains a neighborhood of each of its points, then each member x of A belongs to some open subset of A and hence $x \in U$. In this case $A = U$ and therefore A is open. On the other hand, if A is open it contains a neighborhood (namely, A) of each of its points. ∎

The foregoing theorem evidently implies that a set is open iff it is a neighborhood of each of its points.

The **neighborhood system** of a point is the family of all neighborhoods of the point.

2 THEOREM *If* $\mathfrak{U}$ *is the neighborhood system of a point, then finite intersections of members of* $\mathfrak{U}$ *belong to* $\mathfrak{U}$, *and each set which contains a member of* $\mathfrak{U}$ *belongs to* $\mathfrak{U}$.

PROOF If U and V are neighborhoods of a point x, there are open neighborhoods U_0 and V_0 contained in U and V respectively. Then $U \cap V$ contains the open neighborhood $U_0 \cap V_0$ and is hence a neighborhood of x. Thus the intersection of two (and hence of any finite number of) members of $\mathfrak{U}$ is a member. If a set U contains a neighborhood of a point x it contains an open neighborhood of x and is consequently itself a neighborhood. ∎

3 *Notes* Fréchet [1] first considered abstract spaces. The concept of a topological space developed during the following years, accompanied by a good deal of experimentation with definitions and fundamental processes. Much of the development of the theory may be found in Hausdorff's classic work [1] and, a little later, in the early volumes of *Fundamenta Mathematicae*. There are actually two fundamental concepts which have grown out of these researches: that of a topological space and that of a uniform space (chapter 7). The latter notion, which has been formalized relatively recently (A. Weil [1]), owes much to the study of topological groups.

Standard references on general topology include:

Alexandroff and Hopf [1] (the first two chapters), Bourbaki [1], Fréchet [2], Kuratowski [1], Lefschetz [1] (the first chapter), R. L. Moore [1], Newman [1], Sierpinski [1], Tukey [1], Vaidyanathaswamy [1], and G. T. Whyburn [1].

CLOSED SETS

A subset A of a topological space $(X,\mathfrak{J})$ is **closed** iff its relative complement $X \sim A$ is open. The complement of the complement of the set A is again A, and hence a set is open iff its complement is closed. If $\mathfrak{J}$ is the indiscrete topology the complement of X and the complement of the void set are the only closed sets; that is, only the void set and X are closed. It is always true that the space and the void set are closed as well as open, and it may happen, as we have just seen, that these are the only closed sets. If $\mathfrak{J}$ is the discrete topology, then every subset is closed and open. If X is the set of real numbers and $\mathfrak{J}$ the usual topology, then the situation is quite different. A **closed interval** (that is, a set of the form $\{x: a \leqq x \leqq b\}$) is fortunately closed. An open interval is not closed and a **half-open interval** (that is, a set of the form $\{x: a < x \leqq b\}$ or $\{x: a \leqq x < b\}$ where $a < b$) is neither open nor closed. Indeed—(problem 1.J)—the only sets which are both open and closed are the space and the void set.

According to the De Morgan formulae, 0.3, the union (intersection) of the complements of the members of a family of sets is the complement of the intersection (respectively union). Consequently, the union of a finite number of closed sets is necessarily closed and the intersection of the members of an arbitrary family of closed sets is closed. These properties characterize the family of closed sets, as the following theorem indicates. The simple proof is omitted.

4 THEOREM *Let $\mathfrak{F}$ be a family of sets such that the union of a finite subfamily is a member, the intersection of an arbitrary nonvoid subfamily is a member, and $X = \bigcup\{F: F \in \mathfrak{F}\}$ is a member. Then $\mathfrak{F}$ is precisely the family of closed sets in X relative to the topology consisting of all complements of members of $\mathfrak{F}$.*

ACCUMULATION POINTS

The topology of a topological space can be described in terms of neighborhoods of points and consequently it must be possible to formulate a description of closed sets in terms of neighborhoods. This formulation leads to a new classification of points in the following way. A set A is closed iff $X \sim A$ is open, and hence iff each point of $X \sim A$ has a neighborhood which is contained in $X \sim A$, or equivalently, is disjoint from A. Consequently, A is closed iff for each x, if every neighborhood of x intersects A, then $x \varepsilon A$. This suggests the following definition.

A point x is an **accumulation point** (sometimes called **cluster** point or **limit** point) of a subset A of a topological space $(X,\mathfrak{J})$ iff every neighborhood of x contains points of A other than x. Then it is true that each neighborhood of a point x intersects A if and only if x is either a point of A or an accumulation point of A. The following theorem is then clear.

5 THEOREM *A subset of a topological space is closed if and only if it contains the set of its accumulation points.*

If x is an accumulation point of A it is sometimes said, in a pleasantly suggestive phrase, that there are points of A arbitrarily near x. If we pursue this imagery it appears that an indiscrete topological space is really quite crowded, for each point x is an accumulation point of every set other than the void set and the set $\{x\}$. On the other hand, in a discrete topological space, no point is an accumulation point of a set. If X is the set of real numbers with the usual topology a variety of situations can arise. If A is the open interval $(0,1)$, then every point of the closed interval $[0,1]$ is an accumulation point of A. If A is the set of all non-negative rationals with squares less than 2, then the closed interval $[0,\sqrt{2}]$ is the set of accumulation points. If A is the set of all reciprocals of integers, then 0 is the only accumulation point of A, and the set of integers has no accumulation points.

6 THEOREM *The union of a set and the set of its accumulation points is closed.*

PROOF If x is neither a point nor accumulation point of A, then there is an open neighborhood U of x which does not intersect A. Since U is a neighborhood of each of its points, no one of these is an accumulation point of A. Hence the union of the set A and the set of its accumulation points is the complement of an open set. ▮

The set of all accumulation points of a set A is sometimes called the **derived** set of A.

CLOSURE

The **closure** ($\mathfrak{J}$-closure) of a subset A of a topological space $(X,\mathfrak{J})$ is the intersection of the members of the family of all closed sets containing A. The closure of A is denoted by A^-, or by $\bar{A}$. The set A^- is always closed because it is the intersection of closed sets, and evidently A^- is contained in each closed set which contains A. Consequently A^- is the smallest closed set containing A and it follows that A is closed if and only if $A = A^-$. The next theorem describes the closure of a set in terms of its accumulation points.

7 THEOREM *The closure of any set is the union of the set and the set of its accumulation points.*

PROOF Every accumulation point of a set A is an accumulation point of each set containing A, and is therefore a member of each closed set containing A. Hence A^- contains A and all accumulation points of A. On the other hand, according to the preceding theorem, the set consisting of A and its accumulation points is closed and it therefore contains A^-. ∎

The function which assigns to each subset A of a topological space the value A^- might be called the closure function, or closure operator, relative to the topology. This operator determines the topology completely, for a set A is closed iff $A = A^-$. In other words, the closed sets are simply the sets which are fixed under the closure operator. It is instructive to enquire: Under what circumstances is an operator which is defined for all subsets of a fixed set X the closure operator relative to some topology for X? It turns out that four very simple properties serve to describe closure. First, because the void set is closed, the closure of the void set is void; and, second, each set is contained in its closure. Next, because the closure of each set is closed, the closure of the closure of a set is identical with the closure of the set (in the usual algebraic terminology, the closure operator is idempotent). Finally, the closure of the union of two sets is the union of the closures, for $(A \cup B)^-$ is always a closed set containing A and B, and therefore contains A^- and B^- and hence $A^- \cup B^-$; on the other hand, $A^- \cup B^-$ is a closed set containing $A \cup B$ and hence also $(A \cup B)^-$.

A **closure operator** on X is an operator which assigns to each subset A of X a subset A^c of X such that the following four statements, the **Kuratowski closure axioms**, are true.

(a) *If* 0 *is the void set*, $0^c = 0$.
(b) *For each* A, $A \subset A^c$.
(c) *For each* A, $A^{cc} = A^c$.
(d) *For each* A *and* B, $(A \cup B)^c = A^c \cup B^c$.

The following theorem of Kuratowski shows that these four statements are actually characteristic of closure. The topology defined below is the topology **associated** with a closure operator.

8 THEOREM *Let* c *be a closure operator on* X, *let* $\mathfrak{F}$ *be the family of all subsets* A *of* X *for which* $A^c = A$, *and let* $\mathfrak{I}$ *be the family of complements of members of* $\mathfrak{F}$. *Then* $\mathfrak{I}$ *is a topology for* X, *and* A^c *is the* $\mathfrak{I}$*-closure of* A *for each subset* A *of* X.

PROOF Axiom (a) shows that the void set belongs to $\mathfrak{F}$, and (d) shows that the union of two members of $\mathfrak{F}$ is a member of $\mathfrak{F}$. Consequently the union of any finite subfamily (void or not) of $\mathfrak{F}$ is a member of $\mathfrak{F}$. Because of (b), $X \subset X^c$, so that $X = X^c$, and the union of the members of $\mathfrak{F}$ is then X. In view of theorem 1.4, it will follow that $\mathfrak{I}$ is a topology for X if it is shown that the intersection of the members of any non-void subfamily of $\mathfrak{F}$ is a member of $\mathfrak{F}$. To this end, first observe that, if $B \subset A$, then $B^c \subset A^c$, because $A^c = [(A \sim B) \cup B]^c = (A \sim B)^c \cup B^c$. Now suppose that $\mathfrak{A}$ is a non-void subfamily of $\mathfrak{F}$ and that $B = \bigcap\{A: A \in \mathfrak{A}\}$. The set B is contained in each member of $\mathfrak{A}$, and therefore $B^c \subset \bigcap\{A^c: A \in \mathfrak{A}\} = \bigcap\{A: A \in \mathfrak{A}\} = B$. Since $B \subset B^c$, it follows that $B = B^c$ and $B \in \mathfrak{F}$. This shows that $\mathfrak{I}$ is a topology, and it remains to show that A^c is A^-, the $\mathfrak{I}$-closure of A. By definition, A^- is the intersection of all the $\mathfrak{I}$-closed sets, that is, the members of $\mathfrak{F}$, which contain A. By axiom (c), $A^c \in \mathfrak{F}$, and hence $A^- \subset A^c$; since $A^- \in \mathfrak{F}$ and $A^- \supset A$ it follows that $A^- \supset A^c$ and hence $A^- = A^c$. ∎

INTERIOR AND BOUNDARY

There is another operator defined on the family of all subsets of a topological space, which is very intimately related to the closure operator. A point x of a subset A of a topological space is an **interior point** of A iff A is a neighborhood of x, and the set of all interior points of A is the **interior** of A, denoted A^0. (In the usual terminology, the relation "is an interior point of" is the inverse of the relation "is a neighborhood of.") It is convenient to exhibit the connection between this notion and the earlier concepts before considering examples.

9 THEOREM *Let A be a subset of a topological space X. Then the interior A^0 of A is open and is the largest open subset of A. A set A is open if and only if $A = A^0$. The set of all points of A which are not points of accumulation of $X \sim A$ is precisely A^0. The closure of $X \sim A$ is $X \sim A^0$.*

PROOF If a point x belongs to the interior of a set A, then x is a member of some open subset U of A. Every member of U is also a member of A^0, and consequently A^0 contains a neighborhood of each of its points and is therefore open. If V is an open subset of A and $y \in V$, then A is a neighborhood of y and so $y \in A^0$. Hence A^0 contains each open subset of A and it is therefore the largest open subset of A. If A is open, then A is surely identical with the largest open subset of A; hence A is open iff $A = A^0$. If x is a point of A which is not an accumulation point of $X \sim A$, then there is a neighborhood U of x which does not intersect $X \sim A$ and is therefore a subset of A. Then A is a neighborhood of x and $x \in A^0$. On the other hand, A^0 is a neighborhood of each of its points and A^0 does not intersect $X \sim A$, so that no point of A^0 is an accumulation point of $X \sim A$. Finally, since A^0 consists of the points of A which are not accumulation points of $X \sim A$, the complement, $X \sim A^0$, is precisely the set of all points which are either points of $X \sim A$ or accumulation points of $X \sim A$; that is, the complement is the closure $(X \sim A)^-$. ▮

The last statement of the foregoing theorem deserves a little further consideration. For convenience, let us denote the relative complement $X \sim A$ by A'. Then A'', the complement of the complement of A, is again A (we sometimes say $'$ is an operator of period two). The preceding result can then be stated as $A^{0\prime} = A'^{-}$, and, it follows, taking complements, that $A^0 = A'^{-\prime}$. Thus the interior of A is the complement of the closure of the complement of A. If A is replaced by its complement it follows that $A^- = A'^{0\prime}$, so that the closure of a set is the complement of the interior of the complement.*

If X is an indiscrete space the interior of every set except X itself is void. If X is a discrete space, then each set is open and closed and consequently identical with its interior and with its closure. If X is the set of real numbers with the usual topology, then the interior of the set of all integers is void; the interior of a closed interval is the open interval with the same endpoints. The interior of the set of rational numbers is void, and the closure of the interior of this set is consequently void. The closure of the set of rational numbers is the set X of all numbers, and the interior of this set is X again. Thus the interior of the closure of a set may be quite different from the closure of the interior; that is, the interior operator and the closure operator do not generally commute.

There is one other operator which occurs frequently enough to justify its definition. The **boundary** of a subset A of a topological space X is the set of all points x which are interior to neither A nor $X \sim A$. Equivalently, x is a point of the boundary iff each neighborhood of x intersects both A and $X \sim A$. It is clear that the boundary of A is identical with the boundary of $X \sim A$. If X is indiscrete and A is neither X nor void, then the boundary of A is X, while if X is discrete the boundary of every subset is void. The boundary of an interval of real numbers, in the usual topology for the reals, is the set whose only members are the endpoints of the interval, regardless of whether the interval is open, closed, or half-open. The boundary of the

* An amusing and instructive problem suggests itself. From a given subset A of a topological space, how many different sets can be constructed by successive applications, in any order, of closure, complementation and interior? From the remarks in the above paragraph and the fact that $A^{--} = A^-$, this reduces to: how many distinct sets may be formed from a single set A, by alternate applications of complementation and the closure operator? The surprising answer is given in problem 1.E.

set of rationals, or the set of irrationals, is the set of all real numbers.

It is not difficult to discover the relations between boundary, closure, and interior. The following theorem, whose proof we omit, summarizes the facts.

10 THEOREM *Let A be a subset of a topological space X and let $b(A)$ be the boundary of A. Then $b(A) = A^- \cap (X \sim A)^- = A^- \sim A^0$, $X \sim b(A) = A^0 \cup (X \sim A)^0$, $A^- = A \cup b(A)$ and $A^0 = A \sim b(A)$.*

A set is closed if and only if it contains its boundary and is open if and only if it is disjoint from its boundary.

BASES AND SUBBASES

In defining the usual topology for the set of real numbers we began with the family $\mathfrak{B}$ of open intervals, and from this family constructed the topology $\mathfrak{I}$. The same method is useful in other situations and we now examine the construction in detail. A family $\mathfrak{B}$ of sets is a **base for a topology** $\mathfrak{I}$ iff $\mathfrak{B}$ is a subfamily of $\mathfrak{I}$ and for each point x of the space, and each neighborhood U of x, there is a member V of $\mathfrak{B}$ such that $x \in V \subset U$. Thus the family of open intervals is a base for the usual topology of the real numbers, in view of the definition of the usual topology and the fact that open intervals are open relative to this topology.

There is a simple characterization of bases which is frequently used as a definition: A subfamily $\mathfrak{B}$ of a topology $\mathfrak{I}$ is a base for $\mathfrak{I}$ iff each member of $\mathfrak{I}$ is the union of members of $\mathfrak{B}$. To prove this fact, suppose that $\mathfrak{B}$ is a base for the topology $\mathfrak{I}$ and that $U \in \mathfrak{I}$. Let V be the union of all members of $\mathfrak{B}$ which are subsets of U and suppose that $x \in U$. Then there is W in $\mathfrak{B}$ such that $x \in W \subset U$, and consequently $x \in V$. Hence $U \subset V$ and since V is surely a subset of U, $V = U$. To show the converse, suppose $\mathfrak{B} \subset \mathfrak{I}$ and each member of $\mathfrak{I}$ is the union of members of $\mathfrak{B}$. If $U \in \mathfrak{I}$, then U is the union of the members of a subfamily of $\mathfrak{B}$, and for each x in U there is V in $\mathfrak{B}$ such that $x \in V \subset U$. Consequently $\mathfrak{B}$ is a base for $\mathfrak{I}$.

Although this is a very convenient method for the construction of topologies, a little caution is necessary because not every family of sets is the base for a topology. For example, let X consist of the integers 0, 1, and 2, let A consist of 0 and 1, and let B consist of 1 and 2. If $\mathcal{S}$ is the family whose members are X, A, B and the void set, then $\mathcal{S}$ cannot be the base for a topology because: by direct computation, the union of members of $\mathcal{S}$ is always a member, so that if $\mathcal{S}$ were the base of a topology that topology would have to be $\mathcal{S}$ itself, but $\mathcal{S}$ is not a topology because $A \cap B \notin \mathcal{S}$. The reason for this situation is made clear by the following theorem.

11 THEOREM *A family $\mathcal{B}$ of sets is a base for some topology for the set $X = \bigcup\{B: B \in \mathcal{B}\}$ if and only if for every two members U and V of $\mathcal{B}$ and each point x in $U \cap V$ there is W in $\mathcal{B}$ such that $x \in W$ and $W \subset U \cap V$.*

PROOF If $\mathcal{B}$ is a base for some topology, U and V are members of $\mathcal{B}$ and $x \in U \cap V$ then, since $U \cap V$ is open, there is a member of $\mathcal{B}$ to which x belongs and which is a subset of $U \cap V$. To show the converse, let $\mathcal{B}$ be a family with the specified property and let $\mathfrak{J}$ be the family of all unions of members of $\mathcal{B}$. A union of members of $\mathfrak{J}$ is itself a union of members of $\mathcal{B}$ and is therefore a member of $\mathfrak{J}$, and it is only necessary to show that the intersection of two members U and V of $\mathfrak{J}$ is a member of $\mathfrak{J}$. If $x \in U \cap V$, then we may choose U' and V' in $\mathcal{B}$ such that $x \in U' \subset U$ and $x \in V' \subset V$, and then a member W of $\mathcal{B}$ such that $x \in W \subset U' \cap V' \subset U \cap V$. Consequently $U \cap V$ is the union of members of $\mathcal{B}$, and $\mathfrak{J}$ is a topology. ▌

We have just seen that an arbitrary family $\mathcal{S}$ of sets may fail to be the base for any topology. With admirable persistence we vary the question and enquire whether there is a unique topology which is, in some sense, generated by $\mathcal{S}$. Such a topology should be a topology for the set X which is the union of the members of $\mathcal{S}$, and each member of $\mathcal{S}$ should be open relative to the topology; that is, $\mathcal{S}$ should be a subfamily of the topology. This raises the question: Is there a smallest topology for X which contains $\mathcal{S}$? The following simple result will enable us to exhibit this smallest topology.

12 THEOREM *If $\mathcal{S}$ is any non-void family of sets the family of all*

finite intersections of members of $\mathcal{S}$ *is the base for a topology for the set* $X = \bigcup\{S: S \in \mathcal{S}\}$.

PROOF If $\mathcal{S}$ is a family of sets let $\mathcal{B}$ be the family of finite intersections of members of $\mathcal{S}$. Then the intersection of two members of $\mathcal{B}$ is again a member of $\mathcal{B}$ and, applying the preceding theorem, $\mathcal{B}$ is the base for a topology. ▮

A family $\mathcal{S}$ of sets is a **subbase for a topology** $\mathcal{T}$ iff the family of finite intersections of members of $\mathcal{S}$ is a base for $\mathcal{T}$ (equivalently, iff each member of $\mathcal{T}$ is the union of finite intersections of members of $\mathcal{S}$). In view of the preceding theorem every non-empty family $\mathcal{S}$ is the subbase for some topology, and this topology is, of course, uniquely determined by $\mathcal{S}$. It is the smallest topology containing $\mathcal{S}$ (that is, it is a topology containing $\mathcal{S}$ and is a subfamily of every topology containing $\mathcal{S}$).

There will generally be many different bases and subbases for a topology and the most appropriate choice may depend on the problem under consideration. One rather natural subbase for the usual topology for the real numbers is the family of half-infinite open intervals; that is, the family of sets of the form $\{x: x > a\}$ or $\{x: x < a\}$. Each open interval is the intersection of two such sets, and this family is consequently a subbase. The family of all sets of the same form with a rational is a less obvious and more interesting subbase. (See problem 1.J.)

A space whose topology has a countable base has many pleasant properties. Such spaces are said to satisfy the **second axiom of countability.** (The terms **separable** and **perfectly separable** are also used in this connection, but we shall use neither.)

13 THEOREM *If* A *is an uncountable subset of a space whose topology has a countable base, then some point of* A *is an accumulation point of* A.

PROOF Suppose that no point of A is an accumulation point and that $\mathcal{B}$ is a countable base. For each x in A there is an open set containing no point of A other than x, and since $\mathcal{B}$ is a base we may choose B_x in $\mathcal{B}$ such that $B_x \cap A = \{x\}$. There is then a one-to-one correspondence between the points of A and the members of a subfamily of $\mathcal{B}$, and A is therefore countable. ▮

A sharper form of this theorem is stated in problem 1.H.

A set A is **dense** in a topological space X iff the closure of A is X. A topological space X is **separable** iff there is a countable subset which is dense in X. A separable space may fail to satisfy the second axiom of countability. For example, let X be an uncountable set with the topology consisting of the void set and the complements of finite sets. Then every non-finite set is dense because it intersects every non-void open set. On the other hand, suppose that there is a countable base $\mathcal{B}$ and let x be a fixed point of X. The intersection of the family of all open sets to which x belongs must be $\{x\}$, because the complement of every other point is open. It follows that the intersection of those members of the base $\mathcal{B}$ to which x belongs is $\{x\}$. But the complement of this countable intersection is the union of a countable number of finite sets, hence countable, and this is a contradiction. (Less trivial examples occur later.) There is no difficulty in showing that a space with a countable base is separable.

14 THEOREM *A space whose topology has a countable base is separable.*

PROOF Choose a point out of each member of the base, thus obtaining a countable set A. The complement of the closure of A is an open set which, being disjoint from A, contains no non-void member of the base and is hence void. ∎

A family $\mathcal{A}$ is a **cover** of a set B iff B is a subset of the union $\bigcup\{A: A \in \mathcal{A}\}$; that is, iff each member of B belongs to some member of $\mathcal{A}$. The family is an **open** cover of B iff each member of $\mathcal{A}$ is an open set. A **subcover** of $\mathcal{A}$ is a subfamily which is also a cover.

15 THEOREM (LINDELÖF) *There is a countable subcover of each open cover of a subset of a space whose topology has a countable base.*

PROOF Suppose A is a set, $\mathcal{A}$ is an open cover of A, and $\mathcal{B}$ is a countable base for the topology. Because each member of $\mathcal{A}$ is the union of members of $\mathcal{B}$ there is a subfamily $\mathcal{C}$ of $\mathcal{B}$ which also covers A, such that each member of $\mathcal{C}$ is a subset of some member of $\mathcal{A}$. For each member of $\mathcal{C}$ we may select a containing member of $\mathfrak{A}$ and so obtain a countable subfamily $\mathfrak{D}$ of $\mathfrak{A}$. Then $\mathfrak{D}$ is also a cover of A because $\mathfrak{C}$ covers A. Hence $\mathfrak{A}$ has a countable subcover. ▌

A topological space is a **Lindelöf space** iff each open cover of the space has a countable subcover.

Since the second axiom of countability has been mentioned, it seems only proper that the first be stated. This axiom concerns a localized form of the notion of a base. A **base for the neighborhood system** of a point x, or a **local base** at x, is a family of neighborhoods of x such that every neighborhood of x contains a member of the family. For example, the family of open neighborhoods of a point is always a base for the neighborhood system. A topological space satisfies the **first axiom of countability** if the neighborhood system of every point has a countable base. It is clear that each topological space which satisfies the second axiom of countability also satisfies the first; on the other hand, any uncountable discrete topological space satisfies the first axiom (there is a base for the neighborhood system of each point x which consists of the single neighborhood $\{x\}$) but not the second (the cover whose members are $\{x\}$ for all x in X has no countable subcover). The second axiom of countability is therefore definitely more restrictive than the first.

It is worth noticing that, if $U_1, U_2, \cdots, U_n, \cdots$ is a countable local base at x, then a new local base $V_1, V_2, \cdots, V_n, \cdots$ can be found such that $V_n \supset V_{n+1}$ for each n. The construction is simple: let $V_n = \bigcap \{U_k: k \leqq n\}$.

A **subbase for the neighborhood system** of a point x, or a **local subbase** at x, is a family of sets such that the family of all finite intersections of members is a local base. If $U_1, U_2, \cdots, U_n, \cdots$ is a countable local subbase, then $V_1, V_2, \cdots, V_n, \cdots$, where $V_n = \bigcap \{U_k: k \leqq n\}$ is a countable local base. Hence the existence of a countable local subbase at each point implies the first axiom of countability.

RELATIVIZATION; SEPARATION

If $(X,\mathfrak{J})$ is a topological space and Y is a subset of X we may construct a topology $\mathfrak{U}$ for Y which is called the **relative to-**

pology, or the **relativization of $\mathfrak{J}$ to Y.** The relative topology $\mathfrak{U}$ is defined to be the family of all intersections of members of $\mathfrak{J}$ with Y; that is, U belongs to the relative topology $\mathfrak{U}$ iff $U = V \cap Y$ for some $\mathfrak{J}$-open set V. It is not difficult to see that $\mathfrak{U}$ is actually a topology. Each member U of the relative topology $\mathfrak{U}$ is said to be **open in Y,** and its relative complement $Y \sim U$ is **closed in Y.** The $\mathfrak{U}$-closure of a subset of Y is its **closure in Y.** Each subset Y of X is both open and closed in itself, although Y may be neither open nor closed in X. The topological space $(Y,\mathfrak{U})$ is called a **subspace** of the space $(X,\mathfrak{J})$. More formally, an arbitrary topological space $(Y,\mathfrak{U})$ is a subspace of another space $(X,\mathfrak{J})$ iff $Y \subset X$ and $\mathfrak{U}$ is the relativization of $\mathfrak{J}$.

It is worth noticing that, if $(Y,\mathfrak{U})$ is a subspace of $(X,\mathfrak{J})$ and $(Z,\mathfrak{V})$ is a subspace of $(Y,\mathfrak{U})$, then $(Z,\mathfrak{V})$ is a subspace of $(X,\mathfrak{J})$. This transitivity relation will often be used without explicit mention.

Suppose that $(Y,\mathfrak{U})$ is a subspace of $(X,\mathfrak{J})$ and that A is a subset of Y. Then A may be either $\mathfrak{J}$-closed or $\mathfrak{U}$-closed, a point y may be either a $\mathfrak{U}$ or a $\mathfrak{J}$-accumulation point of A, and A has both a $\mathfrak{J}$ and a $\mathfrak{U}$-closure. The relations between these various notions are important.

16 THEOREM *Let $(X,\mathfrak{J})$ be a topological space, let $(Y,\mathfrak{U})$ be a subspace, and let A be a subset of Y. Then:*

(a) *The set A is $\mathfrak{U}$-closed if and only if it is the intersection of Y and a $\mathfrak{J}$-closed set.*
(b) *A point y of Y is a $\mathfrak{U}$-accumulation point of A if and only if it is a $\mathfrak{J}$-accumulation point.*
(c) *The $\mathfrak{U}$-closure of A is the intersection of Y and the $\mathfrak{J}$-closure of A.*

PROOF The set A is closed in Y iff its relative complement $Y \sim A$ is of the form $V \cap Y$ for some $\mathfrak{J}$-open set V, but this is true iff $A = (X \sim V) \cap Y$ for some V in $\mathfrak{J}$. This proves (a), and (b) follows directly from the definition of the relative topology and the definition of accumulation point. The $\mathfrak{U}$-closure of A is the union of A and the set of its $\mathfrak{U}$-accumulation points, and hence by (b) it is the intersection of Y and the $\mathfrak{J}$-closure of A. ∎

If $(Y,\mathfrak{U})$ is a subspace of $(X,\mathfrak{J})$ and Y is open in X, then each set open in Y is also open in X because it is the intersection of an open set and Y. A similar statement, with "closed" replacing "open" everywhere, is also true. However, knowing that a set is open or closed in a subspace generally tells very little about the situation of the set in X. If X is the union of two sets Y and Z and if A is a subset of X such that $A \cap Y$ is open in Y and $A \cap Z$ is open in Z, then one might hope that A is open in X. But this is not always true, for if Y is an arbitrary subset of X and $Z = X \sim Y$, then $Y \cap Y$ and $Y \cap Z$ are open in Y and Z respectively. There is one important case, in which this result does hold. Two subsets A and B are **separated** in a topological space X iff $A^- \cap B$ and $A \cap B^-$ are both void. This definition of separation involves the closure operation in X. However, the apparent dependence on the space X is illusory, for A and B are separated in X if and only if neither A nor B contains a point or an accumulation point of the other. This condition may be restated in terms of the relative topology for $A \cup B$, in view of part (b) of the foregoing theorem, as: both A and B are closed in $A \cup B$ (or equivalently A (or B) is both open and closed in $A \cup B$) and A and B are disjoint. As an example, notice that the open intervals (0,1) and (1,2) are separated subsets of the real numbers with the usual topology and that there is a point, 1, belonging to the closure of both. However, (0,1) is not separated from the closed interval [1,2] because 1, which is a member of [1,2], is an accumulation point of (0,1).

Three theorems on separation will be needed in the sequel.

17 THEOREM *If Y and Z are subsets of a topological space X and both Y and Z are closed or both are open, then $Y \sim Z$ is separated from $Z \sim Y$.*

PROOF Suppose that Y and Z are closed subsets of X. Then Y and Z are closed in $Y \cup Z$ and therefore $Y \sim Z = ((Y \cup Z) \sim Z)$ and $Z \sim Y$ are open in $Y \cup Z$. It follows that both $Y \sim Z$ and $Z \sim Y$ are open in $(Y \sim Z) \cup (Z \sim Y)$, and since they are complements relative to this set both are closed in $(Y \sim Z) \cup$

$(Z \sim Y)$. Consequently $Y \sim Z$ and $Z \sim Y$ are separated. A dual argument applies to the case where both Y and Z are open in X. ∎

18 Theorem *Let X be a topological space which is the union of subsets Y and Z such that $Y \sim Z$ and $Z \sim Y$ are separated. Then the closure of a subset A of X is the union of the closure in Y of $A \cap Y$ and the closure in Z of $A \cap Z$.*

PROOF The closure of a union of two sets is the union of the closures and hence $A^- = (A \cap Y)^- \cup (A \cap Z \sim Y)^-$. Consequently $A^- \cap Y = [(A \cap Y)^- \cap Y] \cup [(A \cap Z \sim Y)^- \cap Y]$. The set $(Z \sim Y)^-$ is disjoint from $Y \sim Z$, hence $(Z \sim Y)^- \subset Z$, and it follows that $(A \cap Z \sim Y)^-$ is a subset of $(A \cap Z)^- \cap Z$. Similarly $A^- \cap Z$ is the union of $(A \cap Z)^- \cap Z$ and a subset of $(A \cap Y)^- \cap Y$. Consequently $A^- = (A^- \cap Y) \cup (A^- \cap Z) = [(A \cap Y)^- \cap Y] \cup [(A \cap Z)^- \cap Z]$, and the theorem is proved. ∎

19 Corollary *Let X be a topological space which is the union of subsets Y and Z such that $Y \sim Z$ and $Z \sim Y$ are separated. Then a subset A of X is closed (open) if $A \cap Y$ is closed (open) in Y and $A \cap Z$ is closed (open) in Z.*

PROOF If $A \cap Y$ and $A \cap Z$ are closed in Y and Z respectively, then, by the preceding theorem, A is necessarily identical with its closure and is therefore closed. If $A \cap Y$ and $A \cap Z$ are open in Y and Z respectively, then $Y \cap X \sim A$ and $Z \cap X \sim A$ are closed in Y and in Z and hence $X \sim A$ is closed and A is open. ∎

CONNECTED SETS

A topological space $(X,\mathfrak{J})$ is **connected** iff X is not the union of two non-void separated subsets. A subset Y of X is connected iff the topological space Y with the relative topology is connected. Equivalently, Y is connected iff Y is not the union of two non-void separated subsets. Another equivalence follows from the discussion of separation: A set Y is connected iff the only subsets of Y which are both open and closed in Y are Y and the void set. From this form it follows at once that any indiscrete space is connected. A discrete space containing more than one point is not connected. The real numbers, with the usual topology, are connected (problem 1.J), but the rationals, with the usual topology of the reals relativized, are not connected. (For any irrational a the sets $\{x: x < a\}$ and $\{x: x > a\}$ are separated.)

20 THEOREM *The closure of a connected set is connected.*

PROOF Suppose that Y is a connected subset of a topological space and that $Y^- = A \cup B$, where A and B are both open and closed in Y^-. Then each of $A \cap Y$ and $B \cap Y$ is open and closed in Y, and since Y is connected, one of these two sets must be void. Suppose that $B \cap Y$ is void. Then Y is a subset of A and consequently Y^- is a subset of A because A is closed in Y^-. Hence B is void, and it follows that Y^- is connected. ∎

There is another version of this theorem which is apparently stronger, which states that, if Y is a connected subset of X and if Z is a set such that $Y \subset Z \subset Y^-$, then Z is connected. However, the stronger form is an immediate consequence of applying the foregoing theorem to Z with the relative topology.

21 THEOREM *Let $\mathcal{A}$ be a family of connected subsets of a topological space. If no two members of $\mathcal{A}$ are separated, then $\bigcup\{A: A \in \mathcal{A}\}$ is connected.*

PROOF Let C be the union of the members of $\mathcal{A}$ and suppose that D is both open and closed in C. Then for each member A of $\mathcal{A}$, $A \cap D$ is open and closed in A, and since A is connected either $A \subset D$ or $A \subset C \sim D$. Now if A and B are members of $\mathcal{A}$ it is impossible that $A \subset D$ and $B \subset C \sim D$, for in this case A and B, being respectively subsets of the separated sets D and $C \sim D$, would be separated. Consequently either every member of $\mathcal{A}$ is a subset of $C \sim D$ and D is void, or every member of $\mathcal{A}$ is a subset of D and $C \sim D$ is void. ∎

A **component** of a topological space is a maximal connected subset; that is, a connected subset which is properly contained in no other connected subset. A component of a subset A is a component of A with the relative topology; that is, a maximal connected subset of A. If a space is connected, then it is its only component. If a space is discrete, then each component consists of a single point. Of course, there are many spaces which are not discrete which have components consisting of a single point—for example, the space of rational numbers, with the (relativized) usual topology.

22 THEOREM *Each connected subset of a topological space is contained in a component, and each component is closed. If A and B are distinct components of a space, then A and B are separated.*

PROOF Let A be a non-void connected subset of a topological space and let C be the union of all connected sets containing A. In view of the preceding theorem, C is surely connected, and if D is a connected set and contains C, then, since $D \subset C$, it follows that $C = D$. Hence C is a component. (If A is void, and the space is not, a set consisting of a single point is contained in a component, and hence so is A.) Each component C is connected and hence, by 1.20, the closure C^- is connected. Therefore C is identical with C^- and C is closed. If A and B are distinct components and are not separated, then their union is connected, by 1.21, which is a contradiction. ∎

It is well to end our remarks on components with a word of caution. If two points, x and y, belong to the same component of a topological space, then they always lie in the same half of a separation of the space. That is, if the space is the union of separated sets A and B, then both x and y belong to A or both x and y belong to B. The converse of this proposition is false. It may happen that two points always lie in the same half of a separation but nevertheless lie in different components. (See problem 1.**P**.)

PROBLEMS

A LARGEST AND SMALLEST TOPOLOGIES

(a) The intersection of any collection of topologies for X is a topology for X.

(b) The union of two topologies for X may not be a topology for X (unless X consists of at most two points).

(c) For any collection of topologies for X there is a unique largest topology which is smaller than each member of the collection, and a unique smallest topology which is larger than each member of the collection.

B TOPOLOGIES FROM NEIGHBORHOOD SYSTEMS

(a) Let $(X,\mathfrak{I})$ be a topological space and for each x in X let $\mathfrak{U}_x$ be the family of all neighborhoods of x. Then:

(i) If $U \in \mathfrak{U}_x$, then $x \in U$.
(ii) If U and V are members of $\mathfrak{U}_x$, then $U \cap V \in \mathfrak{U}_x$.
(iii) If $U \in \mathfrak{U}_x$ and $U \subset V$, then $V \in \mathfrak{U}_x$.
(iv) If $U \in \mathfrak{U}_x$, then there is a member V of $\mathfrak{U}_x$ such that $V \subset U$ and $V \in \mathfrak{U}_y$ for each y in V (that is, V is a neighborhood of each of its points).

(b) If $\mathfrak{U}$ is a function which assigns to each x in X a non-void family $\mathfrak{U}_x$ satisfying (i), (ii), and (iii), then the family $\mathfrak{I}$ of all sets U, such that $U \in \mathfrak{U}_x$ whenever $x \in U$, is a topology for X. If (iv) is also satisfied, then $\mathfrak{U}_x$ is precisely the neighborhood system of x relative to the topology $\mathfrak{I}$.

Note Various methods of describing a topological space have been investigated intensively. Kuratowski's three closure axioms may be replaced by a single condition, as shown by Monteiro [1] and by Iseki [1]. It is also possible to use the notion of separation as a primitive (Wallace [1], Krishna Murti [1] and Szymanski [1]); the notion of derived set may also be used as primitive (for information and references see Monteiro [2] and Ribeiro [3]). The relation between various operations has been studied by Stopher [1].

C TOPOLOGIES FROM INTERIOR OPERATORS

If ^{i} is an operator which carries subsets of X into subsets of X, and $\mathfrak{I}$ is the family of all subsets such that $A^i = A$, under what conditions will $\mathfrak{I}$ be a topology for X and ^{i} the interior operator relative to this topology?

D ACCUMULATION POINTS IN T_1-SPACES

A topological space is a T_1-*space* iff each set which consists of a single point is closed. (We sometimes say, inaccurately, that "points are closed.")

(a) For any set X there is a unique smallest topology $\mathfrak{I}$ such that $(X,\mathfrak{I})$ is a T_1-space.

(b) If X is infinite and $\mathfrak{I}$ is the smallest topology such that $(X,\mathfrak{I})$ is a T_1-space, then $(X,\mathfrak{I})$ is connected.

(c) If $(X,\mathfrak{I})$ is a T_1-space, then the set of accumulation points of each subset is closed. A sharper result (C. T. Yang): A necessary and sufficient condition that the set of accumulation points of each subset be closed is that the set of accumulation points of $\{x\}$ be closed for each x in X.

Note There is a sequence of successively stronger requirements which may be put upon the topology of a space. A topological space is a T_0-*space* iff for each pair x and y of distinct points, there is a neighborhood of one point to which the other does not belong. In slightly different terminology, the space is a T_0-space iff for distinct points x and y either $x \notin \{y\}^-$ or $y \notin \{x\}^-$. We will define T_2 and T_3-spaces later. The terminology is due to Alexandroff and Hopf [1].

E KURATOWSKI CLOSURE AND COMPLEMENT PROBLEM

If A is a subset of a topological space, then at most 14 sets can be constructed from A by complementation and closure. There is a subset of the real numbers (with the usual topology) from which 14 different sets can be so constructed. (First notice that if A is the closure of an open set, then A is the closure of the interior of A; that is, for such sets $A = A'^{-'-}$ where $'$ denotes complementation.)

F EXERCISE ON SPACES WITH A COUNTABLE BASE

If the topology of a space has a countable base, then each base contains a countable subfamily which is also a base.

G EXERCISE ON DENSE SETS

If A is dense in a topological space and U is open, then $U \subset (A \cap U)^-$.

H ACCUMULATION POINTS

Let X be a space, each subspace of which is Lindelöf, let A be an uncountable subset, and let B be the subset consisting of all points x of A such that each neighborhood of x contains uncountably many points of A. Then $A \sim B$ is countable, and consequently each neighborhood of a point of B contains uncountably many points of B.

Note The accumulation points of a set A may be classified according to the least cardinal number of the intersection of A and a neighborhood of the point. If there is also a cardinal number restriction on a base for the topology then several inequalities result. Theorems 1.13, 1.14, and 1.15 all have generalizations applying to spaces with a base of a given cardinal.

I THE ORDER TOPOLOGY

Let X be a set, linearly ordered by a relation $<$ which is anti-symmetric (it is false that $x < x$). The *order topology* (the $<$ order topology) has a subbase consisting of all sets of the form: $\{x: x < a\}$ or $\{x: a < x\}$ for some a in X.

(a) The order topology for X is the smallest topology in which order is continuous, in the following sense: if a and b are members of X and $a < b$, then there are neighborhoods U of a and V of b such that, whenever $x \in U$ and $y \in V$, then $x < y$.

(b) Let Y be a subset of a set X which is linearly ordered by $<$. Then Y is linearly ordered by $<$, but the $<$ order topology for Y may not be the relativized $<$ order topology for X.

(c) If X, with the order topology, is connected, then X is order-complete (that is, each non-void set with an upper bound has a supremum).

(d) If there are points a and b in X such that $a < b$ and there is no point c such that $a < c < b$, then X is not connected. Such an ordering is said to have a *gap*. Show that X is connected relative to the order topology iff X is order-complete and there are no gaps.

J PROPERTIES OF THE REAL NUMBERS

Let R be the set of real numbers with the usual topology.

(a) An additive subgroup of the reals which contains more than one member is either dense in R or has a smallest positive element. In particular, the set of rational numbers is dense in R.

(b) The usual topology for the reals is identical with the order topology. The usual topology has a countable base.

(c) A closed subgroup of R is either countable or identical with R. A connected subgroup is either $\{0\}$ or R and an open subgroup is necessarily identical with R.

(d) (A. P. Morse) A *proper* interval is a half-open, open, or closed interval which contains more than one point. If $\mathfrak{A}$ is an arbitrary family of proper intervals, then there is a countable subfamily $\mathfrak{B}$ of $\mathfrak{A}$ such that $\bigcup\{B: B \in \mathfrak{B}\} = \bigcup\{A: A \in \mathfrak{A}\}$. (Observe that a disjoint family of proper intervals is countable, and show that all but a countable number of points of $\bigcup\{A: A \in \mathfrak{A}\}$ are interior points of members of $\mathfrak{A}$.)

(e) The family $\mathfrak{S}$ of all proper intervals is a subbase for the discrete topology $\mathfrak{I}$ for R. The space $(R,\mathfrak{I})$ is not a Lindelöf space, although every cover by members of $\mathfrak{S}$ has a countable subcover. (Contrast with the Alexander theorem 5.6.)

Note Further properties of the real numbers are stated in the next problem.

K HALF-OPEN INTERVAL SPACE

Let X be the set of real numbers and let $\mathfrak{J}$ be the topology for X which has for a base the family $\mathfrak{B}$ of all half-open intervals $[a,b) = \{x: a \leqq x < b\}$ where a and b are real numbers. A $\mathfrak{J}$-accumulation point of a set is called an *accumulation point from the right*, and accumulation points from the left are similarly defined.

(a) Members of the base $\mathfrak{B}$ are both open and closed. The space $(X,\mathfrak{J})$ is not connected.

(b) The space $(X,\mathfrak{J})$ is separable but $\mathfrak{J}$ has no countable base. (For every x in X each base must contain a set whose infimum is x.)

(c) Each subspace of $(X,\mathfrak{J})$ is a Lindelöf space. (See 1.J(d).)

(d) If A is a set of real numbers then the set of all points of A which are not accumulation points from the right is countable. More generally, the set of points of A which are not accumulation points from both the right and the left is countable. (See 1.**H**.)

(e) Every subspace of $(X,\mathfrak{J})$ is separable.

L HALF-OPEN RECTANGLE SPACE

Let Y be $X \times X$, where X is the space of the preceding problem, and let $\mathfrak{U}$ be the topology which has as a base the family of all $A \times B$, where A and B are members of the topology $\mathfrak{J}$ of the preceding example.

(a) The space $(Y,\mathfrak{U})$ is separable.

(b) The space $(Y,\mathfrak{U})$ contains a subspace which is not separable. (For example, $\{(x,y): x + y = 1\}$.)

(c) The space $(Y,\mathfrak{U})$ is not a Lindelöf space. (If each open cover of Y has a countable subcover, then every closed subspace has the same property. Consider $\{(x,y): x + y = 1\}$.)

Note The spaces described in 1.**K** and 1.**L** are among the stock counter-examples of general topology. We enumerate other pathological features in 4.I. P. R. Halmos first observed that the product (in a sense to be made specific in chapter 3) of Lindelöf spaces may fail to be a Lindelöf space.

M EXAMPLE (THE ORDINALS) ON 1ST AND 2ND COUNTABILITY

Let Ω' be the set of all ordinals less than or equal to the first uncountable ordinal Ω, let X be $\Omega' \sim \{\Omega\}$, and let ω be the set of all non-negative integers, each with the order topology.

(a) ω is discrete and satisfies the 2nd axiom of countability.

(b) X satisfies the first but not the second axiom of countability.

(c) Ω' satisfies neither axiom of countability; if U is a separable subspace of Ω', then U is itself countable.

N COUNTABLE CHAIN CONDITION

A topological space satisfies the *countable chain condition* iff each disjoint family of open sets is countable. A separable space satisfies the countable chain condition, but not conversely. (Consider an uncountable set with the topology consisting of the void set and the complements of countable sets.) There are more complicated examples (see the Helly space of 5.M) which satisfy the first countability axiom and are separable, but fail to satisfy the second axiom of countability.

O THE EUCLIDEAN PLANE

The Euclidean plane is the set of all pairs of real numbers and the *usual topology* for the plane has a base which consists of all cartesian products $A \times B$ where A and B are open intervals with rational endpoints. This base is countable and the plane is consequently separable.

(a) The usual topology of the plane has a base which consists of all open discs, $\{(x,y): (x - a)^2 + (y - b)^2 < r^2\}$, where a, b, and r are rational numbers.

(b) Let X be the set of all points in the plane with at least one irrational coordinate, and let X have the relative topology. Then X is connected.

P EXAMPLE ON COMPONENTS

Let X be the following subset of the Euclidean plane, with the usual topology relativized. For each positive integer n let $A_n = \{1/n\} \times [0,1]$, where $[0,1]$ is the closed interval, and let X be the union of the sets A_n, with $(0,0)$ and $(0,1)$ adjoined. Then $\{(0,0)\}$ and $\{(0,1)\}$ are components of X, but each open and closed subset of X contains neither or both of the points.

Q THEOREM ON SEPARATED SETS

If X is a connected topological space, Y is a connected subset and $X \sim Y = A \cup B$, where A and B are separated, then $A \cup Y$ is connected.

R FINITE CHAIN THEOREM FOR CONNECTED SETS

Let $\mathfrak{A}$ be a family of connected subsets of a topological space satisfying the condition: if A and B belong to $\mathfrak{A}$, then there is a finite sequence $A_0, A_1, \cdots A_n$, of members of $\mathfrak{A}$ such that $A_0 = A$, $A_n = B$, and, for each i, the sets A_i and A_{i+1} are not separated. Then $\bigcup\{A: A \in \mathfrak{A}\}$ is connected. From this fact deduce 1.21.

S LOCALLY CONNECTED SPACES

A topological space is *locally connected* iff for each point x and each neighborhood U of x the component of U to which x belongs is a neighborhood of x.

(a) Each component of an open subset of a locally connected space is open.

(b) A topological space is locally connected iff the family of open connected subsets is a base for the topology.

(c) If points x and y of a locally connected space X belong to different components, then there are separated subsets A and B of X such that $x \in A$, $y \in B$, and $X = A \cup B$.

Note For many other properties of locally connected spaces and for generalizations, see G. T. Whyburn [1] and R. L. Wilder [1].

T THE BROUWER REDUCTION THEOREM

The usual statement of the theorem is as follows. Let X be a topological space satisfying the second axiom of countability. A property P of subsets of X is called *inductive* iff whenever each member of a countable nest of closed sets has P, then the intersection has P. A set A is *irreducible* with respect to P iff no proper closed subset of A has P. Then: If a closed subset A of X possesses an inductive property P, there is an irreducible closed subset of A which possesses P.

The theorem can be stated more formally in terms of a family of sets (the family of all sets possessing P).

(a) State and prove the theorem in this form. Assume that the topological space is such that every subspace is a Lindelöf space.

(b) If $(X,\mathfrak{T})$ is an arbitrary topological space can any result of this general sort be affirmed? (See 0.25.)

Chapter 2

MOORE-SMITH CONVERGENCE

INTRODUCTION

This chapter is devoted to the study of Moore-Smith convergence. It will turn out that the topology of a space can be described completely in terms of convergence, and the major part of the chapter is devoted to this description. We shall also characterize those notions of convergence which can be described as convergence relative to some topology. This project is similar in purpose to the theory of Kuratowski closure operators; it yields a useful and intuitively natural way of specifying certain topologies. However, the importance of convergence theory extends beyond this particular application, for the fundamental constructions of analysis are limit processes. We are interested in developing a theory which will apply to convergence of sequences, of double sequences, to summation of sequences, to differentiation and integration. The theory which we develop here is by no means the only possible theory, but it is unquestionably the most natural.

Sequential convergence furnishes the pattern on which the theory is developed, and we therefore list a few definitions and theorems on sequences to indicate this pattern. These will be particular cases of the theorems proved later.

A **sequence** is a function on the set ω of non-negative integers. A sequence of real numbers is a sequence whose range is a subset of the set of real numbers. The value of a sequence S at n is denoted, interchangeably, by S_n or $S(n)$. A sequence S is in a set A iff $S_n \varepsilon A$ for each non-negative integer n, and S is eventually in A iff there is an integer m such that $S_n \varepsilon A$ whenever $n \geqq m$. A sequence of real numbers converges to a number s relative to the usual topology iff it is eventually in each neighborhood of s. Using these definitions it turns out that, if A is a set of real numbers, then a point s belongs to the closure of A iff there is a sequence in A which converges to s, and s is an accumulation point of A iff there is a sequence in $A \sim \{s\}$ which converges to s.

We shall want to construct subsequences of a sequence. A sequence S may converge to no point and yet, by a proper construction, a sequence may be obtained from it which converges. We wish to select an integer N_i, for each i in ω, such that S_{N_i} converges. Restated, we want to find a sequence N of integers so that the composition $S \circ N(i) = S_{N_i} = S(N(i))$ converges. If no other requirement is made this is easy enough; if $N_i = 0$ for each i, then $S \circ N$ converges to S_0 since $S \circ N(i) = S_0$ for each i. Of course, an additional condition must be imposed so that the behavior of a subsequence is related to the behavior of the sequence for large integers. The usual condition is that N be strictly monotonically increasing; that is, if $i > j$, then $N_i > N_j$. This condition is unnecessarily stringent, and we impose instead the requirement that, as i becomes large, N_i also becomes large. Formally, then, T is a subsequence of a sequence S iff there is a sequence N of non-negative integers such that $T = S \circ N$ (equivalently, $T_i = S_{N_i}$ for each i) and for each integer m there is an integer n such that $N_i \geqq m$ whenever $i \geqq n$.

The set of points to which the subsequences of a given sequence converge satisfy a condition obtained by weakening the requirement of convergence. A sequence S is frequently in a set A iff for each non-negative integer m there is an integer n such that $n \geqq m$ and $S_n \varepsilon A$. This is precisely the same thing as saying that S is not eventually in the complement of A; intuitively, a sequence is frequently in A if it keeps returning to A. A point s is a cluster point of a sequence S iff S is frequently in each neighborhood of s. Then, if a sequence of real numbers is eventually in a set so is every subsequence, and consequently if a sequence converges to a point so does every subsequence. Each cluster point of a sequence is a limit point of a subsequence, and conversely.

The definitions and statements above are phrased so as to be applicable to any topological space, but unfortunately the theorems, in this generality, are false. (See the problems at the end of this chapter.) This unhappy situation is remedied by noticing that very few of the properties of the integers are used in proving theorems on sequences of real numbers. It is almost evident (although we have not given the proofs) that we need only certain properties of the ordering. Strictly speaking, convergence of sequences involves not only the function S on the non-negative integers ω, but also the ordering, $\geqq$, of ω. For convenience, in the work on convergence, we modify slightly the definition of sequence and agree that a sequence is an ordered pair $(S,\geqq)$ where S is a function on the integers, and we discuss convergence of the pair $(S,\geqq)$. (It will turn out that convergence of the pair $(S,\leqq)$ is also meaningful, but quite different.) Mention of the order will be omitted if no confusion is likely, and convergence of a sequence S will always mean convergence of the pair $(S,\geqq)$.

It is also convenient to have a bound variable (dummy variable) notation for sequences, and accordingly, if S is a function on the non-negative integers ω, $\{S_n, n \in \omega, \geqq\}$ is defined to be the pair $(S,\geqq)$. If A is a subset of ω, then convergence of $\{S_n, n \in A, \geqq\}$ will also be meaningful and will be related to the convergence of $(S,\geqq)$.

After this lengthy introduction the notion of convergence is almost self-evident, lacking a single fact. Which properties of the order $\geqq$ are used? These properties are listed below, and by using them the usual arguments of sequential convergence, with small modifications, are valid.

1 *Notes* E. H. Moore's study of unordered summability of sequences [1] led to the theory of convergence (Moore and Smith [1]). The generalization of the notion of subsequence which we will use is also due to Moore [2]. Garrett Birkhoff [3] applied Moore-Smith convergence to general topology; the form in which we give the theory is approximately that of J. W. Tukey [1]. See McShane [1] for an extremely readable expository account.

The problems at the end of the chapter contain a brief discussion of another theory of convergence and appropriate references.

DIRECTED SETS AND NETS

A binary relation $\geqq$ **directs** a set D if D is non-void and

(a) if m, n and p are members of D such that $m \geqq n$ and $n \geqq p$, then $m \geqq p$;
(b) if $m \in D$, then $m \geqq m$; and
(c) if m and n are members of D, then there is p in D such that $p \geqq m$ and $p \geqq n$.

We say that m follows n in the order $\geqq$ and that n precedes m iff $m \geqq n$. In the usual language of relations (see chapter 0) the condition (a) states that $\geqq$ is transitive on D, or partially orders D, and (b) states that $\geqq$ is reflexive on D. The condition (c) is special in character.

There are several natural examples of sets directed by relations. The real numbers as well as the set ω of non-negative integers are directed by $\geqq$. Observe that 0 is a member of ω which follows every other member in the order $\leqq$. It is also noteworthy that the family of all neighborhoods of a point in a topological space is directed by $\subset$ (the intersection of two neighborhoods is a neighborhood which follows both in the ordering $\subset$). The family of all finite subsets of a set is, on the other hand, directed by $\supset$. Any set is directed by agreeing that $x \geqq y$ for all members x and y, so that each element follows both itself and every other element.

A **directed set** is a pair $(D,\geqq)$ such that $\geqq$ directs D. (This is sometimes called a **directed system.**) A **net** is a pair $(S,\geqq)$ such that S is a function and $\geqq$ directs the domain of S. (A net is sometimes called a directed set.) If S is a function whose domain contains D and D is directed by $\geqq$, then $\{S_n, n \in D, \geqq\}$ is the net $(S|D,\geqq)$ where $S|D$ is S restricted to D. A net $\{S_n, n \in D, \geqq\}$ is **in** a set A iff $S_n \in A$ for all n; it is **eventually in** A iff there is an element m of D such that, if $n \in D$ and $n \geqq m$, then $S_n \in A$. The net is **frequently in** A iff for each m in D there is n in D such that $n \geqq m$ and $S_n \in A$. If $\{S_n, n \in D, \geqq\}$

is frequently in A, then the set E of all members n of D such that $S_n \in A$ has the property: for each $m \in D$ there is $p \in E$ such that $p \geqq m$. Such subsets of D are called **cofinal.** Each cofinal subset E of D is also directed by $\geqq$ because for elements m and n of E there is p in D such that $p \geqq m$ and $p \geqq n$, and there is then an element q of E which follows p. We have the following obvious equivalence: a net $\{S_n, n \in D, \geqq\}$ is frequently in a set A iff a cofinal subset of D maps into the set A, and this is the case iff the net is not eventually in the complement of A.

A net $(S, \geqq)$ in a topological space $(X, \mathfrak{I})$ **converges** to s relative to $\mathfrak{I}$ iff it is eventually in each $\mathfrak{I}$-neighborhood of s. The notion of convergence depends on the function S, the topology $\mathfrak{I}$, and the ordering $\geqq$. However, in cases where no confusion is likely to result we may omit all mention of $\mathfrak{I}$ or of $\geqq$ or of both and simply say "the net S (or the net $\{S_n, n \in D\}$) converges to s." If X is a discrete space (every subset is open), then a net S converges to a point s iff S is eventually in $\{s\}$: that is, from some point on S is constantly equal to s. On the other hand, if X is indiscrete (the only open sets are X and the void set), then every net in X converges to every point of X. Consequently a net may converge to several different points.

It is easy to describe the accumulation points of a set, the closure of a set, and in fact the topology of a space in terms of convergence. The arguments are slight variants of those usually given for sequences of real numbers.

2 THEOREM *Let X be a topological space. Then:*

(a) *A point s is an accumulation point of a subset A of X if and only if there is a net in $A \sim \{s\}$ which converges to s.*
(b) *A point s belongs to the closure of a subset A of X if and only if there is a net in A converging to s.*
(c) *A subset A of X is closed if and only if no net in A converges to a point of $X \sim A$.*

PROOF If s is an accumulation point of A, then for each neighborhood U of s there is a point S_U of A which belongs to $U \sim \{s\}$. The family $\mathfrak{U}$ of all neighborhoods of s is directed by $\subset$, and if U and V are neighborhoods of s such that $V \subset U$, then $S_V \in V$

$\subset U$. The net $\{S_U, U \in \mathfrak{U}, \subset\}$, therefore converges to s. On the other hand, if a net in $A \sim \{s\}$ converges to s, then this net has values in every neighborhood of s and $A \sim \{s\}$ surely intersects each neighborhood of s. This establishes the statement (a). To prove (b), recall that the closure of a set A consists of A together with all the accumulation points of A. For each accumulation point s of A there is, by the preceding, a net in A converging to s; for each point s of A any net whose value at every element of its domain is s converges to s. Therefore each point of the closure of A has a net in A converging to it. Conversely, if there is a net in A converging to s, then every neighborhood of s intersects A and s belongs to the closure of A. Proposition (c) is now obvious. ∎

We have noticed that, in general, a net in a topological space may converge to several different points. There are spaces in which convergence is unique in the sense that, if a net S converges to a point s and also to a point t, then $s = t$. A topological space is a **Hausdorff space** (T_2-space, or **separated** space) iff whenever x and y are distinct points of the space there exists disjoint neighborhoods of x and y.

3 THEOREM *A topological space is a Hausdorff space if and only if each net in the space converges to at most one point.*

PROOF If X is a Hausdorff space and s and t are distinct points of X, then there are disjoint neighborhoods U and V of s and t respectively. Since a net cannot be eventually in each of two disjoint sets it is clear that no net in X converges to both s and t. To establish the converse assume that X is not a Hausdorff space and that s and t are distinct points such that every neighborhood of s intersects every neighborhood of t. Let $\mathfrak{U}_s$ be the family of neighborhoods of s and $\mathfrak{U}_t$ the family of neighborhoods of t; then both $\mathfrak{U}_s$ and $\mathfrak{U}_t$ are directed by $\subset$. We order the cartesian product $\mathfrak{U}_s \times \mathfrak{U}_t$ by agreeing that $(T,U) \geqq (V,W)$ iff $T \subset V$ and $U \subset W$. Clearly the cartesian product is directed by $\geqq$. For each (T,U) in $\mathfrak{U}_s \times \mathfrak{U}_t$ the intersection $T \cap U$ is non-void, and hence we may select a point $S_{(T,U)}$ from $T \cap U$. If $(V,W) \geqq (T,U)$, then $S_{(V,W)} \in V \cap W \subset T \cap U$ and consequently the net $\{S_{(T,U)}, (T,U) \in \mathfrak{U}_s \times \mathfrak{U}_t, \geqq\}$ converges to both s and t. ∎

If $(X,\mathfrak{J})$ is a Hausdorff space and a net $\{S_n, n \in D, \geq\}$ in X converges to s we write $\mathfrak{J}$-lim $\{S_n, n \in D, \geq\} = s$. When no confusion seems possible this will be abbreviated: lim $\{S_n: n \in D\} = s$ or $\lim_n S_n = s$. The use of "limit" should be restricted to nets in a Hausdorff space so that the usual rule concerning substitution of equals for equals may remain valid. If lim $\{S_n: n \in D\} = s$ and lim $\{S_n: n \in D\} = t$, then $s = t$, since we always use equality in the sense of identity. As a matter of fact we shall occasionally use the notation $\lim_n S_n = s$ to mean S converges to s in cases where the space is not Hausdorff.

The device used in the preceding proof is often useful. If $(D,\geq)$ and $(E,\succ)$ are directed sets, then the cartesian product $D \times E$ is directed by $\gg$, where $(d,e) \gg (f,g)$ iff $d \geq f$ and $e \succ g$. The directed set $(D \times E,\gg)$ is the **product directed set.** We also want to define the product of a family of directed sets. Suppose for each a in a set A we are given a directed set $(D_a,>_a)$. The cartesian product $\times\{D_a: a \in A\}$ is the set of all functions d on A such that d_a $(= d(a))$ is a member of D_a for each a in A. The product directed set is $(\times\{D_a: a \in A\},\geq)$ where, if d and e are members of the product $d \geq e$ iff $d_a >_a e_a$ for each a in A. The **product order** is $\geq$. Of course, it must be verified that the product directed set is, in fact, directed. If d and e are members of the cartesian product $\times\{D_a: a \in A\}$, then for each a there is a member f_a of D_a which follows both d_a and e_a in the order $>_a$, and consequently the function f whose value at a is f_a follows both d and e in the order $\geq$. An important special case of the product directed set is that in which all coordinate sets D_a are identical and all relations $>_a$ are identical. In this case $\times\{D: a \in A\}$ is simply the set D^A of all functions on A to D, which is directed by the convention that d follows e iff $d(a)$ follows $e(a)$ for each member a of A. This is, for example, precisely the usual ordering of the set of all real valued functions on the set of real numbers.

The next result on limits is related to the closure axiom: $A^{--} = A^-$. It is important because it replaces an iterated limit by a single limit. The situation is as follows: Consider the class of all functions S such that $S(m,n)$ is defined whenever m belongs to a directed set D and n belongs to a directed set E_m. We want to find a net R with values in this domain such that $S \circ R$ converges to $\lim_m \lim_n S(m,n)$ whenever S is a function to a topological space and this iterated limit exists. It is interesting to notice that the solution of this problem requires Moore-Smith convergence, for, considering double sequences, no sequence whose range is a subset of $\omega \times \omega$ can have this property. The construction which yields a solution to the problem is a variant of the diagonal process. Let F be the product directed set $D \times \times\{E_n: n \in D\}$, and for each point (m,f) of F let $R(m,f) = (m,f(m))$. Then R is the required net.

4 THEOREM ON ITERATED LIMITS *Let D be a directed set, let E_m be a directed set for each m in D, let F be the product $D \times \times\{E_m: m \in D\}$, and for (m,f) in F let $R(m,f) = (m,f(m))$. If $S(m,n)$ is a member of a topological space for each m in D and each n in E_m, then $S \circ R$ converges to* $\lim_m \lim_n S(m,n)$ *whenever this iterated limit exists.*

PROOF Suppose $\lim_m \lim_n S(m,n) = s$ and that U is an open neighborhood * of s. We must find a member (m,f) of F such that, if $(p,g) \geq (m,f)$, then $S \circ R(p,g) \in U$. Choose m in D so that $\lim_n S(p,n) \in U$ for each p following m and then, for each such p, choose a member $f(p)$ of E_p such that $S(p,n) \in U$ for all n following $f(p)$ in E_p. If p is a member of D which does not follow m let $f(p)$ be an arbitrary member of E_p. If $(p,g) \geq (m,f)$, then $p \geq m$, hence $\lim_n S(p,n) \in U$, and (since $g(p) \geq f(p)$) $S \circ R\,(p,g) = S(p,g(p)) \in U$. ∎

SUBNETS AND CLUSTER POINTS

Following the pattern discussed in the introduction to the chapter we now define the generalization of subsequence and prove the hoped-for theorems.

* The existence of an open neighborhood of s is essential to the proof. The iterated limit theorem, the fact that the family of open neighborhoods of a point is a local base, and the closure axiom "$A^{--} = A^-$" are intimately related. Convergence has been studied in spaces with a structure less restrictive than a topology. See Ribeiro [1].

A net $\{T_m, m \in E\}$ is a **subnet** of a net $\{S_n, n \in D\}$ iff there is a function N on E with values in D such that

(a) $T = S \circ N$, or equivalently, $T_i = S_{N_i}$ for each i in E; and
(b) for each m in D there is n in E with the property that, if $p \geq n$, then $N_p \geq m$.

Since there seems to be no possibility of confusion we omit specific mention of the orderings involved. The second condition states, intuitively, "as p becomes large so does N_p." From this condition it is immediately clear that, if S is eventually in a set A, then the subnet $S \circ N$ of S is also eventually in A. This is a very important fact and the definition of subnet is designed to obtain precisely this result. Notice that each cofinal subset E of D is directed by the same ordering, and that $\{S_n, n \in E\}$ is a subnet of S. (Let N be the identity function on E, and the second condition of the definition becomes the requirement that E be cofinal.) This is a standard way of constructing subnets, and it is unfortunate that this simple variety of subnet is not adequate for all purposes. (2.E.)

There is a special sort of subnet which is adequate for almost all purposes. Suppose N is a function on the directed set E to the directed set D such that N is isotone ($N_i \geq N_j$ if $i \geq j$) and the range of N is cofinal in D. Then clearly $S \circ N$ is a subnet of S for each net S. The subnet constructed in the proof of the following lemma is of this sort (as remarked by K. T. Smith).

5 LEMMA *Let S be a net and $\mathfrak{A}$ a family of sets such that S is frequently in each member of $\mathfrak{A}$, and such that the intersection of two members of $\mathfrak{A}$ contains a member of $\mathfrak{A}$. Then there is a subnet of S which is eventually in each member of $\mathfrak{A}$.*

PROOF The intersection of any two members of $\mathfrak{A}$ contains a member of $\mathfrak{A}$ and therefore $\mathfrak{A}$ is directed by $\subset$. Let $\{S_n, n \in D\}$ be a net which is frequently in each member of $\mathfrak{A}$ and let E be the set of all pairs (m,A) such that $m \in D$, $A \in \mathfrak{A}$, and $S_m \in A$. Then E is directed by the product ordering for $D \times \mathfrak{A}$, for if (m,A) and (n,B) are members of E there is C in $\mathfrak{A}$ such that $C \subset A \cap B$ and p in D such that p follows both m and n and

$S_p \in C$; then $(p,C) \in E$ and (p,C) follows both (m,A) and (n,B). For (m,A) in E let $N(m,A) = m$. Then N is clearly isotone, and the range of N is cofinal in D ($\{S_n, n \in D\}$ is frequently in each member of $\mathcal{A}$). Consequently $S \circ N$ is a subnet of S. Finally, if A is a member of $\mathcal{A}$, if m is an arbitrary member of D such that $S_m \in A$, and if (n,B) is a member of E which follows (m,A), then $S \circ N(n,B) = S_n \in B \subset A$; it follows that $S \circ N$ is eventually in A. ▮

We now apply this lemma to convergence in a topological space. A point s of the space is a **cluster point** of a net S iff S is frequently in every neighborhood of s. A net may have one, many, or no cluster points. For example, if ω is the set of non-negative integers, then $\{n, n \in \omega\}$ is a net which has no cluster point relative to the usual topology for the real numbers. The other sort of extreme occurs if S is a sequence whose range is the set of all rational numbers (such a sequence exists because the set of rationals is countable). It is easy to see that this sequence is frequently in each open interval, and consequently every real number is a cluster point. If a net converges to a point, then this point is surely a cluster point, but it is possible that a net may have a single cluster point and fail to converge to this point. For example, consider the sequence $-1, 1, -1, 2, -1, 3, -1 \cdots$, constructed by alternating -1 and the sequence of positive integers. Then -1 is the unique cluster point of the sequence, but the sequence fails to converge to -1.

6 THEOREM *A point s in a topological space is a cluster point of a net S if and only if some subnet of S converges to s.*

PROOF Let s be a cluster point of S and let $\mathfrak{U}$ be the family of all neighborhoods of s. Then the intersection of two members of $\mathfrak{U}$ is again a member of $\mathfrak{U}$, and S is frequently in each member of $\mathfrak{U}$. Consequently the preceding lemma applies and there is a subnet of S which is eventually in each member of $\mathfrak{U}$, that is, converges to s. If s is not a cluster point of S, then there is a neighborhood U of s such that S is not frequently in U, and therefore S is eventually in the complement of U. Then each subnet of S is eventually in the complement of U and hence cannot converge to s. ▮

The following is a characterization of cluster points in terms of closure.

7 THEOREM *Let $\{S_n, n \in D\}$ be a net in a topological space and for each n in D let A_n be the set of all points S_m for $m > n$. Then s is a cluster point of $\{S_n, n \in D\}$ if and only if s belongs to the closure of A_n for each n in D.*

PROOF If s is a cluster point of $\{S_n, n \in D\}$, then for each n, A_n intersects each neighborhood of s because $\{S_n, n \in D\}$ is frequently in each neighborhood. Therefore s is in the closure of each A_n. If s is not a cluster point of $\{S_n, n \in D\}$ there is a neighborhood U of s such that $\{S_n, n \in D\}$ is not frequently in U. Hence for some n in D, if $m \geqq n$, then $S_m \notin U$, so that U and A_n are disjoint. Consequently s is not in the closure of A_n. ▮

SEQUENCES AND SUBSEQUENCES

It is of some interest to know when a topology can be described in terms of sequences alone, not only because it is a convenience to have a fixed domain for all nets, but also because there are properties of sequences which fail to generalize. The most important class of topological spaces for which sequential convergence is adequate are those satisfying the first countability axiom: the neighborhood system of each point has a countable base. That is, for each point x of the space X there is a countable family of neighborhoods of x such that every neighborhood of x contains some member of the family. In this case we may replace "net" by "sequence" in almost all of the preceding theorems.

It should be noticed that a sequence may have subnets which are not subsequences.

8 THEOREM *Let X be a topological space satisfying the first axiom of countability. Then:*

(a) *A point s is an accumulation point of a set A if and only if there is a sequence in $A \sim \{s\}$ which converges to s.*
(b) *A set A is open if and only if each sequence which converges to a point of A is eventually in A.*

(c) *If s is a cluster point of a sequence S there is a subsequence of S converging to s.*

PROOF Suppose that s is an accumulation point of a subset A of X, and that $U_0, U_1, \cdots U_n \cdots$ is a sequence which is a base for the neighborhood system of s. Let $V_n = \bigcap\{U_i: i = 0, 1, \cdots, n\}$. Then the sequence $V_0, V_1, \cdots, V_n \cdots$ is also a base for the neighborhood system of s and, moreover, $V_{n+1} \subset V_n$ for each n. For each n select a point S_n from $V_n \cap (A \sim \{s\})$, thus obtaining a sequence $\{S_n, n \in \omega\}$ which evidently converges to s. This establishes half of (a), and the converse is obvious. If A is a subset of X which is not open, then there is a sequence in $X \sim A$ which converges to a point of A. Such a sequence surely fails to be eventually in A, and part (b) of the theorem follows. Finally, suppose that s is a cluster point of a sequence S and that $V_0, V_1 \cdots$ is a sequence which is a base for the neighborhood system of s such that $V_{n+1} \subset V_n$ for each n. For every non-negative integer i, choose N_i such that $N_i \geqq i$ and S_{N_i} belongs to V_i. Then surely $\{S_{N_i}, i \in \omega\}$ is a subsequence of S which converges to s. ∎

*CONVERGENCE CLASSES

It is sometimes convenient to define a topology by specifying what nets converge to which points. For example, if $\mathfrak{F}$ is a family of functions each on a fixed set X to a topological space Y it is natural to specify that a net $\{f_n, n \in D\}$ converges to a function g iff $\{f_n(x), n \in D\}$ converges to $g(x)$ for each x in X. (This sort of convergence is discussed in some detail in chapter 3.) Having made such a specification the question naturally arises: Is there a topology for $\mathfrak{F}$ such that this convergence is convergence relative to the topology? An affirmative answer would enable us to use the machinery developed for topological spaces to investigate the structure of $\mathfrak{F}$.

The problem may be formally phrased as follows. If $\mathcal{C}$ is a class consisting of pairs (S,s), where S is a net in X and s a point, when is there a topology $\mathfrak{J}$ for X such that $(S,s) \in \mathcal{C}$ iff S converges to s relative to the topology $\mathfrak{J}$? From the preceding discussion of convergence we know several properties which $\mathcal{C}$ must possess if such a topology exists. We shall say that $\mathcal{C}$ is a **convergence class** for X iff it satisfies the conditions listed below.* For convenience, we say that S converges ($\mathcal{C}$) to s or that $\lim_n S_n \equiv s$ ($\mathcal{C}$) iff $(S,s) \in \mathcal{C}$.

(a) *If S is a net such that $S_n = s$ for each n, then S converges ($\mathcal{C}$) to s.*
(b) *If S converges ($\mathcal{C}$) to s, then so does each subnet of S.*
(c) *If S does not converge ($\mathcal{C}$) to s, then there is a subnet of S, no subnet of which converges ($\mathcal{C}$) to s.*
(d) *(Theorem 2.4 on iterated limits) Let D be a directed set, let E_m be a directed set for each m in D, let F be the product $D \times \times\{E_m: m \in D\}$ and for (m,f) in F let $R(m,f) = (m,f(m))$. If* $\lim_m \lim_n S(m,n) \equiv s$ ($\mathcal{C}$), *then $S \circ R$ converges ($\mathcal{C}$) to s.*

It has previously been shown that convergence in a topological space satisfies (a), (b), and (d). Statement (c) is easily established, in this case, by the argument: If a net $\{S_n, n \in D\}$ fails to converge to a point s, then it is frequently in the complement of a neighborhood of s, and hence for a cofinal subset E of D, $\{S_n, n \in E\}$ is in the complement. But clearly $\{S_n, n \in E\}$ is a subnet, no subnet of which converges to s.

We now show that every convergence class is actually derived from a topology.

9 THEOREM *Let $\mathcal{C}$ be a convergence class for a set X, and for each subset A of X let A^c be the set of all points s such that, for some net S in A, S convergences ($\mathcal{C}$) to s. Then ^{c} is a closure operator, and $(S,s) \in \mathcal{C}$ if and only if S converges to s relative to the topology associated with ^{c}.*

PROOF It is first shown that ^{c} is a closure operator. (See 1.8.) Since a net is a function on a directed set, and the set is non-void by definition, $(0)^c$ is void. In view of condition (a) on constant nets, for each member s of a set A there is a net S which converges ($\mathcal{C}$) to s, and hence $A \subset A^c$. If $s \in A^c$, then because of the definition of the operator ^{c} $s \in (A \cup B)^c$ and consequently

* The first three of these, with "net" replaced by "sequence," are Kuratowski's modification of the Fréchet axioms for limit space. See Kuratowski [1].

$A^c \subset (A \cup B)^c$ for each set B. Therefore $A^c \cup B^c \subset (A \cup B)^c$. To show the opposite inclusion suppose that $\{S_n, n \in D\}$ is a net in $A \cup B$, and suppose that $\{S_n, n \in D\}$ converges ($\mathfrak{C}$) to s. If $D_A = \{n: n \in D \text{ and } S_n \in A\}$, and $D_B = \{n: n \in D \text{ and } S_n \in B\}$, then $D_A \cup D_B = D$. Hence either D_A or D_B is cofinal in D, and consequently either $\{S_n, n \in D_A\}$ or $\{S_n, n \in D_B\}$ is a subnet of $\{S_n, n \in D\}$ which also converges ($\mathfrak{C}$) to s by virtue of condition (b). Hence $s \in A^c \cup B^c$ and we have shown that $A^c \cup B^c = (A \cup B)^c$. It must now be shown that $A^{cc} = A^c$, and condition (d) is precisely what is needed. If $\{T_m, m \in D\}$ is a net in A^c which converges ($\mathfrak{C}$) to t, then for each m in D there are a directed set E_m and a net $\{S(m,n), n \in E_m\}$ which converge ($\mathfrak{C}$) to T_m. Condition (d) then exhibits a net which converges ($\mathfrak{C}$) to t and consequently $t \in A^c$. Hence $A^{cc} = A^c$.

The more delicate part of the proof, that of showing that convergence ($\mathfrak{C}$) is identical with convergence relative to the topology $\mathfrak{J}$ associated with the operator ^{c}, remains. First, suppose $\{S_n, n \in D\}$ converges ($\mathfrak{C}$) to s and S does not converge to s relative to $\mathfrak{J}$. Then there is an open neighborhood U of s such that $\{S_n, n \in D\}$ is not eventually in U. Hence there is a cofinal subset E of D such that $S_n \in X \sim U$ for n in E. Since $\{S_n, n \in E\}$ is a subnet of $\{S_n, n \in D\}$ this subnet in $X \sim U$ converges ($\mathfrak{C}$) to s by condition (b). Hence $X \sim U \neq (X \sim U)^c$, and U is not open relative to $\mathfrak{J}$, which is a contradiction.

Finally, suppose that a net P converges to a point r relative to the topology $\mathfrak{J}$ and fails to converge ($\mathfrak{C}$). Then by (c) there is a subnet $\{T_m, m \in D\}$, no subnet of which converges ($\mathfrak{C}$) to r, and a contradiction results if we construct such a subnet. For each m in D let $B_m = \{n: n \in D \text{ such that } n \geqq m\}$ and let A_m be the set of all T_n for n in B_m. Because $\{T_m, m \in D\}$ converges relative to $\mathfrak{J}$ to r, r must lie in the closure of each A_m. Consequently, for each m in D there are a directed set E_m and a net $\{U(m,n), n \in E_m\}$ in A_m, such that the composition $\{T \circ U(m,n), n \in E_m\}$ converges ($\mathfrak{C}$) to r. Condition (d) on convergence classes now applies. If $R(m,f) = (m,f(m))$ for each (m,f) in $D \times \mathsf{X}\{E_m, m \in D\}$, then $T \circ U \circ R$ converges ($\mathfrak{C}$) to r. Moreover, if $p \geqq m$, then $U \circ R(p,f) = U(p,f(p)) \in B_m$; that is, $U \circ R(p,f) \geqq m$. It follows that $T \circ U \circ R$ is a subnet of T, and the theorem follows. ▌

The preceding theorem sets up a one-to-one correspondence between the topologies for a set X and the convergence classes on it. This correspondence is order inverting in the following sense. If $\mathcal{C}_1$ and $\mathcal{C}_2$ are convergence classes and $\mathfrak{J}_1$ and $\mathfrak{J}_2$ are the associated topologies, then $\mathcal{C}_1 \subset \mathcal{C}_2$ if and only if $\mathfrak{J}_2 \subset \mathfrak{J}_1$. (This fact is immediately evident from the definition of convergence.) We also notice that the intersection $\mathcal{C}_1 \cap \mathcal{C}_2$ is a convergence class in view of the four characteristic properties of such classes. It is easy to see that the topology associated with $\mathcal{C}_1 \cap \mathcal{C}_2$ is the smallest topology which is larger than each of $\mathfrak{J}_1$ and $\mathfrak{J}_2$, and dually, the convergence class of $\mathfrak{J}_1 \cap \mathfrak{J}_2$ is the smallest convergence class which is larger than each of $\mathcal{C}_1$ and $\mathcal{C}_2$.

PROBLEMS

A EXERCISE ON SEQUENCES

Let X be a countable set with a topology consisting of the void set together with all sets whose complements are finite. What sequences converge to what points?

B EXAMPLE: SEQUENCES ARE INADEQUATE

Let Ω' be the set of ordinal numbers which are less than or equal to the first uncountable ordinal Ω, and let the topology be the order topology. Then Ω is an accumulation point of $\Omega' \sim \{\Omega\}$, but no sequence in $\Omega' \sim \{\Omega\}$ converges to Ω.

C EXERCISE ON HAUSDORFF SPACES: DOOR SPACES

A topological space is a *door space* iff every subset is either open or closed. A Hausdorff door space has at most one accumulation point, and if x is a point which is not an accumulation point, then $\{x\}$ is open. (If U is an arbitrary neighborhood of an accumulation point y, then $U \sim \{y\}$ is open.)

D EXERCISE ON SUBSEQUENCES

Let N be a sequence of non-negative integers such that no integer occurs more than a finite number of times; that is, for each m, the set $\{i: N_i = m\}$ is finite. Then if $\{S_n, n \in \omega\}$ is any sequence, $\{S_{N_i}, i \in \omega\}$ is a subsequence. If $\{S_n, n \in \omega\}$ is a sequence in a topological space, and N is an arbitrary sequence of non-negative integers, then $\{S_{N_i}, i \in \omega\}$ is either a subsequence of $\{S_n, n \in \omega\}$ or else has a cluster point.

E EXAMPLE: COFINAL SUBSETS ARE INADEQUATE

Let X be the set of all pairs of non-negative integers with the topology described as follows: For each point (m,n) other than $(0,0)$ the set $\{(m,n)\}$ is open. A set U is a neighborhood of $(0,0)$ iff for all except a finite number of integers m the set $\{n: (m,n) \notin U\}$ is finite. (Visualizing X in the Euclidean plane, a neighborhood of $(0,0)$ contains all but a finite number of the members of all but a finite number of columns.)

(a) The space X is a Hausdorff space.

(b) Each point of X is the intersection of a countable family of closed neighborhoods.

(c) The space is a Lindelöf space; that is, each open cover has a countable subcover.

(d) No sequence in $X \sim \{(0,0)\}$ converges to $(0,0)$. (If a sequence S in $X \sim \{(0,0)\}$ converges to $(0,0)$, then it is eventually in the complement of each column, and the sequence has only a finite number of values in each column.)

(e) There is a sequence S in $X \sim \{(0,0)\}$ with $(0,0)$ as a cluster point, and S restricted to any cofinal subset of the integers fails to converge.

Note This example is due to Arens [1].

F MONOTONE NETS

Let X be an order-complete chain; that is, X is linearly ordered by a relation $>$, such that each non-void subset of X which has an upper bound has a supremum. Let X have the order topology (1.I). A net $(S,\succ)$ in X is monotone increasing (decreasing) iff whenever $m \succ n$, then $S_m \geqq S_n (S_n \geqq S_m)$.

(a) Each monotone increasing net in X whose range is bounded (there is x in X such that $x \geqq S_n$ for all n) converges to the supremum of its range.

(b) If X is the set of all real numbers with the usual order or if X is the set of all ordinal numbers less than the first uncountable ordinal, then each monotone increasing (decreasing) net whose range has an upper (lower) bound converges to the supremum (infimum) of its range.

G INTEGRATION THEORY, JUNIOR GRADE

Let f be a real-valued function whose domain includes a set A, let $\mathcal{A}$ be the family of all finite subsets of A, and for each F in $\mathcal{A}$ let $S_F = \sum\{f(a): a \in F\}$. Then $\mathcal{A}$ is directed by $\supset$ and $\{S_F, F \in \mathcal{A}, \supset\}$

is a net. If this net converges f is *summable* over A and the number to which the net converges is the *unordered sum* of f over A, denoted $\sum\{f(a): a \in A\}$ or simply $\sum_A f$.

(a) If f is non-negative (non-positive), then f is summable iff there is an upper bound (lower bound) for the sums over finite subsets of A. (Use the preceding problem on monotone nets.)

(b) Let $A_+ = \{a: f(a) \geq 0\}$ and $A_- = \{a: f(a) < 0\}$. Then f is summable over A iff it is summable over both A_+ and A_-. If f is summable over A, then $\sum_A f = \sum_{A_+} f + \sum_{A_-} f$.

(c) A function f is summable over A iff $|f|$ is summable over A, where $|f|(a) = |f(a)|$.

(d) If f is summable on a set A, then f is zero outside some countable subset of A. (If f is different from zero at every point of some uncountable subset, then, for some positive integer n, $\{a: f(a) \geq 1/n\}$ is uncountable.)

(e) If f and g are summable over A and r and s are real numbers, then $rf + sg$ is summable over A and $\sum_A(rf + sg) = r\sum_A f + s\sum_A g$.

(f) If f is summable over A, and B and C are disjoint subsets of A, then f is summable over each of B and C and $\sum_{B \cup C} f = \sum_B f + \sum_C f$.

(g) If x is a sequence of real numbers, then the *ordered sum* ("sum of the series") is the limit of the sequence S_n where $S_n = \sum\{x_i: i = 0, 1, \cdots, n\}$. In other words, the ordered sum is the limit $\{S_F, F \in \mathcal{B}\}$, where $\mathcal{B}$ is the family of all sets which are of the form $\{m: m \leq n\}$ for some n. This is a subnet of the net defining the unordered sum. The sequence x is *absolutely summable* iff the sequence $|x|$, where $|x|_n = |x_n|$, has an ordered sum. The unordered sum of x over the integers exists iff the sequence is absolutely summable, and in this case, the unordered and ordered sums are equal.

(h) (*Fubinito*) Let f be a real-valued function on a cartesian product $A \times B$. Then:

(i) If f is summable over $A \times B$, then $\sum_{A \times B} f = \sum\{\sum\{f(a,b): b \in B\}: a \in A\}$. (The latter is one of the two *iterated* sums.)

(ii) If, for each member a of A, $f(a,b)$ is either non-negative for all b or non-positive for all b, if $F(a) = \sum\{f(a,b): b \in B\}$, and if F is summable over A, then f is summable over $A \times B$.

(iii) In general, both iterated sums may exist and f may fail to be summable. In fact, if both A and B are countably infinite and F and G are arbitrary real functions on A and on B respectively, then there is f on $A \times B$ such that $\sum\{f(a,b): a \in A\} = G(b)$ and $\sum\{f(a,b): b \in B\} = F(a)$ for all b in B and all a in A.

Notes The results stated in this problem are those which are needed to develop measure theory using unordered summation instead of absolutely convergent series. All the results except (d), (g) and (h,ii) can be established in a much more general situation; in chapter 7 the problem will be reexamined using the notion of completeness. The order-theoretic treatment above gives some insight into more complicated examples of integration.

Historically, unordered summation was the forerunner of Moore-Smith convergence. (Moore [1].)

H INTEGRATION THEORY, UTILITY GRADE

Let f be a bounded real-valued function on the closed interval of real numbers $[a,b]$. A *subdivision* S of $[a,b]$ is a finite family of closed intervals, covering $[a,b]$, such that any two intervals have at most one point in common. The length of an interval I is denoted $|I|$, and for a subdivision S the *mesh*, $\|S\|$, is the maximum of $|I|$ for I in S. We direct the family of subdivisions in two different ways:

(i) $S \geqq S'$ iff S is a *refinement* of S', in the sense that each member of S is a subset of a member of S'; and
(ii) $S \gg S'$ iff $\|S\| \leqq \|S'\|$.

Let $M_f(I)$ be the supremum of f on I, and let $m_f(I)$ be the infimum. The *upper* and *lower Darboux* sums corresponding to the subdivision S are defined to be $D_f(S) = \sum\{|I|M_f(I): I \in S\}$ and $d_f = \sum\{|I|m_f(I): I \in S\}$ respectively. The Riemann sums are more complicated. A *choice function* for a subdivision S is a function c on S such that $c(I) \in I$ for each I in S. The set of all pairs (S,c), such that S is a subdivision and c is a choice function for S, is ordered in two ways: $(S,c) \succ (S',c')$ iff $S \geqq S'$ and $(S,c) \succ\succ (S',c')$ iff $S \gg S'$. For a pair (S,c) the *Riemann sum* is $R_f(S,c) = \sum\{|I|f(c(I)): I \in S\}$.

The basic computation is made in terms of the ordering by refinement.

(a) The nets $(D_f,\geqq)$ and $(d_f,\geqq)$ are monotonically decreasing and increasing respectively, and hence converge.

(b) $d_f(S) \leqq R_f(S,c) \leqq D_f(S)$ for all subdivisions S and all choice functions c.

(c) For each positive number e there is a $\succ$-cofinal subset of the set of pairs (S,c) such that $R_f(S,c) + e \geqq D_f(S)$. (There is also a dual proposition.)

(d) The net $(R_f,\succ)$ converges iff $\lim (D_f,\geqq) = \lim (d_f,\geqq)$. If $(R_f,\succ)$ converges, then $\lim (R_f,\succ) = \lim (D_f,\geqq) = \lim (d_f,\geqq)$.

(e) The net $(R_f,\succ)$ is a subnet of $(R_f,\succ\succ)$.

(f) The net $(R_f,\succ\succ)$ converges iff $\lim(D_f,\geqq) = \lim(d_f,\geqq)$. If $(R_f,\succ\succ)$ converges $\lim(R_f,\succ\succ) = \lim(R_f,\succ)$.

Notes The Riemann integral of f is usually defined to be the limit of $(R_f,\succ\succ)$. The advantage of considering refinement as well as mesh is, here, essentially a matter of technique. If instead of considering finite subdivisions and length of intervals we consider countable subdivisions and let $|I|$ be the Lebesgue measure of I, the net $(R_f,\succ)$ converges to the usual Lebesgue integral of f, while $(R_f,\succ\succ)$ may not. Further, a definition of the refinement type may be used to integrate certain functions whose values lie in a vector space. (See Hille [1], chapter 3.) An integral of the Darboux type requires that the range of the function to be integrated be partially ordered. The Daniell integral and various generalizations (Bourbaki [2], McShane [2] and [3], and M. H. Stone [1]) are essentially of this sort. There is another standard way of introducing an integral, via a completion process with respect to a metric, which has many advantages (Halmos [1]).

I MAXIMAL IDEALS IN LATTICES

A *lattice* is a non-void set X with a reflexive partial ordering $\geqq$ such that for every pair x and y of members of X there is a (unique) smallest element $x \vee y$ which is greater than each of x and y and a (unique) largest element $x \wedge y$ which is smaller than each. The elements $x \vee y$ and $x \wedge y$ are respectively the *join* and the *meet* of x and y. The lattice is *distributive* iff $x \wedge (y \vee z) = (x \wedge y) \vee (x \wedge z)$ and $x \vee (y \wedge z) = (x \vee y) \wedge (x \vee z)$ for all x, y, and z in X. A subset A of X is an *ideal* (a *dual ideal*) iff whenever $y \geqq x$ and $y \in A$, then $x \in A$, and if y and z belong to A so does $y \vee z$ (respectively, whenever $x \geqq y$ and $y \in A$, then $x \in A$, and if $y \in A$ and $z \in A$, then $y \wedge z \in A$).

Let A and B be disjoint subsets of a distributive lattice X such that A is an ideal and B is a dual ideal. Then there are disjoint sets A' and B' such that A' is an ideal containing A, B' is a dual ideal containing B, and $A' \cup B' = X$.

The proof of this proposition is broken down into a sequence of lemmas.

(a) The family of all ideals which contain A and are disjoint from B contains a maximal member A'. (See 0.25.) Similarly there is a dual ideal B' which contains B, is disjoint from A', and is maximal with respect to these properties.

(b) The smallest ideal which contains A' and a member c of X is $\{x : x \leqq c \text{ or } x \leqq c \vee y \text{ for some } y \text{ in } A'\}$. Since A' is maximal, if c

does not belong to either A' or B, then $c \vee x \in B$ for some x in A'. (If $z \geq x \in B$, then $z \in B$.)

(c) If c belongs to neither A' nor B', then there is x in A' and y in B' such that $c \vee x \in B'$ and $c \wedge y \in A'$. Then $(c \vee x) \wedge y = (c \wedge y) \vee (x \wedge y)$ belongs to both A' and B'.

Notes This theorem is due to M. H. Stone [2]; it is the best form of one of the basic facts about ordered sets. It is used in the next two problems and it is the fact underlying the most important results on compactness (chapter 5). An application of some form of the maximal principle seems to be essential to its proof. It has been stated in the literature that this theorem (or, more precisely, a corollary to the theorem which occurs in problem 2. **K**) implies the axiom of choice, but I do not know whether this is the case. Finally, the definition of distributivity which is given above is unduly restrictive. Either of the two equalities implies the other (Birkhoff [1]).

J UNIVERSAL NETS

A net in a set X is said to be *universal* iff for each subset A of X the net is eventually in A or eventually in $X \sim A$.

(a) If a universal net is frequently in a set it is eventually in the set. Hence a universal net in a topological space converges to each of its cluster points.

(b) If a net is universal, then each subnet is also universal. If S is a universal net in X and f is a function on X to Y, then $f \circ S$ is a universal net in Y.

(c) *Lemma* If S is a net in X, then there is a family $\mathcal{C}$ of subsets of X such that: S is frequently in each member of $\mathcal{C}$, the intersection of two members of $\mathcal{C}$ belongs to $\mathcal{C}$, and for each subset A of X either A or $X \sim A$ belongs to $\mathcal{C}$. (Either show that there is a family $\mathcal{C}$ maximal with respect to the first two listed properties and demonstrate that it possesses the third, or apply 2.I, letting $\mathcal{A}$ be the family of all sets A such that S is eventually in $X \sim A$, $\mathcal{B}$ the family of all B such that S is eventually in B, and let the ordering be $\subset$.)

(d) There is a universal subnet of each net in X. (Use the preceding result and 2.5.)

K BOOLEAN RINGS: THERE ARE ENOUGH HOMOMORPHISMS

A *Boolean ring* is a ring $(R,+,\cdot)$ such that $r \cdot r = r$ and $r + r = 0$ for each r in R. The field of integers modulo 2 is denoted I_2.

(a) A Boolean ring is commutative. (Observe that $(r + s) \cdot (r + s) = r + s$.)

(b) If $(R,+,\cdot)$ is a Boolean ring, then multiplication of members of R by members of I_2 can be defined so that R is an algebra over I_2.

(c) The *symmetric difference* $A \Delta B$ of two sets A and B is defined to be $(A \cup B) \sim (A \cap B)$. If $\mathcal{A}$ is the family of all subsets of a set X, then $(\mathcal{A},\Delta,\cap)$ is a Boolean ring with unit.

(d) Let X be a set and let $I_2{}^X$ be the family of all functions on X to I_2. Define addition and multiplication of functions pointwise (that is, $(f+g)(x) = f(x) + g(x)$ and $(f\cdot g)(x) = f(x)\cdot g(x)$). Then $(I_2{}^X,+,\cdot)$ is a Boolean ring with unit and is isomorphic to $(\mathcal{A},\Delta,\cap)$ where $\mathcal{A}$ is the family of all subsets of X.

(e) The *natural ordering* of a Boolean ring is defined by agreeing that $r \geqq s$ iff $r\cdot s = s$. The relation $\geqq$ partially orders R in such a way that the least element which follows both r and s is $r \vee s = r + s + r\cdot s$ and the greatest element which precedes both r and s is $r \wedge s = r\cdot s$. Each of $\vee$ and $\wedge$ are associative operations and the following distributive laws hold: $r \wedge (s \vee t) = (r \wedge s) \vee (r \wedge t)$ and $r \vee (s \wedge t) = (r \vee s) \wedge (r \vee t)$.

(f) Recall that S is an ideal in a Boolean ring $(R,+,\cdot)$ iff S is an additive subgroup and $r\cdot s \in S$ whenever $r \in R$ and $s \in S$; the ideal S is maximal iff $R \neq S$ and no ideal other than R properly contains S. There is a one-to-one correspondence between maximal ideals in R and homomorphisms into I_2 which are not identically zero. (The kernel of such a homomorphism is a maximal ideal.)

(g) A necessary and sufficient condition that S be an ideal in a Boolean ring is that $r \vee s \in S$ whenever r and s are members of S and $t \in S$ whenever t precedes a member of S in the natural order (that is, $t \leqq$ some member of S). A subset T of R is called a *dual ideal* iff $r \wedge s \in T$ whenever r and s are members of T and $t \in T$ whenever t follows a member of T. If $r \in R$, then $\{s: r \geqq s\}$ is an ideal and $\{s: s \geqq r\}$ is a dual ideal. If S is an ideal, T is a disjoint dual ideal, and $S \cup T = R$, then the function which is zero on S and one on T is a homomorphism of R into I_2. (In a Boolean ring of sets an ideal is frequently called an $\cap$-ideal and a dual ideal a $\cup$-ideal.)

(h) *Theorem* If S is an ideal in a Boolean ring and T is a dual ideal which is disjoint from S, then there is a homomorphism of the ring into I_2 which is zero on S and one on T. In particular, if r is a non-zero member of the ring there is a homomorphism h of the ring such that $h(r) = 1$. (In other words, there are enough homomorphisms to distinguish members of the ring. A proof of this theorem may be based on 2.I.)

(i) If X is a topological space and $\mathcal{B}$ is the family of all subsets of X which are both open and closed, then $(\mathcal{B},\Delta,\cap)$ is a Boolean algebra.

(j) Not all Boolean algebras are isomorphic to an algebra of all subsets of a set. (Show by example that there are countable Boolean algebras.)

Note This investigation is completed in 5.S.

L FILTERS

A theory of convergence has been built on the concept of filter. A *filter* $\mathfrak{F}$ in a set X is a family of non-void subsets of X such that

(i) the intersection of two members of $\mathfrak{F}$ always belongs to $\mathfrak{F}$; and
(ii) if $A \in \mathfrak{F}$ and $A \subset B \subset X$, then $B \in \mathfrak{F}$.

In the terminology of the previous problem a filter is a proper dual ideal in the Boolean ring of all subsets of X. A filter $\mathfrak{F}$ *converges* to a point x in a topological space X iff each neighborhood of x is a member of $\mathfrak{F}$ (that is, the neighborhood system of x is a subfamily of $\mathfrak{F}$).

(a) A subset U is open iff U belongs to every filter which converges to a point of U.

(b) A point x is an accumulation point of a set A iff $A \sim \{x\}$ belongs to some filter which converges to x.

(c) Let ϕ_x be the collection of all filters which converge to a point x. Then $\bigcap \{\mathfrak{F}: \mathfrak{F} \in \phi_x\}$ is the neighborhood system of x.

(d) If $\mathfrak{F}$ is a filter converging to x and $\mathfrak{G}$ is a filter which contains $\mathfrak{F}$, then $\mathfrak{G}$ converges to x.

(e) A filter in X is an *ultrafilter* iff it is properly contained in no filter in X. If $\mathfrak{F}$ is an ultrafilter in X and the union of two sets is a member of $\mathfrak{F}$, then one of the two sets belongs to $\mathfrak{F}$. In particular, if A is a subset of X, then either A or $X \sim A$ belongs to $\mathfrak{F}$. (Problem 2.I again.)

(f) One might suspect that filters and nets lead to essentially equivalent theories. Grounds for this suspicion may be found in the following facts:

(i) If $\{x_n, n \in D\}$ is a net in X, then the family $\mathfrak{F}$ of all sets A such that $\{x_n, n \in D\}$ is eventually in A is a filter in X.
(ii) Let $\mathfrak{F}$ be a filter in X and let D be the set of all pairs (x,F) such that $x \in F$ and $F \in \mathfrak{F}$. Direct D by agreeing that $(y,G) \geqq (x,F)$ iff $G \subset F$, and let $f(x,F) = x$. Then $\mathfrak{F}$ is precisely the family of all sets A such that the net $\{f(x,F), (x,F) \in D\}$ is eventually in A.

Notes The definition of filter is due to H. Cartan; his treatment of convergence is given in full in Bourbaki [1]. Proposition (c) is a remark of W. H. Gottschalk; (f) is part of the folklore of the subject.

Chapter 3

PRODUCT AND QUOTIENT SPACES

It is the purpose of this chapter to investigate two methods of constructing new topological spaces from old. One of these involves assigning a standard sort of topology to the cartesian product of spaces, thus building a new space from those originally given. For example, the Euclidean plane is the product space of the real numbers (with the usual topology) with itself, and Euclidean n-space is the product of the real numbers n times. In chapter 4 arbitrary cartesian products of the real numbers will serve as standard spaces with which one may compare other topological spaces.

The second method of constructing a new space from a given one depends on dividing the given space X into equivalence classes, each of which is a point of the newly constructed space. Roughly speaking, we "identify" the points of certain subsets of X, so obtaining a new set of points, which is then assigned the "quotient" topology. For example, the equivalence classes of real numbers modulo the integers are assigned a topology so that the resulting space is a "copy" of the unit circle in the plane.

Both of these methods of constructing spaces are motivated by making certain functions continuous. We therefore begin by defining continuity and proving a few simple propositions about it.

CONTINUOUS FUNCTIONS

For convenience we review some of the terminology and a few elementary propositions about functions (chapter 0). The words "function," "map," "mapping," "correspondence," "operator," and "transformation" are synonymous. A function f is said to be on X iff its domain is X. It is to Y, or into Y, iff its range is a subset of Y and it is onto Y if its range is Y. The value of f at a point x is $f(x)$ and $f(x)$ is also called the image under f of x. If B is a subset of Y, then the inverse under f of B, $f^{-1}[B]$, is $\{x: f(x) \in B\}$. The inverse under f of the intersection (union) of the members of a family of subsets of Y is the intersection (union) of the inverses of the members; that is, if Z_c is a subset of Y for each member c of a set C, then $f^{-1}[\bigcap\{Z_c: c \in C\}] = \bigcap\{f^{-1}[Z_c]: c \in C\}$, and similarly for unions. If $y \in Y$, then $f^{-1}[\{y\}]$, the inverse of the set whose only member is y, is abbreviated $f^{-1}[y]$. The image $f[A]$ of a subset A of X is the set of all points y such that $y = f(x)$ for some x in A. The image of the union of a family of subsets of X is the union of the images, but, in general, the image of the intersection is not the intersection of the images. A function f is one to one iff no two distinct points have the same image, and in this case f^{-1} is the function inverse to f. (Notice that the notation is arranged so that, roughly speaking, square brackets occur in the designations of subsets of the range and domain of a function, and parentheses in the designation of members. For example, if f is one to one onto Y and $y \in Y$, then $f^{-1}(y)$ is the unique point x of X such that $f(x) = y$, and $f^{-1}[y] = \{x\}$.)

A map f of a topological space $(X,\mathfrak{J})$ into a topological space $(Y,\mathfrak{U})$ is **continuous** iff the inverse of each open set is open. More precisely, f is continuous with respect to $\mathfrak{J}$ and $\mathfrak{U}$, or $\mathfrak{J}$-$\mathfrak{U}$ continuous, iff $f^{-1}[U] \in \mathfrak{J}$ for each U in $\mathfrak{U}$. The concept depends on the topology of both the range and the domain space, but we follow the usual practice of suppressing all mention of the topologies when confusion is unlikely. There are one or two propositions about continuity which are quite important, although almost self-evident. First, if f is a continuous function on X to Y and g is a continuous function on Y to Z, then the composition $g \circ f$ is a continuous function on X to Z, for $(g \circ f)^{-1}[V] = f^{-1}[g^{-1}[V]]$ for each subset V of Z, and using first the continuity of g, then that of f, it follows that if V is open so is $(g \circ f)^{-1}[V]$. If f is a continuous function on X to Y, and A is a subset of X, then the restriction of f to A, $f|A$, is also continuous with respect to the relative topology for A, for if U is open in Y, then $(f|A)^{-1}[U] = A \cap f^{-1}[U]$, which is open in A. A function f such that $f|A$ is continuous is **continuous on** A. It may happen that f is continuous on A but fails to be continuous on X.

The following is a list of conditions, each equivalent to continuity; it is useful because it is frequently necessary to prove functions continuous.

1 THEOREM *If X and Y are topological space and f is a function on X to Y, then the following statements are equivalent.*

(a) *The function f is continuous.*
(b) *The inverse of each closed set is closed.*
(c) *The inverse of each member of a subbase for the topology for Y is open.*
(d) *For each x in X the inverse of every neighborhood of f(x) is a neighborhood of x.*
(e) *For each x in X and each neighborhood U of f(x) there is a neighborhood V of x such that* $f[V] \subset U$.
(f) *For each net S (or* $\{S_n, n \in D\}$*) in X which converges to a point s, the composition* $f \circ S$ ($\{f(S_n), n \in D\}$) *converges to* $f(s)$.
(g) *For each subset A of X the image of the closure is a subset of the closure of the image; that is,* $f[A^-] \subset f[A]^-$.
(h) *For each subset B of Y,* $f^{-1}[B]^- \subset f^{-1}[B^-]$.

PROOF (a) ↔ (b): This is a simple consequence of the fact that the inverse of a function preserves relative complements; that is, $f^{-1}[Y \sim B] = X \sim f^{-1}[B]$ for every subset B of Y.

(a) ↔ (c): If f is continuous then the inverse of a member of a subbase is open because each subbase member is open. Conversely, since each open set V in Y is the union of finite intersections of subbase members, $f^{-1}[V]$ is the union of finite intersections of the inverses of subbase members; if these are open, then the inverse of each open set is open.

(a) → (d): If f is continuous, $x \in X$, and V is a neighborhood of $f(x)$, then V contains an open neighborhood W of $f(x)$ and

$f^{-1}[W]$ is an open neighborhood of x which is a subset of $f^{-1}[V]$; consequently $f^{-1}[V]$ is a neighborhood of x.

(d) $\rightarrow$ (e): Assuming (d), if U is a neighborhood of $f(x)$, then $f^{-1}[U]$ is a neighborhood of x such that $f[f^{-1}[U]] \subset U$.

(e) $\rightarrow$ (f): Assuming (e), let S be a net in X which converges to a point s. Then if U is a neighborhood of $f(s)$ there is a neighborhood V of s such that $f[V] \subset U$, and since S is eventually in V, $f \circ S$ is eventually in U.

(f) $\rightarrow$ (g): Assuming (f), let A be a subset of X and s a point of the closure A. Then there is a net S in A which converges to s, and $f \circ S$ converges to $f(s)$, which is therefore a member of $f[A]^-$. Hence $f[A^-] \subset f[A]^-$.

(g) $\rightarrow$ (h): Assuming (g), if $A = f^{-1}[B]$, then $f[A^-] \subset f[A]^- \subset B^-$ and hence $A^- \subset f^{-1}[B^-]$. That is, $f^{-1}[B]^- \subset f^{-1}[B^-]$.

(h) $\rightarrow$ (b): Assuming (h), if B is a closed subset of Y, then $f^{-1}[B]^- \subset f^{-1}[B^-] = f^{-1}[B]$ and $f^{-1}[B]$ is therefore closed. ∎

There is also a localized form of continuity which is useful.* A function f on a topological space X to a topological space Y is **continuous at a point** x iff the inverse under f of each neighborhood of $f(x)$ is a neighborhood of x. It is easy to give characterizations of the form of 3.1(e) and 3.1(f) for continuity at a point. Evidently f is continuous iff it is continuous at each point of its domain.

A **homeomorphism,** or **topological transformation,** is a continuous one-to-one map of a topological space X onto a topological space Y such that f^{-1} is also continuous. If there exists a homeomorphism of one space onto another, the two spaces are said to be **homeomorphic** and each is a **homeomorph** of the other. The identity map of a topological space onto itself is always a homeomorphism, and the inverse of a homeomorphism is again a homeomorphism. It is also evident that the composition of two homeomorphisms is a homeomorphism. Consequently the collection of topological spaces can be divided into equivalence classes such that each topological space is homeomorphic to every member of its equivalence class and to these spaces only. Two topological spaces are **topologically equivalent** iff they are homeomorphic.

* If f is defined on a subset A of a topological space, then continuity at points of the closure A^- may also be defined (see 3.D); several useful propositions result.

Two discrete spaces, X and Y, are homeomorphic iff there is a one-to-one function on X onto Y, that is, iff X and Y have the same cardinal number. This is true because every function on a discrete space is continuous, regardless of the topology of the range space. It is also true that two indiscrete spaces (the only open sets are the space and the void set) are homeomorphic iff there is a one-to-one map of one onto the other, because each function into an indiscrete space is continuous regardless of the topology of the domain space. In general, it may be quite difficult to discover whether two topological spaces are homeomorphic. The set of all real numbers, with the usual topology, is homeomorphic to the open interval (0,1), with the relative topology, for the function whose value at a member x of (0,1) is $(2x - 1)/x(x - 1)$ is easily proved to be a homeomorphism. However, the interval (0,1) is not homeomorphic to $(0,1) \cup (1,2)$, for if f were a homeomorphism (or, in fact, just a continuous function) on (0,1) with range $(0,1) \cup (1,2)$, then $f^{-1}[(0,1)]$ would be a proper open and closed subset of (0,1), and (0,1) is connected. This little demonstration was achieved by noticing that one of the spaces is connected, the other is not, and the homeomorph of a connected space is again connected. A property which when possessed by a topological space is also possessed by each homeomorph is a **topological invariant.** The proof that two spaces are not homeomorphic usually depends on exhibiting a topological invariant which is possessed by one but not by the other. A property which is defined in terms of the members of the space and the topology turns out, automatically, to be a topological invariant. Besides connectedness, the property of having a countable base for the topology, having a countable base for the neighborhood system of each point, being a T_1 space or being a Hausdorff space, are all topological invariants. Formally, topology is the study of topological invariants.*

PRODUCT SPACES

There is a standard way of topologizing the cartesian product of a collection of topological spaces. The construction is ex-

* A *topologist* is a man who doesn't know the difference between a doughnut and a coffee cup.

tremely important and consequently we examine minutely the properties of the topology introduced. Let X and Y be topological spaces and let $\mathfrak{B}$ be the family of all cartesian products $U \times V$ where U is an open subset of X and V is an open subset of Y. The intersection of two members of $\mathfrak{B}$ is a member of $\mathfrak{B}$, because $(U \times V) \cap (R \times S) = (U \cap R) \times (V \cap S)$, and consequently $\mathfrak{B}$ is the base for a topology for $X \times Y$ by theorem 1.11. This topology is called the **product topology** for $X \times Y$. A subset W of $X \times Y$ is open relative to the product topology if and only if for each member (x,y) of W there are open neighborhoods U of x and V of y such that $U \times V \subset W$. The spaces X and Y are **coordinate spaces,** and the functions P_0 and P_1 which carry a point (x,y) of $X \times Y$ into x and into y respectively are the **projections** into the coordinate spaces. These projections are continuous functions, for if U is open in X, $P_0^{-1}[U]$ is $U \times Y$, which is open. Continuity of the projections actually serves to characterize the product topology in the following sense. Suppose $\mathfrak{J}$ is a topology for $X \times Y$ such that each of the projections is continuous. Then if U is open in X and V is open in Y the set $U \times V$ is open relative to $\mathfrak{J}$, for $U \times V = P_0^{-1}[U] \cap P_1^{-1}[V]$ and this set is open relative to $\mathfrak{J}$ because the projections are continuous. Consequently $\mathfrak{J}$ is larger than the product topology and the product topology is therefore the smallest topology for which the projections into coordinate spaces are continuous.

There is no difficulty in extending this definition of product topology to cartesian product of any finite number of coordinate spaces. If each of $X_0, X_1, \cdots X_{n-1}$ is a topological space, then a base for the product topology for $X_0 \times X_1 \times \cdots \times X_{n-1}$ is the family of all products $U_0 \times U_1 \times \cdots U_{n-1}$ where each U_i is open in X_i. In particular, if each X_i is the set of real numbers with the usual topology, then the product space is **Euclidean n-space** E_n. The members of E_n are real-valued functions on the set $0, 1, \cdots n - 1$, the value of the function x at the integer i being x_i $(= x(i))$.

The product topology for the cartesian product of an arbitrary family of topological spaces will now be defined. Suppose we are given a set X_a for each member a of an index set A. The cartesian product $\times\{X_a: a \in A\}$ is defined to be the set of all functions x on A such that $x_a \in X_a$ for each a in A. The set X_a is the a-th coordinate set and the projection P_a of the product into the a-th coordinate set is defined by $P_a(x) = x_a$. Suppose that a topology $\mathfrak{J}_a$ is given for each coordinate set. The construction * of the product topology is motivated by the requirement that each projection P_a is to be continuous. In order to attain continuity of the projections it is necessary and sufficient that each set of the form $P_a^{-1}[U]$ be open, where U is an open subset of X_a. The family of all sets of this form is a subbase for a topology; it is clearly the smallest topology such that projections are continuous. The **product topology** is this topology. The members of the defining subbase are of the form $\{x: x_a \in U\}$ where U is open in X_a; they are, intuitively, cylinders over open sets in the coordinate spaces. It is sometimes said that elements of the subbase consist of sets obtained by "restricting the a-th coordinate to lie in an open subset of the a-th coordinate space." A base for the product topology is the family of all finite intersections of subbase elements. A member U of this base is of the form $\bigcap\{P_a^{-1}[U_a]: a \in F\} = \{x: x_a \in U_a$ *for each a in* $F\}$ where F is a finite subset of A and U_a is open in X_a for each a in F. It is to be emphasized that these are *finite* intersections. It is not true that $\times\{U_a: a \in A\}$ is always open relative to the product topology if U_a is open in X_a for each a. The **product space** is the cartesian product with the product topology.

The projections of a product space into the coordinate spaces have another very useful property. A function f on a topological space X to another space Y is **open (interior)** iff the image of each open set is open; that is, if U is open in X, then $f[U]$ is open in Y.

2 THEOREM *The projection of a product space into each of its coordinate spaces is open.*

PROOF Let P_c be the projection of $\times\{X_a: a \in A\}$ into X_c. In order to show that P_c is open it is sufficient to show that the image of a neighborhood of a point x in the product is a neighborhood of $P_c(x)$, and it may be assumed that the neighborhood

* This description of the product topology is due to N. Bourbaki.

in the product space is a member of the defining base for the product topology. Suppose that $x \in V = \{y: y_a \in U_a \textit{ for } a \textit{ in } F\}$, where F is a finite subset of A and U_a is open in X_a for each a in F. We construct a copy of X_c which contains the point x. For $z \in X_c$ let $f(z)_c = z$, and for $a \neq c$ let $f(z)_a = x_a$. Then $P_c \circ f(z) = z$. If $c \notin F$, then clearly $f[X_c] \subset V$ and $P_c[V] = X_c$ which is open. If $c \in F$, then $f(z) \in V$ iff $z \in U_c$ and $P_c[V] = U_c$. The theorem follows. (As a matter of fact, the function f defined in this proof is a homeomorphism, a fact that is occasionally useful.) ▮

It might be conjectured that the projection of a closed set in a product space is closed. This, however, is easily seen to be false, for in the Euclidean plane the set $\{(x,y): xy = 1\}$ has a non-closed projection on each coordinate space.

There is an extremely useful characterization of continuity of a function whose range is a subset of a product space.

3 THEOREM *A function f on a topological space to a product $\times\{X_a: a \in A\}$ is continuous if and only if the composition $P_a \circ f$ is continuous for each projection P_a.*

PROOF If f is continuous, then $P_a \circ f$ is always continuous because P_a is continuous. If $P_a \circ f$ is continuous for each a, then for each open subset U of X_a the set $(P_a \circ f)^{-1}[U] = f^{-1}[P_a^{-1}[U]]$ is open. It follows that the inverse under f of each member of the defining subbase for the product topology is open, and hence (3.1c) f is continuous. ▮

Convergence in a product space can be described very simply in terms of the projections.

4 THEOREM *A net S in a product space converges to a point s if and only if its projection in each coordinate space converges to the projection of s.*

PROOF Since the projection into each coordinate space is continuous, if $\{S_n, n \in D\}$ is a net in the cartesian product $\times\{X_a: a \in A\}$ which converges to a point s, then the net $\{P_a(S_n), n \in D\}$ surely converges to $P_a(s)$. To show the converse, let $\{S_n, n \in D\}$ be a net such that $\{P_a(S_n), n \in D\}$ converges to s_a for each a in A. Then for each open neighborhood U_a of s_a, $\{P_a(S_n), n \in D\}$

is eventually in U_a, consequently $\{S_n, n \in D\}$ is eventually in $P_a^{-1}[U_a]$, and hence $\{S_n, n \in D\}$ must eventually be in each finite intersection of sets of the form $P_a^{-1}[U_a]$. Since the family of such finite intersections is a base for the neighborhood system of s in the product topology, $\{S_n, n \in D\}$ converges to s. ∎

Convergence in the product topology is called **coordinatewise convergence,** or **pointwise convergence.** The latter term is used most frequently in the case in which all coordinate spaces are identical. In this important special case the cartesian product $\times\{X: a \in A\}$ is simply the set of all functions on A to X, usually denoted X^A. A net $\{F_n, n \in D\}$ in X^A converges to f in the topology of pointwise convergence iff the net $\{F_n(a), n \in D\}$ converges to $f(a)$ for each a in A. This fact makes the terminology, "pointwise convergence," seem reasonable. The product topology is also called the **simple** topology in this case.

It is natural to ask whether the product of topological spaces inherits properties which are possessed by the coordinate spaces. For example, we might ask, in case each coordinate space is a Hausdorff space or satisfies the first or second axiom of countability, whether the product space also has these properties. The following theorems answer these questions.

5 THEOREM *The product of Hausdorff spaces is a Hausdorff space.*

PROOF If x and y are distinct members of the product $\times\{X_a: a \in A\}$, then $x_a \neq y_a$ for some a in A. If each coordinate space is Hausdorff, then there are disjoint open neighborhoods U and V of x_a and y_a respectively and $P_a^{-1}[U]$ and $P_a^{-1}[V]$ are disjoint neighborhoods of x and y in the product. ∎

Recall that an indiscrete topological space is one in which the only open sets are the void set and the space.

6 THEOREM *Let X_a be a topological space satisfying the first axiom of countability for each member a of an index set A. Then the product $\times\{X_a: a \in A\}$ satisfies the first axiom of countability if and only if all but a countable number of the spaces X_a are indiscrete.*

PROOF Suppose that B is a countable subset of A, that X_a is indiscrete for a in $A \sim B$, and that x is a point in the product space. For each a in A choose a countable base $\mathfrak{U}_a$ for the neighborhood system of x_a in X_a. Then $\mathfrak{U}_a = \{X_a\}$ if a is in $A \sim B$. Consider the family of all finite intersections of sets of the form $P_a^{-1}[U]$ for a in A and U in $\mathfrak{U}_a$. This is a countable family because $P_a^{-1}[U] = \times\{X_b: b \in A\}$ if $a \in A \sim B$. But the family of these finite intersections is a base for the neighborhood system of x and consequently the product space satisfies the first axiom of countability.

To prove the converse suppose that B is an uncountable subset of A such that for each a in B there is a neighborhood of x_a in X_a which is a proper subset of X_a, and suppose that there is a countable base $\mathfrak{U}$ for the neighborhood system of x. Each member U of $\mathfrak{U}$ contains a member of the defining base for the product topology, and consequently, except for a finite number of members a of A, $P_a[U] = X_a$. Since B is uncountable, there is a member a of B such that $P_a[U] = X_a$ for every U in $\mathfrak{U}$. But there is an open neighborhood V of x_a which is a proper subset of X_a, and clearly no member of $\mathfrak{U}$ is a subset of $P_a^{-1}[V]$ since each member of $\mathfrak{U}$ projects onto X_a. This is a contradiction. ∎

It is also true that the coordinate spaces inherit certain properties of a product space. If a product space is Hausdorff, so is each coordinate space, and if the product space has a countable local base at each point, then so does each coordinate space. These propositions are easy to establish, and the proofs are omitted.

7 *Notes* Tychonoff defined the product topology and proved the most important properties in two classic papers (Tychonoff [1] and [2]). His results are now among the standard tools of general topology. (See also chapter 5.) Prior to Tychonoff's work a great deal of investigation had been done on the convergence of sequences of functions relative to the topology of pointwise convergence. Many difficulties occur in this work because the topology cannot be completely described by sequential convergence, at least in the most interesting cases. (See problem 3.W.)

QUOTIENT SPACES

We begin by reviewing briefly the considerations which led to the definition of the product topology. If f is a function on a set X with values in a topological space Y, then it is always possible to assign a topology to X such that f is a continuous function. One obvious and uninteresting topology which has this property is the discrete topology; a more interesting topology is the family $\mathfrak{I}$ of all sets of the form $f^{-1}[U]$ for U open in Y. This family is evidently a topology because the inverse of a function preserves unions. Each topology, relative to which f is continuous, contains $\mathfrak{I}$ and consequently $\mathfrak{I}$ is the smallest topology for which f is continuous. If we are given a family of functions, a function f_a for each member a of an index set A, then the topology, a subbase for which is the family of all sets of the form $f_a^{-1}[U]$ for a in A and U open in the range of f_a, has precisely the same properties. This is the method by which the product topology was defined.

It is the purpose of this section to investigate the reciprocal situation. If f is a function on a topological space X with range Y, how may Y be topologized so that f is continuous? If a subset U of Y is open in a topology relative to which f is continuous, then $f^{-1}[U]$ is open in X. On the other hand, the family $\mathfrak{U}$ of all subsets U such that $f^{-1}[U]$ is open in X is a topology for Y because the inverse of an intersection (or union) of members of the family is the intersection (union) of the inverses. The topology $\mathfrak{U}$ is therefore the largest topology for Y such that the function f is continuous; it is called the **quotient topology** for Y (the quotient topology relative to f and the topology of X). A subset B of Y is closed relative to the quotient topology iff $f^{-1}[Y \sim B] = X \sim f^{-1}[B]$ is open in X. Hence B is closed iff $f^{-1}[B]$ is closed.

Without some severe limitation on f very little can be said about the quotient topology. Consequently we consider only functions belonging to one of two dual categories. Recall that f, a function on a topological space with values in another space, is open iff the image of each open set is open. A function f is said to be **closed** iff the image of each closed set is closed. It has already been observed that projection of the Euclidean plane onto its first coordinate space is an open map which is not closed, and subspaces of the plane give examples of maps which are closed but not open, and maps which are neither open nor closed. The subspace $X = \{(x,y): x = 0 \text{ or } y = 0\}$, consisting of the two axes, is mapped onto the reals by the projection $P(x,y) = x$. The image of a small neighborhood of (0,1) is mapped into the single point 0. Consequently P is not an open map on X, but it is easy to verify that it is closed. If (0,0) is removed, leaving $X \sim \{(0,0)\}$, then P on this subspace is neither open nor closed (the image of the closed set $\{(x,y): y = 0 \text{ and } x \neq 0\}$ is not closed).

It is apparent from the definition that the notion of an open or a closed map depends on the topology of the range space. However, if a map f is continuous and either open or closed, then the topology of the range is entirely determined by the map f and the topology of the domain.

8 THEOREM *If f is a continuous map of the topological space $(X,\mathfrak{J})$ onto the space $(Y,\mathfrak{U})$ such that f is either open or closed, then $\mathfrak{U}$ is the quotient topology.*

PROOF If f is an open map and U is a subset of Y such that $f^{-1}[U]$ is open, then $U = f[f^{-1}[U]]$ is open relative to $\mathfrak{U}$. Consequently, if f is open, each set open relative to the quotient topology is open relative to $\mathfrak{U}$, and the quotient topology is smaller than $\mathfrak{U}$. If f is continuous as well as open, then since the quotient topology is the largest for which f is continuous, $\mathfrak{U}$ is the quotient topology. To prove the theorem for a closed function f it is only necessary to replace "open" by "closed" in each of the preceding statements. ▮

If f is a function on a topological space to a product space, then f is continuous iff the composition of f with each projection is continuous. There is an analogue of this proposition for quotient spaces.

9 THEOREM *Let f be a continuous map of a space X onto a space Y and let Y have the quotient topology. Then a function g on Y to a topological space Z is continuous if and only if the composition $g \circ f$ is continuous.*

PROOF If U is open in Z and $g \circ f$ is continuous, then $(g \circ f)^{-1}[U] = f^{-1}[g^{-1}[U]]$, which is open in X, and $g^{-1}[U]$ is therefore open in Y by the definition of the quotient topology. The converse is clear. ∎

It is almost evident that the quotient topology and the properties of open or closed maps have little to do with the range space. In fact, if f is a continuous map of a topological space X onto a space Y with the quotient topology, then a topological copy of Y may be reconstructed from X, its topology, and the family of all sets of the form $f^{-1}[y]$ for y in Y. The construction goes as follows. Let $\mathfrak{D}$ be the family of all subsets of X of the form $f^{-1}[y]$ for y in Y, and let P be the function on X to $\mathfrak{D}$ whose value at x is $f^{-1}[f(x)]$. For each member y of Y let $g(y) = f^{-1}[y]$, so that g is a one-to-one map of Y onto $\mathfrak{D}$. Then $g \circ f = P$, and $f = g^{-1} \circ P$. If $\mathfrak{D}$ is assigned the quotient topology (relative to P) the preceding theorem 3.9 shows the continuity of g (since $g \circ f = P$) and the continuity of g^{-1} (since $g^{-1} \circ P = f$). Consequently g is a homeomorphism.

The preceding remarks show that the range space is essentially extraneous to the discussion, and the remaining theorems of the section will be formulated so as to display this fact. As a preliminary we consider briefly the families of subsets of a fixed set X. A **decomposition (partition)** of X is a disjoint family $\mathfrak{D}$ of subsets of X whose union is X. The **projection (quotient map)** of X onto the decomposition $\mathfrak{D}$ is the function P whose value at x is the unique member of $\mathfrak{D}$ to which x belongs. There is an equivalent way of describing a decomposition. Given $\mathfrak{D}$, define a relation R on X by agreeing that a point x is R related to a point y iff x and y belong to the same member of the decomposition. Formally, the relation R **of the decomposition** $\mathfrak{D}$ is the subset of $X \times X$ consisting of all pairs (x,y) such that x and y belong to the same member of $\mathfrak{D}$, or, briefly, $R = \bigcup\{D \times D: D \in \mathfrak{D}\}$. If P is the projection of X onto $\mathfrak{D}$, then $R = \{(x,y): P(x) = P(y)\}$. The relation R is an equivalence relation: that is, it is reflexive, symmetric, and transitive (see chapter 0). Reciprocally, each equivalence relation on X defines a family of subsets (the equivalence classes) which is a decomposition of X. If R is an equivalence relation on X, then $\boldsymbol{X/R}$ is defined to be the family of equivalence classes. If A is a subset of X, then $R[A]$ is the set of all points which are R relatives of points of A; that is, $R[A] = \{y: (x,y) \in R \textit{ for some } x \textit{ in } A\}$. Equivalently, $R[A] = \bigcup\{D: D \in X/R \textit{ and } D \cap A \textit{ is non-void}\}$. If x is a point of X, then we abbreviate $R[\{x\}]$ as $R[x]$. The set $R[x]$ is the equivalence class to which x belongs, and if P is the projection of X onto the decomposition, then $P(x) = R[x]$.

We assume for the rest of the section that X is a fixed topological space, R is an equivalence relation on X, and that P is the projection of X onto the family X/R of equivalence classes. The **quotient space** is the family X/R with the quotient topology (relative to P). If $\mathcal{A} \subset X/R$, then $P^{-1}[\mathcal{A}] = \bigcup\{A: A \in \mathcal{A}\}$ and hence $\mathcal{A}$ is open (closed) relative to the quotient topology iff $\bigcup\{A: A \in \mathcal{A}\}$ is open (respectively closed) in X.

10 Theorem *Let P be the projection of the topological space X onto the quotient space X/R. Then the following statements are equivalent.*

(a) *P is an open mapping.*
(b) *If A is an open subset of X, then $R[A]$ is open.*
(c) *If A is a closed subset of X, then the union of all members of X/R which are subsets of A is closed.*

If "open" and "closed" are interchanged in (a), (b), *and* (c) *the resulting statements are equivalent.*

PROOF It is first shown that (a) is equivalent to (b). For each subset A of X, the set $R[A] = P^{-1}[P[A]]$. If P is open and A is open, then, since P is continuous, $P^{-1}[P[A]]$ is open. If $P^{-1}[P[A]]$ is open for each open set A, then, since by the definition of the quotient topology $P[A]$ is open, P is an open mapping. To prove (b) equivalent to (c) notice that the union of all members of X/R which are subsets of A is $X \sim R[X \sim A]$, and this set is closed for each closed A iff $R[X \sim A]$ is open whenever $X \sim A$ is open. A proof of the dual proposition is obtained by interchanging "open" and "closed" throughout. ▌

If X is a Hausdorff space or satisfies one of the axioms of countability it is natural to ask whether the quotient space X/R necessarily inherits these properties. Without some drastic restriction the answer is "no." For example, if X is the set of real numbers with the usual topology and R is the set of pairs (x,y) such that $x - y$ is rational, then the quotient space X/R is indiscrete, and the projection P of X onto X/R is open. Consequently an open map may carry a Hausdorff space into a non-Hausdorff space. An example of a closed map which carries a Hausdorff space onto a non-Hausdorff space or carries a space satisfying the first axiom of countability onto a space which fails to satisfy the axiom, is slightly more complex but not difficult. (3.**R**, 4.**G**.) There is an additional hypothesis which is sometimes useful. It is sometimes assumed that R, which is a set of ordered pairs, is closed in the product space $X \times X$. This condition may be restated: if x and y are members of X which are not R related, then there is a neighborhood W of (x,y) in the product space $X \times X$ which is disjoint from R. Such a neighborhood W contains a neighborhood of the form $U \times V$, where U and V are neighborhoods of x and y respectively, and $U \times V$ is disjoint from R iff there is no point of U which is R related to a point of V. That is, R is closed in $X \times X$ iff, whenever x and y are points of X which are not R related, then there are neighborhoods U and V of x and y respectively such that no point of U is R related to a point of V. Equivalently, no member of X/R intersects both U and V.

11 THEOREM *If the quotient space X/R is Hausdorff, then R is closed in the product space $X \times X$.*

If the projection P of a space X onto the quotient space X/R is open, and R is closed in $X \times X$, then X/R is a Hausdorff space.

PROOF If X/R is a Hausdorff space and $(x,y) \notin R$, then $P(x) \neq P(y)$ and there are disjoint open neighborhoods U of $P(x)$ and V of $P(y)$. The sets $P^{-1}[U]$ and $P^{-1}[V]$ are open, and since their images under P are disjoint, no point of $P^{-1}[U]$ is R related to a point of $P^{-1}[V]$. Therefore $P^{-1}[U] \times P^{-1}[V]$ is a neighborhood of (x,y) which is disjoint from R, and R is closed. The first statement of the theorem is proved. Suppose now that P is open, R is closed in $X \times X$, and $P(x)$ and $P(y)$ are distinct members of X/R. Then x is not R related to y and, since R is closed, there are open neighborhoods U and V of x and y respectively such that no point of U is R related to a point of V. Hence the images of U and V are disjoint, and since P is open they are open neighborhoods of $P(x)$ and $P(y)$ respectively. ∎

Closed maps have been studied rather extensively under a different name. A decomposition $\mathfrak{D}$ of a topological space X is **upper semi-continuous** iff for each D in $\mathfrak{D}$ and each open set U containing D there is an open set V such that $D \subset V \subset U$ and V is the union of members of $\mathfrak{D}$. (See problem 3.F for the origin of the term "upper semi-continuous.")

12 THEOREM *A decomposition $\mathfrak{D}$ of a topological space X is upper semi-continuous if and only if the projection P of X onto $\mathfrak{D}$ is closed.*

PROOF According to theorem 3.10, P is a closed map iff for each open subset U of X it is true that the union V of all members of $\mathfrak{D}$ which are subsets of U is an open set. If P is closed, $D \in \mathfrak{D}$ and $D \subset U$, then V is the required open set and hence $\mathfrak{D}$ is upper semi-continuous. To prove the converse suppose that $\mathfrak{D}$ is upper semi-continuous and that U is an open subset of X. Let V be the union of all members of $\mathfrak{D}$ which are subsets of U. If $x \in V$, then $x \in D \subset U$ for some D in $\mathfrak{D}$. Hence by upper semi-continuity there is an open set W, the union of members of $\mathfrak{D}$, such that $D \subset W \subset U$. Then W is a subset of V and consequently V is a neighborhood of x. The set V is open because it is a neighborhood of each of its points, and it follows from 3.10 that P is a closed map. ∎

If A and B are disjoint closed subsets of X one may define the decomposition $\mathfrak{D}$ of X whose members are A, B, and all sets $\{x\}$ for x in $X \sim (A \cup B)$. The quotient space of this decomposition is sometimes called "the space obtained by identifying all points of A and identifying all points of B." It is very easy to verify that $\mathfrak{D}$ is upper semi-continuous, and if X is Hausdorff the relation $R = \bigcup\{D \times D: D \in \mathfrak{D}\}$ is closed in $X \times X$. One might suppose that with this simple construction the quotient space might inherit pleasant properties of the space X. Unfortunately, even in this case X may be Hausdorff or satisfy the first or second countability axiom and the corresponding proposition for the quotient space be false.

13 *Note* The notion of upper semi-continuous collection was introduced by R. L. Moore in the late twenties; open mappings were first studied intensively by Aronszajn a little later (Aronszajn [2]). Many of the results of the preceding section will be found in Whyburn [2].

PROBLEMS

A CONNECTED SPACES

The image under a continuous map of a connected space is connected.

B THEOREM ON CONTINUITY

Let A and B be subsets of a topological space X such that $X = A \cup B$, and $A \sim B$ and $B \sim A$ are separated. If f is a function on X which is continuous on A and continuous on B, then f is continuous on X. (See 1.19.)

C EXERCISE ON CONTINUOUS FUNCTIONS

If f and g are continuous functions on a topological space X with values in a Hausdorff space Y, then the set of all points x in X such that $f(x) = g(x)$ is closed. Consequently, if f and g agree on a dense subset of X ($f(x) = g(x)$ for x belonging to a dense subset of X), then $f = g$.

D CONTINUITY AT A POINT; CONTINUOUS EXTENSION

Let f be defined on a subset X_0 of a topological space X with values in a Hausdorff space Y; then f *is continuous at* x iff x belongs to the closure of X_0 and for some member y of the range the inverse of each neighborhood of y is the intersection of X_0 and a neighborhood of x.

(a) A function f is continuous at x iff $x \in \overline{X_0}$ and whenever S and T are nets in X_0 converging to x then $f \circ S$ and $f \circ T$ converge to the same point of Y.

(b) Let C be the set of points at which f is continuous and let f' be the function on C whose value at a point x is the member y of the range space which is given by the definition of continuity at a point (more precisely, the graph of f' is the intersection of $C \times Y$ with the closure of the graph of f). The function f' has the property: If U is open in X, then $f'[U] \subset f[U]^-$. The function f' is continuous, provided Y has the property: The family of closed neighborhoods of each point of Y is a base for the neighborhood system of the point. (Such topological spaces are called *regular*. The requirement that Y be regular is essential here, as shown by Bourbaki and Dieudonné [1].)

E EXERCISE ON REAL-VALUED CONTINUOUS FUNCTIONS

Let f and g be real-valued functions on a topological space, let f and g be continuous with respect to the usual topology for the real numbers, and let a be a fixed real number.

(a) The function af, whose value at x is $af(x)$, is continuous. (Show that the function which carries the real number r into ar is continuous, and use the fact that the composition of continuous functions is continuous.)

(b) The function $|f|$, whose value at x is $|f(x)|$, is continuous.

(c) If $F(x) = (f(x), g(x))$, then F is continuous relative to the usual topology for the Euclidean plane. (Verify that F followed by projection into a coordinate space is continuous.)

(d) The functions $f + g$, $f - g$, and $f \cdot g$ are continuous, and if g is never zero, then f/g is continuous. (First show that $+$, $-$, and $\cdot$ are continuous functions on the Euclidean plane to the space of real numbers. (See also 3.S.))

(e) The functions $\max [f,g] = [(f + g) + |f - g|]/2$ and $\min [f,g] = [(f + g) - |f - g|]/2$ are continuous.

F UPPER SEMI-CONTINUOUS FUNCTIONS

A real-valued function f on a topological space X is *upper semi-continuous* iff the set $\{x: f(x) \geqq a\}$ is closed for each real number a. The *upper* topology $\mathcal{U}$ for the set R of real numbers consists of the void set, R, and all sets of the form $\{t: t < a\}$ for a in R. If $\{S_n, n \in D\}$ is a net of real numbers, then $\limsup \{S_n: n \in D\}$ is defined to be $\lim \{\sup \{S_m: m \in D \text{ and } m \geqq n\}: n \in D\}$ where this limit is taken relative to the usual topology for the real numbers.

(a) A net $\{S_n, n \in D\}$ of real numbers converges to s relative to $\mathcal{U}$ iff $\limsup \{S_n: n \in D\} \geqq s$.

(b) If f is a real-valued function on X, then f is upper semi-continuous iff f is continuous relative to the upper topology $\mathcal{U}$, and this is the case iff $\limsup \{f(x_n): n \in D\} \leqq f(x)$ whenever $\{x_n, n \in D\}$ is a net in X converging to a point x.

(c) If f and g are upper semi-continuous and t is a non-negative real number, then $f + g$ and tf are upper semi-continuous.

(d) If F is a family of upper semi-continuous functions such that $i(x) = \inf \{f(x): f \in F\}$ exists for each x in X, then i is upper semi-continuous. (Observe that $\{x: i(x) \geqq a\} = \bigcap \{\{x: f(x) \geqq a\}: f \in F\}$.)

(e) If f is a bounded real-valued function on X, then there is a smallest upper semi-continuous function f^- such that $f^- \geqq f$. If $\mathfrak{U}$ is the family of neighborhoods of a point x and $S_V = \sup \{f(y): y \in V\}$, then $f^-(x) = \lim \{S_V, V \in \mathfrak{U}, \subset\}$.

(f) A real-valued function g is called *lower semi-continuous* iff $-g$ is upper semi-continuous. If f is a bounded real-valued function, let $f_- = -(-f)^-$ and let the *oscillation* of f, Q_f, be defined by $Q_f(x) = f^-(x) - f_-(x)$ for x in X. Then Q_f is upper semi-continuous, and f is continuous iff $Q_f(x) = 0$ for all x in X.

(g) Let f be a non-negative real valued function on X, let R have the usual topology, and let $G = \{(x,t): 0 \leqq t \leqq f(x)\}$ have the relativized product topology of $X \times R$. Let $\mathfrak{D}$ be the decomposition of G into "vertical slices"; that is, sets of the form $(\{x\} \times R) \cap G$. If the decomposition $\mathfrak{D}$ is upper semi-continuous, then f is upper semi-continuous. (The converse is also true but the simplest proof requires theorem 5.12.)

G EXERCISE ON TOPOLOGICAL EQUIVALENCE

(a) Any two open intervals of real numbers, with the relativized usual topology for the reals, are homeomorphic.

(b) Any two closed intervals are homeomorphic, and any two half-open intervals are homeomorphic.

(c) No open interval is homeomorphic to a closed or half-open interval, and no closed interval is homeomorphic to a half-open interval.

(d) The subspace $\{(x,y): x^2 + y^2 = 1\}$ of the Euclidean plane is not homeomorphic to a subspace of the space of real numbers.

(Certain of the foregoing spaces have one or more points x such that the complement of $\{x\}$ is connected.)

H HOMEOMORPHISMS AND ONE-TO-ONE CONTINUOUS MAPS

Given two topological spaces X and Y, a one-to-one continuous map of Y onto X and a one-to-one continuous map of X onto Y: then X and Y are not necessarily homeomorphic. (Let the space X consist of a countable number of disjoint half-open intervals and a countable number of *isolated* points (points x such that $\{x\}$ is open). Let Y consist of a countable number of open intervals and a countable number of isolated points. Observe that a countable number of half-open intervals can be mapped in a one-to-one continuous way onto an open interval. I believe this example is due to R. H. Fox.)

I CONTINUITY IN EACH OF TWO VARIABLES

Let X and Y be topological spaces, $X \times Y$ the product space, and let f be a function on $X \times Y$ to another topological space. Then $f(x,y)$

is **continuous in** x iff for each y the function $f(\ ,y)$ whose value at x is $f(x,y)$, is continuous. Similarly, $f(x,y)$ is continuous in y iff for each $x \in X$, the function $f(x,\ )$ such that $f(x,\ )(y) = f(x,y)$, is continuous. If f is continuous on the product space, then $f(x,y)$ is continuous in x and in y, but the converse is false. (The classical example is the real-valued function f on the Euclidean plane such that $f(x,y) = xy/(x^2 + y^2)$ and $f(0,0) = 0$.)

J EXERCISE ON EUCLIDEAN n-SPACE

A subset A of Euclidean n-space E_n is *convex* iff for every pair x and y of points of A and every real number t, such that $0 \leq t \leq 1$, the point $tx + (1 - t)y$ is a member of A. (We define $(tx + (1 - t)y)_i = tx_i + (1 - t)y_i$.) Then any two non-void open convex subsets of E_n are homeomorphic. What of closed convex subsets?

K EXERCISE ON CLOSURE, INTERIOR AND BOUNDARY IN PRODUCTS

Let X and Y be topological spaces and let $X \times Y$ be the product space. For each set C let C^b be the boundary of C. Then, if A and B are subsets of X and Y respectively,

(a) $(A \times B)^- = A^- \times B^-$,

(b) $(A \times B)^0 = A^0 \times B^0$, and

(c) $(A \times B)^b = (A \times B)^- \sim (A \times B)^0 = ((A^b \cup A^0) \times (B^b \cup B^0)) \sim (A^0 \times B^0) = (A^b \times B^b) \cup (A^b \times B^0) \cup (A^0 \times B^b) = (A^b \times B^-) \cup (A^- \times B^b)$.

L EXERCISE ON PRODUCT SPACES

Suppose that, for each member a of an index set A, X_a is a topological space. Let B and C be disjoint subsets of A such that $A = B \cup C$. Then the product space $\times\{X_b: b \in B\} \times \times\{X_c: c \in C\}$ is homeomorphic to the product space $\times\{X_a: a \in A\}$. For each fixed topological space X the product X^A is homeomorphic to $X^B \times X^C$ and $(X^B)^C$ is homeomorphic to $X^{B \times C}$, all spaces being given the product topology.

M PRODUCT OF SPACES WITH COUNTABLE BASES

The product topology has a countable base iff the topology of each coordinate space has a countable base and all but a countable number of the coordinate spaces are indiscrete.

N EXAMPLE ON PRODUCTS AND SEPARABILITY

Let Q be the closed unit interval and let X be the product space Q^Q. Let A be the subset of X consisting of characteristic functions of points;

more precisely, $x \in A$ iff for some q in Q, $x(q) = 1$ and x is zero on $Q \sim \{q\}$.

(a) The space X is separable. (The set of all x in X with finite range (sometimes called step functions) are dense in X. There is also a countable subset of this set which is dense in X.)

(b) The set A, with the relative topology, is discrete and not separable.

(c) There is a single accumulation point x of A in X, and if U is a neighborhood of x, then $A \sim U$ is finite.

O PRODUCT OF CONNECTED SPACES

The product of an arbitrary family of connected topological spaces is connected. (Fix a point x in the product, and let A be the set of all points y such that there is a connected subset to which both x and y belong. Show that A is dense.)

P EXERCISE ON T_1-SPACES

The product of T_1-spaces is a T_1-space. If $\mathfrak{D}$ is a decomposition of a topological space, then the quotient space is T_1 iff the members of $\mathfrak{D}$ are closed.

Q EXERCISE ON QUOTIENT SPACES

The projection of a topological space X into the quotient space X/R is a closed map iff, for each subset A of X, $R[A]^- \subset R[A^-]$. The projection is open iff $R[A^0] \subset R[A]^0$ for each subset A. ($^-$ and ^{0} are the closure and interior respectively.)

R EXAMPLE ON QUOTIENT SPACES AND DIAGONAL SEQUENCES

Let X be the Euclidean plane with the usual topology, let A be the set of all points (x,y) with $y = 0$, and let the decomposition $\mathfrak{D}$ consist of A and all sets $\{(x,y)\}$ with $(x,y) \notin A$. Then $\mathfrak{D}$, with the quotient topology, has the following properties.

(a) The projection of X onto the quotient space is closed.

(b) There is a countable number of neighborhoods of A whose intersection is $\{A\}$.

(c) For each non-negative integer m the sequence $\{(m,1/(n+1)), n \in \omega\}$ converges, in the quotient space, to A. If $\{N_n, n \in \omega\}$ is a subsequence of the sequence of non-negative integers, then the sequence $\{(n,1/(N_n+1)), n \in \omega\}$ does not converge to A. (The latter might be called a diagonal of the original family of sequences.)

(d) The quotient space does not satisfy the first axiom of countability.

Note This example is due to R. S. Novosad.

S TOPOLOGICAL GROUPS

A triple $(G,\cdot,\mathfrak{J})$ is a *topological group* iff $(G,\cdot)$ is a group, $(G,\mathfrak{J})$ is a topological space, and the function whose value at a member (x,y) of $G \times G$ is $x\cdot y^{-1}$ is continuous relative to the product topology for $G \times G$. When confusion is unlikely all mention of the group operation $\cdot$ and the topology $\mathfrak{J}$ is suppressed, and we say "G is a topological group." If X and Y are subsets of G, then $X\cdot Y$ is the set of all points z of G such that $z = x\cdot y$ for some x in X and some y in Y. If x is a point of G we abbreviate $\{x\}\cdot Y$ and $Y\cdot\{x\}$ to $x\cdot Y$ and $Y\cdot x$ respectively, and Y^{-1} is defined to be $\{x: x^{-1} \in Y\}$.

(a) If X, Y, and Z are subsets of G, then $(X\cdot Y)\cdot Z = X\cdot(Y\cdot Z)$ and $(X\cdot Y)^{-1} = Y^{-1}\cdot X^{-1}$.

(b) Let $(G,\cdot)$ be a group and $\mathfrak{J}$ a topology for G. Then $(G,\cdot,\mathfrak{J})$ is a topological group iff for each x and y in G and each neighborhood W of $x\cdot y^{-1}$ there are neighborhoods U of x and V of y such that $U\cdot V^{-1} \subset W$. Equivalently, $(G,\cdot,\mathfrak{J})$ is a topological group iff i and m are continuous, where $i(x) = x^{-1}$ and $m(x,y) = x\cdot y$.

(c) If G is a topological group, then i, where $i(x) = x^{-1}$, is a homeomorphism of G onto G. For each a in G both L_a and R_a (called the *left* and *right translations* by a) are homeomorphisms, where $L_a(x) = a\cdot x$ and $R_a(x) = x\cdot a$.

It is very important to notice that the topology of a topological group is determined by the neighborhood system of a member of the group. This fact (precisely stated below) permits the "localization" of many notions.

(d) If G is a topological group and $\mathfrak{U}$ is the neighborhood system of the identity, then a subset A of G is open iff $x^{-1}\cdot A \in \mathfrak{U}$ for each x in A or equivalently if $A\cdot x^{-1} \in \mathfrak{U}$ for each x in A. The closure of the subset A is $\bigcap \{U\cdot A: U \in \mathfrak{U}\} = \bigcap\{A\cdot U: U \in \mathfrak{U}\}$. (Notice that $x \in U\cdot A$ iff $(U^{-1}\cdot x) \cap A$ is not void.)

(e) The family $\mathfrak{U}$ of neighborhoods of the identity e of a topological group has the properties:

(i) if U and V belong to $\mathfrak{U}$, then $U \cap V \in \mathfrak{U}$;
(ii) if $U \in \mathfrak{U}$ and $U \subset V$, then $V \in \mathfrak{U}$;
(iii) if $U \in \mathfrak{U}$, then for some $V \in \mathfrak{U}$, $V\cdot V^{-1} \subset U$; and
(iv) for each U in $\mathfrak{U}$ and each x in G, $x\cdot U\cdot x^{-1} \in \mathfrak{U}$.

On the other hand, given a group G and a non-void family $\mathfrak{U}$ of non-void subsets satisfying these four propositions there is a unique topology $\mathfrak{J}$ for G such that $(G,\cdot,\mathfrak{J})$ is a topological group and $\mathfrak{U}$ is the neighborhood system of the identity element.

(f) Every group, with the discrete topology or with the indiscrete topology, is a topological group If G is the set of real numbers, then $(G,+,\mathfrak{J})$, where $\mathfrak{J}$ is the usual topology, is a topological group and $(G \sim \{0\},\cdot,\mathfrak{J})$ is also a topological group. If G is the set of all integers, p is a prime and $\mathfrak{U}$ is the family of all subsets U of G such that for some positive integer k every integral multiple of p^k belongs to U, then $\mathfrak{U}$ is the neighborhood system of 0 relative to a topology $\mathfrak{J}$ such that $(G,+,\mathfrak{J})$ is a topological group.

(g) A topological group is a Hausdorff space whenever it is T_0-space. (That is, if x and y are distinct elements there is either a neighborhood of x to which y does not belong or the reverse. Observe that if $x \notin U\cdot y$, then $x\cdot y^{-1} \notin U$, and if $V^{-1}\cdot V \subset U$, then $V\cdot x \cap V\cdot y$ is void.)

(h) If U is open and X is an arbitrary subset of a topological group, then $U\cdot X$ and $X\cdot U$ are open. However, both X and Y may be closed subsets and $X\cdot Y$ may fail to be closed. (Consider the Euclidean plane with the usual addition with $X = Y = \{(x,y): y = 1/x^2\}$.)

(i) A cartesian product $\times\{G_a: a \in A\}$ of groups is a group under the operation: $(x\cdot y)_a = x_a\cdot y_a$ for each a in A. The product, with the product topology, is a topological group and the projection into each coordinate space is a continuous open homomorphism.*

Note Bourbaki [1], Pontrjagin [1], and Weil [2] are standard references on topological groups; see also Chevalley [1].

T SUBGROUPS OF A TOPOLOGICAL GROUP

(a) A subgroup of a topological group is, with the relative topology, a topological group.

(b) The closure of a subgroup is a subgroup and the closure of an invariant subgroup is invariant (invariant = normal = distinguished).

(c) Every subgroup with non-void interior is open and closed. A subgroup H is either closed or $H^- \sim H$ is dense in H^-.

(d) The smallest subgroup which contains a fixed open subset of a topological group is both open and closed.

(e) The component of the identity in a topological group is an invariant subgroup.

* Some authors use the term "representation" to mean continuous homomorphism, and the term "homomorphism" to mean a continuous homomorphism which is an open map onto its range.

(f) A discrete (with the relative topology) normal subgroup of a connected topological group is a subset of the center. (For a fixed member h of the subgroup H consider the map of G into H which carries x into $x^{-1} \cdot h \cdot x$.)

U FACTOR GROUPS AND HOMOMORPHISMS

Let G be a topological group, H a subgroup, G/H the family of left cosets (sets of the form $x \cdot H$ for some x in G). Then G/H with the quotient topology is a *homogeneous space*. If H is an invariant subgroup, then G/H is a group, called the factor group or quotient group.

(a) The projection of a topological group G onto the homogeneous quotient space G/H is open and continuous. (Show that the union of all left cosets which intersect an open set U is $U \cdot H$ and apply 3.10.)

(b) If H is an invariant subgroup, then G/H, with the quotient topology, is a topological group and the projection is a continuous, open homomorphism.

(c) The map of the homogeneous space which carries an element A into $a \cdot A$, where a is a fixed member of G, is a homeomorphism.

(d) If f is a homomorphism of a topological group G into another group H, then f is continuous iff the inverse of a neighborhood of the identity element of H is a neighborhood of the identity of G.

(e) If f is a continuous homomorphism of the topological group G into a topological group J, then the map of G onto $f[G]$, where $f[G]$ has the quotient topology, is a continuous open homomorphism, and the identity map of $f[G]$, with the quotient topology, into J is continuous. Hence each continuous homomorphism may be "factored" into a continuous open homomorphism followed by a continuous one-to-one homomorphism. If f is a continuous open homomorphism of G onto J, then J is topologically isomorphic to G/K where K is the kernel of f.

(f) If $J \subset H \subset G$ and J and H are invariant subgroups of G, then H/J is a subgroup of G/J, the quotient topology for H/J is the relative quotient topology for G/J, and the map of G/J into G/H which carries A into $A \cdot H$ is continuous and open, and hence $(G/J)/(H/J)$ is topologically isomorphic to G/H.

V BOX SPACES

A base for the *box* topology for the cartesian product $\times\{X_a: a \in A\}$ is the family of all sets $\times\{U_a: a \in A\}$ where U_a is open in X_a for each a in A. Hence the cartesian product of open sets is open relative to the box topology.

(a) Projection into each coordinate space is, relative to the box topology, continuous and open.

(b) Let Y be the cartesian product of the real numbers an infinite number of times; that is, $Y = R^A$, where R is the set of real numbers and A is an infinite set. With the box topology Y does not satisfy the first countability axiom, and the component of Y to which a point y belongs is the set of all points x such that $\{a: x_a \neq y_a\}$ is finite. (Let x and y be points of Y whose coordinates differ for an infinite set $a_0, a_1, \cdots, a_p \cdots$ of members of A. Let Z be the set of all z in Y such that for some k, $p| z(a_p) - x(a_p) |/| x(a_p) - y(a_p) | < k$ for all p. Then Z is open and closed, $x \in Z$ and $y \notin Z$.)

(c) Prove the results of (b) for the product of an infinite number of connected, Hausdorff topological groups, each of which contains at least two points. Show first that the product of topological groups is, with the box topology, a topological group.

W FUNCTIONALS ON REAL LINEAR SPACES

Let $(X,+,\cdot)$ be a real linear space. A real-valued linear function on X is called a *linear functional*. The set Z of all linear functionals on X is, with the natural definition of addition and scalar multiplication, a real linear space. It is clear that Z is a subset of the product $R^X = \times\{R: x \in X\}$, where R is the set of real numbers. The relativized product topology for Z is called the *weak** or w^*-topology (the *simple* topology). (The space Z is a subgroup of R^X, which is a topological group according to 3.S(i); however, the following results do not require the propositions on topological groups.)

The following propositions characterize w^*-dense subspaces of Z and w^*-continuous linear functionals.

(a) If $f, g_1, \cdots, g_n$ are members of Z and $f(x) = 0$ whenever $g_i(x) = 0$ for each i, then there are real numbers $a_1, \cdots, a_n$ such that $f = \sum\{a_i g_i: i = 1, \cdots, n\}$. (Consider the map G of X into E^n defined by $(G(x))_i = g_i(x)$. Show that there is an induced map F (see chapter 0) such that $f = F \circ G$.)

(b) *Density lemma* Let Y be a linear subspace of Z such that for each non-zero member x of X there is g in Y such that $g(x) \neq 0$. Then Y is w^*-dense in Z. (To show that $f \in Y^-$ it is necessary to prove that for each finite subset $x_1, \cdots, x_n$ of X there is a member of Y which approximates f at each of $x_1, \cdots, x_n$. Show there is g in Y such that $g(x_i) = f(x_i)$ for each i, $i = 1, \cdots, n$.)

(c) *Evaluation theorem* A linear functional F on Z is w^*-continuous iff it is an evaluation; that is, iff for some x in X it is true that

$F(g) = g(x)$ for all g in Z. (If F is w^*-continuous, then for some $x_1, \cdots, x_n$ in X and some positive real numbers $r_1, \cdots, r_n$ it is true that $|F(g)| < 1$ whenever $|g(x_i)| < r_i$ for each i. Show that, if $g(x_i) = 0$ for each i, then $F(g) = 0$.)

Notes The concept of the product topology grew out of the study of sequential convergence relative to the w^*-topology. The latter has been studied extensively (see, for example, Banach [1]). There were several awkward situations which arose in this study, which have been somewhat clarified by further topological developments. One might define the sequential closure of a set to be the union of the set and all limit points of sequences in the set, and agree that a set is sequentially closed iff it is identical with its sequential closure. Then it is not hard to see that a set may be sequentially closed relative to the w^*-topology but may fail to be w^*-closed. This is not a serious criticism if sequential convergence is the object under study. However, the really damaging fact is that the sequential closure of a set may fail to be sequentially closed; that is, sequential closure is not a Kuratowski closure operator. Because of this the machinery of general topology does not apply to the sequential closure operator, and *ad hoc* arguments are necessary for each conclusion. See Banach [1; 208 ff] for further discussion and examples.

X REAL LINEAR TOPOLOGICAL SPACES

A *real linear topological space* (r.l.t.s) is a quadruple $(X,+,\cdot,\mathfrak{J})$ such that $(X,+,\cdot)$ is a real linear space, $(X,+,\mathfrak{J})$ is a topological group, and the scalar multiplication, $\cdot$, is a continuous function on $X \times$ (*real numbers*) to X. Recall that a subset K of a real linear space is convex iff, whenever $0 \leqq t \leqq 1$ and x and y are members of K, then $t\cdot x + (1 - t)\cdot y \in K$.

(a) The function which, for a fixed real number a, $a \neq 0$, carries each member x of a real linear topological space into $a\cdot x$ is a homeomorphism.

(b) The cartesian product of real linear topological spaces is, with addition and scalar multiplication defined coordinate-wise, and with the product topology, a r.l.t.s.

(c) If Y is a linear subspace of the r.l.t.s. X, then Y, with the relative topology, is a r.l.t.s., and X/Y, with the quotient topology, is a r.l.t.s.

(d) Let K be a convex subset of a r.l.t.s. X and f a linear functional on X. Then f is continuous on K iff, for each real number t, the set $f^{-1}[t] \cap K$ is closed in K. (If $\{x_n, n \in D\}$ is a net in K, converging to a member x of K such that $\{f(x_n), n \in D\}$ fails to converge to $f(x)$, choose for n in a cofinal subset of D a point y_n on the segment from x_n to x such that $f(y_n)$ is a constant different from $f(x)$.)

(e) If f is a real-valued linear function (that is, a linear functional) on a r.l.t.s. X, then f is continuous iff $\{x: f(x) = 0\}$ is closed.

Notes The concept of a linear topological space is relatively recent (Kolmogoroff [1] and v. Neumann [1]); it is a notion which grew out of the study of the weak and weak* topologies for a Banach space and its adjoint. Much of the elementary theory of linear topological spaces is a direct application of the theory of topological groups; the results which distinguish the theory from that of topological groups all depend on convexity arguments. (This is a perfectly normal state of affairs; the chief use of the scalar multiplication, which is the only distinguishing feature, is in convexity arguments.) The few results on r.l.t. spaces which occur in the problems of this book do not constitute an adequate introduction to the theory because we do not list the propositions on convexity which are essential to a serious study. The following are suggested as reading references: Bourbaki [3], Nachbin [1], and Nakano [1]. The first of these contains a study of linear topological spaces over a topologized (not necessarily commutative) field.

Chapter 4

EMBEDDING AND METRIZATION

The development of general topology has followed an evolutionary pattern which occurs frequently in mathematics. One begins by observing similarities and recurring arguments in several situations which superficially seem to bear little resemblance to each other. We then attempt to isolate the concepts and methods which are common to the various examples, and if the analysis has been sufficiently penetrating we may find a theory containing many or all of our examples, which in itself seems worthy of study. It is in precisely this way, after much experimentation, that the notion of a topological space was developed. It is a natural product of a continuing consolidation, abstraction, and extension process. Each such abstraction, if it is to contain the examples from which it was derived in more than a formal way, must be tested to find whether we have really found the central ideas involved. This testing is usually done by comparing the abstractly constructed object with the objects from which it derived. In this case we want to find whether a topological space, at least under some reasonable restrictions, must necessarily be one of the particular concrete spaces from which the notion is derived. The "standard" examples with which we compare spaces are cartesian products of unit intervals and metric spaces. In this chapter the elementary properties of metric and pseudo-metric spaces are developed, and necessary and sufficient conditions are given under which a space is a copy of a metric space or of a subspace of the cartesian product of intervals.

A word of caution: the notion of a topological space by no means includes all of the properties which metric spaces possess. In chapter 6 a different and more penetrating abstraction of the concept of a metric space is made.

EXISTENCE OF CONTINUOUS FUNCTIONS

In this section we prove four lemmas, all part of a program to construct real-valued continuous functions on topological spaces. For the moment we are concerned with a rather special sort of topological space. A space is **normal** * iff for each disjoint pair of closed sets, A and B, there are disjoint open sets U and V such that $A \subset U$ and $B \subset V$. A **T_4-space** is a normal space which is T_1 ($\{x\}$ is closed for each x). If it is agreed that a set U is a **neighborhood of a set** A iff A is a subset of the interior U^0 of U, then the definition of normality can be stated: a space is normal iff disjoint closed sets have disjoint neighborhoods. There is another restatement of the condition which is also suggestive. A family of neighborhoods of a set is a **base for the neighborhood system** of the set iff every neighborhood of the set contains a member of the family. If W is a neighborhood of a closed subset A of a normal space X, then there are disjoint open sets U and V such that $A \subset U$ and $X \sim W^0 \subset V$, and hence the arbitrary neighborhood W of A contains the closed neighborhood U^-. Consequently the family of closed neighborhoods of a closed set A is a base for the neighborhood system of A if the space is normal. The converse is also true, for if A and B are disjoint closed sets and W is a closed neighborhood of A which is contained in $X \sim B$, then W^0 and $X \sim W$ are disjoint open neighborhoods of A and B respectively.

Every discrete space and every indiscrete space is normal and consequently a normal space need not be Hausdorff and may fail to satisfy the first or second axiom of countability. However, a T_4-space (T_1 and normal) is surely a Hausdorff space. A closed

* This nomenclature is an excellent example of the time-honored custom of referring to a problem we cannot handle as abnormal, irregular, improper, degenerate, inadmissible, and otherwise undesirable. A brief discussion of the abnormalities of the class of normal spaces occurs in the problems at the end of the chapter.

subset of a normal space is, with the relative topology, normal. However, subspaces, products, and quotient spaces of normal spaces may not be normal. (See 4.E, 4.F.)

There is a condition which for T_1-spaces is intermediate to Hausdorff and normal, and under certain circumstances implies normality. A topological space is **regular** iff for each point x and each neighborhood U of x there is a closed neighborhood V of x such that $V \subset U$; that is, the family of closed neighborhoods of each point is a base for the neighborhood system of the point. An equivalent statement: for each point x and each closed set A, if $x \notin A$, then there are disjoint open sets U and V such that $x \in U$ and $A \subset V$. A regular space which is also T_1 is called a **T_3-space.** Recall that a Lindelöf space is a topological space such that each open cover has a countable subcover.

1 LEMMA (TYCHONOFF) *Each regular Lindelöf space is normal.*

PROOF Suppose A and B are closed disjoint subsets of X. Because X is regular, for each point of A there is a neighborhood whose closure fails to intersect B and consequently the family $\mathfrak{U}$ of all open sets whose closures do not intersect B is a cover of A. Similarly, the family $\mathfrak{V}$ of all open sets whose closures do not intersect A is a cover of B, and $\mathfrak{U} \cup \mathfrak{V} \cup \{X \sim (A \cup B)\}$ is a cover of X. There is then a sequence $\{U_n, n \in \omega\}$ of members of $\mathfrak{U}$ which covers A, and a sequence $\{V_n, n \in \omega\}$ of members of $\mathfrak{V}$ which covers B. Let $U_n' = U_n \sim \bigcup\{V_p^-: p \leq n\}$ and let $V_n' = V_n \sim \bigcup\{U_p^-: p \leq n\}$. Since $U_n' \cap V_m$ is void for $m \leq n$, it follows that $U_n' \cap V_m'$ is void for $m \leq n$. Applying the same argument with the roles of U and V interchanged, $U_n' \cap V_m'$ is void for all m and n and consequently $\bigcup\{U_n': n \in \omega\}$ is disjoint from $\bigcup\{V_n': n \in \omega\}$. Finally, $V_p^- \cap A$ and $U_p^- \cap B$ are void for all p and hence the open disjoint sets $\bigcup\{U_n': n \in \omega\}$ and $\bigcup\{V_n': n \in \omega\}$ contain A and B respectively. ∎

In particular, a regular topological space satisfying the second axiom of countability is always normal.

We now begin the construction of continuous real-valued functions. If A and B are disjoint closed sets, we want to construct a continuous real-valued function which is zero on A and one on B, with all values in the closed interval [0,1]. Instead of constructing the function f directly we construct sets which correspond (approximately) to sets of the form $\{x: f(x) < t\}$. The two following lemmas show the relation between a family of subsets and a real-valued function.

2 LEMMA *Suppose that for each member t of a dense subset D of the positive reals F_t is a subset of a set X such that*

(a) *if $t < s$, then $F_t \subset F_s$; and*
(b) $\bigcup\{F_t: t \in D\} = X$.

For x in X let $f(x) = \inf\{t: x \in F_t\}$. Then $\{x: f(x) < s\} = \bigcup\{F_t: t \in D$ and $t < s\}$ and $\{x: f(x) \leqq s\} = \bigcap\{F_t: t \in D$ and $t > s\}$ for each real number s.

PROOF The calculation is direct. The set $\{x: f(x) < s\} = \{x: \inf\{t: x \in F_t\} < s\}$, and since the infimum is less than s iff some member of $\{t: x \in F_t\}$ is less than s, the set $\{x: f(x) < s\}$ is the set of all x such that for some t, $t < s$ and $x \in F_t$; that is, $\bigcup\{F_t: t \in D \text{ and } t < s\}$. This establishes the first equality. To prove the second notice that $\inf\{t: x \in F_t\} \leqq s$ if for each u greater than s there is $t < u$ such that $x \in F_t$. Conversely, if for each t in D such that $t > s$ it is true that $x \in F_t$, then $\inf\{t: x \in F_t\} \leqq s$ because D is dense in the set of positive numbers. Consequently the set of all x such that $f(x) = \inf\{t: x \in F_t\} \leqq s$ is $\{x: \text{if } t \in D \text{ and } t > s, \text{ then } x \in F_t\} = \bigcap\{F_t: t \in D \text{ and } t > s\}$. ∎

3 LEMMA *Suppose that for each member t of a dense subset D of the positive reals F_t is an open subset of a topological space X such that*

(a) *if $t < s$, then the closure of F_t is a subset of F_s; and*
(b) $\bigcup\{F_t: t \in D\} = X$.

Then the function f such that $f(x) = \inf\{t: x \in F_t\}$ is continuous.

PROOF According to 3.1 a function is continuous if the inverse of each member of some subbase for the topology of the range space is open, and the family of all sets of the form $\{t: t < s\}$ or $\{t: t > s\}$, for real numbers s, is a subbase for the usual topology for the set of real numbers. Consequently, to show f continuous it is sufficient to show that $\{x: f(x) < s\}$ is open and $\{x: f(x) \leqq s\}$

is closed for each real s. In view of the previous lemma the first of these, $\{x: f(x) < s\}$, is the union of open sets F_t and is therefore open. With reference again to the previous lemma, $\{x: f(x) \leqq s\} = \bigcap\{F_t: t \in D \text{ and } t > s\}$, and the proof will be complete if we show this set is identical with $\bigcap\{F_t^-: t \in D \text{ and } t > s\}$. Since $F_t \subset F_t^-$ for each t, surely $\bigcap\{F_t: t \in D \text{ and } t > s\} \subset \bigcap\{F_t^-: t \in D \text{ and } t > s\}$. On the other hand, for each t in D with $t > s$ there is r in D such that $s < r < t$, and hence such that $F_r^- \subset F_t$. The reverse inclusion follows. ∎

The principal result of the section is now easily proved.

4 Lemma (Urysohn) *If A and B are disjoint closed subsets of a normal space X, then there is a continuous function f on X to the interval* [0,1] *such that f is zero on A and one on B.*

PROOF Let D be the set of positive dyadic rational numbers (that is, the set of all numbers of the form $p2^{-q}$, where p and q are positive integers). For t in D and $t > 1$ let $F(t) = X$, let $F(1) = X \sim B$ and let $F(0)$ be an open set containing A such that $F(0)^-$ is disjoint from B. For t in D and $0 < t < 1$ write t in the form $t = (2m + 1)2^{-n}$ and choose, inductively on n, $F(t)$ to be an open set containing $F(2m2^{-n})^-$ and such that $F(t)^- \subset F((2m + 2)2^{-n})$. This choice is possible because X is normal. Let $f(x) = \inf\{t: x \in F(t)\}$. The previous lemma shows that f is continuous. The function f is zero on A because $A \subset F(t)$ for each t in D, and f is one on B because $F(t) \subset X \sim B$ for $t \leqq 1$ and $F(t) = X$ for $t > 1$. ∎

EMBEDDING IN CUBES

The cartesian product of closed unit intervals, with the product topology, is called a **cube.** A cube is then the set Q^A of all functions on a set A to the closed unit interval Q, with the topology of pointwise, or coordinate-wise, convergence. The cube is used as a standard sort of space, and we want to describe those topological spaces which are homeomorphic to subspaces of cubes. The device used to accomplish this description is simple but noteworthy; it will be used again in other connections.

Suppose that F is a family of functions such that each member

f of F is on a topological space X to a space Y_f (the range may be different for different members of the family). There is then a natural mapping of X into the product $\times\{Y_f: f \in F\}$ which is defined by mapping a point x of X into the member of the product whose f-th coordinate is $f(x)$. Formally, the **evaluation** map e is defined by: $e(x)_f = f(x)$. It turns out that e is continuous if the members of F are continuous and e is a homeomorphism if, in addition, F contains "enough functions." A family F of functions on X **distinguishes points** iff for each pair of distinct points x and y there is f in F such that $f(x) \neq f(y)$. The family **distinguishes points and closed sets** iff for each closed subset A of X and each member x of $X \sim A$ there is f in F such that $f(x)$ does not belong to the closure of $f[A]$.

5 EMBEDDING LEMMA *Let F be a family of continuous functions, each member f being on a topological space X to a topological space Y_f. Then:*

(a) *The evaluation map e is a continuous function on X to the product space* $\times\{Y_f: f \in F\}$.
(b) *The function e is an open map of X onto e[X] if F distinguishes points and closed sets.*
(c) *The function e is one to one if and only if F distinguishes points.*

PROOF The map e followed by projection P_f into the f-th coordinate space is continuous because $P_f \circ e(x) = f(x)$. Consequently, by theorem 3.3, e is continuous. To prove statement (b) it is sufficient to show that the image under e of an open neighborhood U of a point x contains the intersection of $e[X]$ and a neighborhood of $e(x)$ in the product. Choose a member f of F such that $f(x)$ does not belong to the closure of $f[X \sim U]$. The set of all y in the product such that $y_f \notin f[X \sim U]^-$ is open, and evidently its intersection with $e[X]$ is a subset of $e[U]$. Hence e is an open map of X onto $e[X]$. Statement (c) is clear. ∎

The preceding lemma reduces the problem of embedding a space topologically in a cube to the problem of finding a "rich" set of continuous real-valued functions on the space. There are topological spaces on which each continuous real-valued function is constant. For example, any indiscrete space has this property. There are less trivial examples; there are regular Hausdorff spaces on which every real continuous function is constant.* A topological space X is called **completely regular** iff for each member x of X and each neighborhood U of x there is a continuous function f on X to the closed unit interval such that $f(x) = 0$ and f is identically one on $X \sim U$. It is clear that the family of all continuous functions on a completely regular space to the unit interval $[0,1]$ distinguishes points and closed sets, in the sense of the preceding lemma. (The converse statement is also true, but will not be needed here.) If a completely regular space is T_1 ($\{x\}$ is closed for each x), then the family of continuous functions on the space to $[0,1]$ also distinguishes points. A completely regular T_1-space is called a **Tychonoff** space. If X is a Tychonoff space and F is the family of all continuous functions on X to $[0,1]$, then the embedding lemma 4.5 shows that the evaluation map of X into the cube Q^F is a homeomorphism. Thus each Tychonoff space is homeomorphic to a subspace of a cube. This fact is actually characteristic of Tychonoff spaces, as will presently be demonstrated.

Each normal T_1-space is a Tychonoff space in view of Urysohn's lemma 4.4. Each completely regular space is regular, for if U is a neighborhood of x and f is a continuous function which is zero at x and one on $X \sim U$, then $V = \{y: f(y) < 1/2\}$ is an open set whose closure is contained in $\{y: f(y) \leq 1/2\}$, which is a subset of U. For T_1-spaces there is a hierarchy of so-called separation axioms: Hausdorff, regular, completely regular, and normal. Except for normality, these properties are hereditary, in the sense that each subspace of a space X enjoys the property if X does. The product of spaces of each of these types is again of the same type, excepting, again, normality. The proofs of these facts are left as problems (4.**H**) except for the following, which is needed now.

6 THEOREM *The product of Tychonoff spaces is a Tychonoff space.*

PROOF For convenience, let us agree that a continuous function f on a topological space X to the closed unit interval is *for a*

* See Hewitt [1] and Novak [1]. For other facts on separation axioms see van Est and Freudenthal [1].

pair (x,U) iff x is a point, U is a neighborhood of x, $f(x) = 0$, and f is identically one on $X \sim U$. If $f_1, \cdots, f_n$ are functions for $(x,U_1), \cdots, (x,U_n)$, where n is a positive integer, and if $g(x) = \sup \{f_i(x): i = 1, \cdots, n\}$, then g is a function for $(x, \bigcap \{U_i: i = 1, \cdots, n\})$. Consequently the space is completely regular if for each x and each neighborhood U of x belonging to some subbase for the topology there is a function for (x,U). If X is the product $\times \{X_a: a \in A\}$ of Tychonoff spaces and $x \in X$ let U_a be a neighborhood of x_a in X_a. If f is a function for (x_a,U_a), then $f \circ P_a$, where P_a is the projection into the a-th coordinate space, is a function for $(x,P_a^{-1}[U_a])$. The family of sets of the form $P_a^{-1}[U_a]$ is a subbase for the product topology and hence the product space is completely regular. Since the product of T_1-spaces is a T_1-space the theorem follows. ∎

7 EMBEDDING THEOREM *In order that a topological space be a Tychonoff space it is necessary and sufficient that it be homeomorphic to a subspace of a cube.*

PROOF The closed unit interval is a Tychonoff space, and hence a cube, being a product of unit intervals, is a Tychonoff space by 4.6. Each subspace of a cube is therefore a Tychonoff space. It has already been observed that if X is a Tychonoff space and F the set of all continuous functions on X to the closed unit interval Q, then (by the embedding lemma 4.5) the evaluation map is a homeomorphism of X into the cube Q^F. ∎

METRIC AND PSEUDO-METRIC SPACES

There are many topological spaces in which the topology is derived from a notion of distance. A **metric** for a set X is a function d on the cartesian product $X \times X$ to the non-negative reals such that for all points x, y, and z of X,

(a) $d(x,y) = d(y,x)$,
(b) (*triangle inequality*) $d(x,y) + d(y,z) \geq d(x,z)$,
(c) $d(x,y) = 0$ if $x = y$, and
(d) if $d(x,y) = 0$, then $x = y$.

The last of these conditions is inessential for many purposes. A function d which satisfies only (a) and (b) and (c) is called a **pseudo-metric** (sometimes an **écart,** although "écart" is also used in a slightly different sense). All of the definitions of this section will be made for pseudo-metrics, it being understood that the same definitions are to hold with "pseudo-metric" replaced by "metric."

A **pseudo-metric** space is a pair (X,d) such that d is a pseudo-metric for X. For members x and y of X the number $d(x,y)$ is the **distance** (if confusion seems possible the d-distance) from x to y. If r is a positive number the set $\{y: d(x,y) < r\}$ is the **open sphere** of d-radius r about x, or briefly the open r-sphere about x, and $\{y: d(x,y) \leq r\}$ is the **closed r-sphere** about x. The intersection of two open spheres may not be a sphere. However, if $d(x,y) < r$ and $d(x,z) < s$, then each point w such that $d(w,x) < \min [r - d(x,y), s - d(x,z)]$ is a member of both the open r-sphere about y and the open s-sphere about z because of the triangle inequality. Consequently the intersection of two open spheres contains an open sphere about each of its points, and hence the family of all open spheres is the base for a topology for X (see 1.11). This topology is the **pseudo-metric topology** for X. Observe that each closed sphere is closed relative to the pseudo-metric topology.

Let X be a set and define $d(x,y)$ to be zero if $x = y$ and one otherwise. Then d is a metric for X and the open 1-sphere about each point x is $\{x\}$; hence $\{x\}$ is open relative to the metric topology and the space is discrete. The closed 1-sphere about each point is X and it follows that the closure of an open r-sphere may be different from the closed r-sphere. If d is defined to be zero for all pairs (x,y) in $X \times X$, then d is not a metric, but is a pseudo-metric. Then the open r-sphere about each point is the entire space, and the pseudo-metric topology for X is the indiscrete topology. If X is the set of all real numbers and $d(x,y) = |x - y|$ then d is a metric for X; it is called the **usual metric** for the real numbers. The usual metric topology is fortunately the usual topology for the reals.

The distance from a point x to a non-void subset A of a pseudo-metric space is defined to be $D(A,x) = \inf \{d(x,y): y \in A\}$.

8 THEOREM *If A is a fixed subset of a pseudo-metric space, then the distance from a point x to A is a continuous function of x relative to the pseudo-metric topology.*

PROOF Since $d(x,z) \leqq d(x,v) + d(y,z)$ it follows, taking lower bounds for z in A, that $D(A,x) \leqq d(x,y) + D(A,y)$. The same inequality holds with x and y interchanged and hence $|D(A,x) - D(A,y)| \leqq d(x,y)$. Consequently, if y is in the open r-sphere about x, then $|D(A,x) - D(A,y)| < r$ and continuity follows. ∎

9 THEOREM *The closure of a set A in a pseudo-metric space is the set of all points which are zero distance from A.*

PROOF Since $D(A,x)$ is continuous in x the set $\{x: D(A,x) = 0\}$ is closed and contains A and hence contains the closure A^- of A. If $y \notin A^-$, then there is a neighborhood of y, which may be taken to be an open r-sphere, which does not intersect A. Consequently $D(A,y) \geqq r$ and hence $\{x: D(A,x) = 0\} \subset A^-$. Therefore $A^- = \{x: D(A,x) = 0\}$. ∎

10 THEOREM *Each pseudo-metric space is normal.*

PROOF Let A and B be disjoint closed subsets of a pseudo-metric space X, and let $D(A,x)$ and $D(B,x)$ be the distance from x to A and B respectively. Let $U = \{x: D(A,x) - D(B,x) < 0\}$ and let $V = \{x: D(A,x) - D(B,x) > 0\}$. The function $D(A,x) - D(B,x)$ is continuous in x and therefore U and V are open. Clearly U is disjoint from V, and using 4.9 it follows that $A \subset U$ and $B \subset V$. ∎

11 THEOREM *Every pseudo-metric space satisfies the first axiom of countability. The second is satisfied if and only if the space is separable.*

PROOF A set is open relative to the pseudo-metric topology iff it contains an open sphere about each of its points. Therefore the family of open spheres about a point x is a base for the neighborhood system of x. Since each open sphere about x contains a sphere with rational radius there is a countable base for the neighborhood system and the space satisfies the first axiom of countability. Each space which satisfies the second axiom of countability is separable, so it remains to prove that a separable pseudo-metric space has a countable base for its topology. Let Y be a countable dense subset and let $\mathfrak{U}$ be the family of all open spheres with rational radii about members of Y. Surely $\mathfrak{U}$ is countable. If U is a neighborhood of a point x there is, for some positive r, an open r-sphere about x which is contained in U. Let s be a positive rational number less than r, let y be a point of Y such that $d(x,y) < s/3$, and let V be the open $2s/3$ sphere about y. Then $x \in V \subset U$ and hence $\mathfrak{U}$ is a base for the topology. ∎

12 THEOREM *A net* $\{S_n, n \in D\}$ *in a pseudo-metric space* (X,d) *converges to a point* s *if and only if* $\{d(S_n,s), n \in D\}$ *converges to zero.*

PROOF A net $\{S_n, n \in D\}$ converges to s iff the net is eventually in each open r-sphere about s, but this is true iff $\{d(S_n,s), n \in D\}$ is eventually in each open r-sphere about 0 in the space of real numbers with the usual metric. ∎

The **diameter** of a subset A of a pseudo-metric space (X,d) is sup $\{d(x,y)\colon x \in A \text{ and } y \in A\}$. If this supremum does not exist the diameter is said to be infinite. It is interesting to notice that the property of having a finite diameter is not a topological invariant.

13 THEOREM *Let* (X,d) *be a pseudo-metric space, and let* $e(x,y) = \min [1,d(x,y)]$. *Then* (X,e) *is a pseudo-metric space whose topology is identical with that of* (X,d).

Consequently each pseudo-metric space is homeomorphic to a pseudo-metric space of diameter at most one.

PROOF To prove that e is a pseudo-metric it is sufficient to show that if a, b, and c are non-negative numbers such that $a + b \geqq c$, then $\min [1,a] + \min [1,b] \geqq \min [1,c]$, for the latter inequality becomes the triangle inequality for e if we set $a = d(x,y)$, $b = d(y,z)$ and $c = d(x,z)$. If either $\min [1, a]$ or $\min [1,b]$ is one the inequality is surely correct since $\min [1,c] \leqq 1$. If neither of these is one the inequality $a + b \geqq c \geqq \min [1,c]$ completes the proof. Consequently e is a pseudo-metric for X. The family of all open r-spheres, for r less than one, is a base for the pseudo-metric topology. Since this family is the same whether d or e is used as pseudo-metric, the two pseudo-metric topologies are identical. Clearly the e-diameter of X is at most one. ▮

The product of uncountably many topological spaces does not generally satisfy the first axiom of countability (see 3.6) and consequently one cannot expect to find a pseudo-metric for the product of arbitrarily many pseudo-metric spaces such that the pseudo-metric topology is the product topology. For countable products the situation is pleasant. Because of the previous theorem we restrict our attention to spaces of diameter at most one.

14 THEOREM *Let* $\{(X_n,d_n), n \in \omega\}$ *be a sequence of pseudo-metric spaces, each of diameter at most one, and define d by:* $d(x,y) = \sum\{2^{-n}d_n(x_n,y_n): n \in \omega\}$. *Then d is a pseudo-metric for the cartesian product, and the pseudo-metric topology is the product topology.*

PROOF The simple proof that d is a pseudo-metric is omitted. (Problem 2.G on summability contains the necessary machinery.) To show the two topologies identical, first observe that, if V is a 2^{-p}-sphere about a point x of the product and $U = \{y: d_n(x_n,y_n) < 2^{-p-n-2}$ *for* $n \leq p + 2\}$, then $U \subset V$, for if $y \in U$, then $d(x,y) < \sum\{2^{-p-n-2}: n = 0, \cdots, p + 2\} + \sum\{2^{-n}: n = p + 3, \cdots\} < 2^{-p-1} + 2^{-p-1} = 2^{-p}$. But U is a neighborhood of x in the product topology and it follows that each set which is open relative to the pseudo-metric topology is open relative to the product topology. To show the converse consider a member U of the defining subbase of the product topology. Then U is of the form $\{x: x_n \in W\}$ where W is open in X_n. For x in U there is an open r-sphere about x_n which is a subset of W, and since $d(x,y) \geq 2^{-n}d_n(x_n,y_n)$ the open $r2^{-n}$-sphere about x is a subset of U. Therefore each member of the defining subbase, and consequently each member of the product topology, is open relative to the pseudo-metric topology. ▮

If (X,d) and (Y,e) are pseudo-metric spaces and f is a map of X onto Y, then f is an **isometry** (a d-e isometry) iff $d(x,y) = e(f(x),f(y))$ for all points x and y of X. Every isometry is a continuous open map (relative to the two pseudo-metric topologies) because the image of each open r-sphere about x is an open r-sphere about $f(x)$. The composition of two isometrics is again an isometry and if an isometry is one to one the inverse is also an isometry. On a metric space an isometry is necessarily one to one and an isometry of a metric space onto a metric space is a homeomorphism. The collection of all metric spaces is divided into equivalence classes of mutually isometric spaces. Each property which, when possessed by a metric space, is also possessed by each isometric metric space, is a **metric invariant.** A metric invariant is not necessarily a topological invariant (for example, consider the property of being of infinite diameter).

Each pseudo-metric space differs but little, in one sense, from a metric space. In making this statement precise it is convenient to agree that the **distance between two subsets,** A and B, of a pseudo-metric space is $D(A,B) = \text{dist}\,(A,B) = \inf \{d(x,y): x \in A \textit{ and } y \in B\}$. It is generally not true that D is a pseudo-metric, for the space X is zero distance from every non-void subset and the triangle inequality fails. However, D is actually a metric for the members of the decomposition which we want to consider. For a pseudo-metric space (X,d) let $\mathfrak{D}$ be the family of all sets of the form $\{x\}^-$. Because of 4.9, $\{x\}^-$ is precisely the set of all points y such that $d(x,y) = 0$, and the decomposition $\mathfrak{D}$ is the quotient X/R where R is the relation $\{(x,y): d(x,y) = 0\}$.

15 THEOREM *Let (X,d) be a pseudo-metric space, let $\mathfrak{D}$ be the family of all sets $\{x\}^-$ for x in X, and for members A and B of $\mathfrak{D}$ let $D(A,B) =$ dist (A,B). Then $(\mathfrak{D},D)$ is a metric space whose topology is the quotient topology for $\mathfrak{D}$, and the projection of X onto $\mathfrak{D}$ is an isometry.*

PROOF A point u is a member of $\{x\}^-$ iff $d(u,x) = 0$, and this is true iff $x \in \{u\}^-$. If $u \in \{x\}^-$ and $v \in \{y\}^-$, then $d(u,v) \leq d(u,x) + d(x,y) + d(y,v) = d(x,y)$. Consequently, since in this case it is also true that $x \in \{u\}^-$ and $y \in \{v\}^-$, $d(u,v) = d(x,y)$. It follows that for members A and B of $\mathfrak{D}$, $D(A,B)$ is identical with $d(x,y)$ for every x in A and every y in B. Therefore $(\mathfrak{D},D)$ is a metric space and the projection of X onto $\mathfrak{D}$ is an isometry. If U is an open set in X and $x \in U$, then, for some $r > 0$, U contains an open r-sphere about x, and hence contains $\{x\}^-$. The projection of X onto $\mathfrak{D}$ is therefore an open map relative to the quotient topology for $\mathfrak{D}$, by 3.10. The projection is also open relative to the metric topology derived from D and hence, by 3.8, these two topologies are identical. ∎

METRIZATION

Given a topological space $(X,\mathfrak{I})$, it is natural to ask whether there is a metric for X such that $\mathfrak{I}$ is the metric topology. Such a metric **metrizes** the topological space and the space is said to be **metrizable.** Similarly, a topological space is **pseudo-metrizable** iff there is a pseudo-metric such that the topology is the pseudo-metric topology. A pseudo-metric is a metric if and only if the topology is T_1 (that is, iff $\{x\}$ is closed for each point x) and it follows that a space is metrizable if and only if it is T_1 and pseudo-metrizable. The theorems of this section are stated for metrizable spaces; the corresponding theorems for pseudo-metrizable spaces will be self-evident.

The two principal theorems of the section give necessary and sufficient conditions that a topological space be, respectively, metrizable and separable, and metrizable. The first of these is the classical metrization theorem of Urysohn; all of the pieces of its proof are already available and it is simply a matter of fitting the facts together. The second theorem has been proved only recently (its history is given in the notes at the end of the section). It turns out that a mild variant of Urysohn's procedure proves the sufficiency of the conditions imposed, but the necessity requires a new sort of construction. A further study of the concepts introduced here is made in the last section of chapter 5. Finally, the entire problem of metrization is approached from a different point of view in chapter 7; however, the results obtained there do not include the theorems of this section.

The pattern for a proof of metrizability is very simple. According to 4.14 the product of countably many pseudo-metric spaces is pseudo-metrizable. According to the embedding lemma 4.5, if F is a family of continuous functions on a T_1-space X, where a member f of F maps X into a space Y_f, then the evaluation map of X into $\times\{Y_f: f \in F\}$ is a homeomorphism whenever

F distinguishes points and closed sets (that is, if A is a closed subset of X and x is a member of $X \sim A$, then $f(x) \notin f[A]^-$ for some member f of F). The problem of metrizing a T_1-space X then reduces to that of finding a countable family of continuous functions, each on X to some pseudo-metrizable space, such that F distinguishes points and closed sets. (A pseudo-metrizable T_1-space is necessarily metrizable.)

For convenience, let Q^ω denote the product of the closed unit interval with itself countably many times; that is Q^ω is the set of all functions on the non-negative integers to the closed unit interval Q, with the product topology.

16 METRIZATION THEOREM (URYSOHN) *A regular T_1-space whose topology has a countable base is homeomorphic to a subspace of the cube Q^ω and is hence metrizable.*

PROOF In view of the remarks preceding the theorem it is sufficient to show that there is a countable family of continuous functions on X to Q which distinguishes points and closed sets. Let $\mathfrak{B}$ be a countable base for the topology of X and let $\mathfrak{A}$ be the set of all pairs (U,V) such that U and V belong to $\mathfrak{B}$ and $U^- \subset V$. Surely $\mathfrak{A}$ is countable. For each pair (U,V) in $\mathfrak{A}$ choose a continuous function f on X to Q such that f is zero on U and one on $X \sim V$ (such a function exists because of the Tychonoff lemma 4.1 and the Urysohn lemma 4.4) and let F be the family of functions so obtained. Then F is countable and it remains to be proved that F distinguishes points and closed sets. If B is closed and $x \in X \sim B$ choose a member V of $\mathfrak{B}$ such that $x \in V \subset X \sim B$ and choose U in $\mathfrak{B}$ such that $x \in U^- \subset V$. Then $(U,V) \in \mathfrak{A}$, and if f is the corresponding member of F, then $f(x) = 0 \notin \{1\} = f[B]^-$. ∎

It is easy to describe the class of topological spaces to which the foregoing metrization theorem applies.

17 THEOREM *If X is a T_1-space, then the following are equivalent:*

(a) *X is regular and there is a countable base for its topology.*
(b) *X is homeomorphic to a subspace of the cube Q^ω.*
(c) *X is metrizable and separable.*

PROOF The previous theorem shows that (a) → (b). The cube Q^ω is metrizable, by 4.14, and satisfies the second axiom of countability (3.**M**). Hence each subspace is metrizable and satisfies the second axiom of countability and is therefore separable. Hence (b) → (c). (Caution: it is not true that a subspace of a separable space is necessarily separable.) Finally (c) → (a), for if X is metrizable and separable, then it is surely regular and by 4.11 it satisfies the second axiom of countability. ▮

The metrization theorem for spaces which are not necessarily separable depends heavily on the ideas which we have already exploited. A brief discussion of methodology will indicate where the procedure used so far can be improved. The construction of a metric for X is accomplished by finding a family of mappings of X into pseudo-metrizable spaces. But observe: so far the only space which has been used as the range space is the unit interval Q. Stated in slightly different form, if f is a function on X to Q, then one may construct a pseudo-metric for X by letting $d(x,y) = | f(x) - f(y) |$. The Urysohn metrization is accomplished by using a countable number of pseudo-metrics of this sort, and the problem is to generalize this construction. If F is a family of functions on X to Q, then a possible candidate for a pseudo-metric is the sum: $\sum\{| f(x) - f(y) | : f \in F\}$. This sum must be continuous in x and y in order that the identity map of X into the pseudo-metric space (X,d) be continuous, and a condition much weaker than finiteness of the family F will ensure continuity. It is sufficient, to obtain continuity, that for each point x of X there be a neighborhood U of x such that all but a finite number of the members of F vanish on U; in other words, a sort of local finiteness suffices. This notion of local finiteness is the key to the problem.

A family $\mathfrak{A}$ of subsets of a topological space is **locally finite** iff each point of the space has a neighborhood which intersects only finitely many members of $\mathfrak{A}$. It follows immediately from the definition that a point is an accumulation point of the union $\bigcup\{A: A \in \mathfrak{A}\}$ iff it is an accumulation point of some member of $\mathfrak{A}$, and hence the closure of the union is the union of the closures; that is, $[\bigcup\{A: A \in \mathfrak{A}\}]^- = \bigcup\{A^-: A \in \mathfrak{A}\}$. It is also evident that the family of all closures of members of $\mathfrak{A}$ is locally finite.

A family $\mathcal{A}$ is **discrete** if each point of the space has a neighborhood which intersects at most one member of $\mathcal{A}$. A discrete family is locally finite, and if $\mathcal{A}$ is discrete, then the family of closures of members of $\mathcal{A}$ is also discrete. Finally, a family $\mathcal{A}$ is **σ-locally finite (σ-discrete)** if and only if it is the union of a countable number of locally finite (respectively, discrete) subfamilies.

The following metrization theorem can now be stated. Its proof is contained in the sequence of lemmas which follows the statement.

18 METRIZATION THEOREM *The following three conditions on a topological space are equivalent.*

(a) *The space is metrizable.*
(b) *The space is T_1 and regular, and the topology has a σ-locally finite base.*
(c) *The space is T_1 and regular, and the topology has a σ-discrete base.*

It is clear that condition (c) implies (b) and it will be proved that (b) implies (a), and finally that (a) implies (c). The first step in the proof is a variant of Tychonoff's lemma, 4.1.

19 LEMMA *A regular space whose topology has a σ-locally finite base is normal.*

PROOF If A and B are disjoint closed subsets of the space X, then there are open covers $\mathcal{U}$ and $\mathcal{V}$ of A and B respectively such that the closure of each member of $\mathcal{U}$ is disjoint from B, the closure of each member of $\mathcal{V}$ is disjoint from A, and both $\mathcal{U}$ and $\mathcal{V}$ are subfamilies of a σ-locally finite base $\mathcal{B}$. It follows that $\mathcal{U} = \bigcup\{\mathcal{U}_n: n \in \omega\}$ and $\mathcal{V} = \bigcup\{\mathcal{V}_n: n \in \omega\}$ where $\mathcal{U}_n$ and $\mathcal{V}_n$ are locally finite families. Let $U_n = \bigcup\{W: W \in \mathcal{U}_n\}$ and let $V_n = \bigcup\{W: W \in \mathcal{V}_n\}$. Then $U_n^- = \bigcup\{W^-: W \in \mathcal{U}_n\}$, and hence U_n^- is disjoint from B and similarly V_n^- is disjoint from A. This is precisely the situation which occurs in the proof of 4.1, and as there, the proof is completed by letting $U_n' = U_n \sim \bigcup\{V_k^-: k \leq n\}$, $V_n' = V_n \sim \bigcup\{U_k^-: k \leq n\}$. The union of the sets U_n' and the union of the sets V_n' are the required disjoint neighborhoods of A and B respectively. ∎

The following lemma now completes the proof that the conditions listed in 4.18 are sufficient for metrizability.

20 Lemma *A regular T_1-space whose topology has a σ-locally finite base is metrizable.*

PROOF It will be shown that there is a countable family D of pseudo-metrics on the space X such that each member of D is continuous on $X \times X$ and such that for each closed subset A of X and each point x of $X \sim A$ there is a member d of D such that the d-distance from x to A is positive. This will prove metrizability, for the map of X into each of the pseudo-metric spaces (X,d) will then be continuous, and 4.5 and 4.14 will apply just as for the Urysohn theorem. The problem is then to construct the family D. Let $\mathcal{B}$ be a σ-locally finite base for the topology of X, and suppose that $\mathcal{B} = \bigcup \{\mathcal{B}_n : n \in \omega\}$ where each $\mathcal{B}_n$ is locally finite. For every ordered pair of integers m and n and for each member U of $\mathcal{B}_m$, let U' be the union of all members of $\mathcal{B}_n$ whose closures are contained in U. Because $\mathcal{B}_n$ is locally finite the closure of U' is a subset of U, and there is a continuous function f_U on X to the unit interval which is one on U' and zero on $X \sim U$ by 4.19 and 4.4. Let $d(x,y) = \sum\{|f_U(x) - f_U(y)| : U \in \mathcal{B}_m\}$. The continuity of d on $X \times X$ is a straightforward consequence of the local finiteness of $\mathcal{B}_m$. Finally, let D be the family of pseudo-metrics so obtained; since one pseudo-metric was constructed for each ordered pair of integers, D is countable. If A is a closed subset of X and $x \in X \sim A$, then for some m and some U in $\mathcal{B}_m$ it is true that $x \in U \subset X \sim A$, and for some n and some V in $\mathcal{B}_n$ it is true that $x \in V$ and $V^- \subset U$. For the pseudo-metric d constructed for this pair it is clear that the d-distance from x to A is at least one. ∎

The most interesting part of the proof of the metrization theorem remains. It must be proved that each metric space has a σ-discrete base. A stronger result than this is true and, since the more potent theorem will be needed later, we introduce a new concept. A cover $\mathcal{B}$ of a set X is a **refinement** of a cover $\mathcal{A}$ iff each member of $\mathcal{B}$ is a subset of a member of $\mathcal{A}$. For example, in a metric space, the family of all open spheres of radius one half is a refinement of the family of all open spheres of radius one. The following theorem states that any open cover of a pseudo-metric space has an open refinement which is σ-discrete. This will imply that each pseudo-metric topology has a σ-discrete base, for one may select a σ-discrete refinement $\mathcal{B}_n$ of the cover consisting of all open spheres of radius $1/n$, and the union of the families $\mathcal{B}_n$ is then a σ-discrete base. This fact completes the proof of the metrization theorem 4.18.

21 THEOREM *Each open cover of a pseudo-metrizable space has an open σ-discrete refinement.*

PROOF Let $\mathcal{U}$ be an open cover of the pseudo-metric space (X,d). The first step in the proof is the decomposition of each member U of $\mathcal{U}$ into "concentric disks." For each positive integer n and each member U of $\mathcal{U}$ let U_n be the set of all members x of U such that dist $[x,X \sim U] \geqq 2^{-n}$. Because of the triangle inequality it is clear that dist $[U_n,X \sim U_{n+1}] \geqq 2^{-n} - 2^{-n-1} = 2^{-n-1}$. Choose a relation $<$ which well orders the family $\mathcal{U}$ (see 0.25h) and for each positive integer n and each member U of $\mathcal{U}$ let $U_n^* = U_n \sim \bigcup \{V_{n+1}: V \in \mathcal{U} \text{ and } V < U\}$. For each U and V in $\mathcal{U}$ and each positive integer n it is true that $U_n^* \subset X \sim V_{n+1}$, or $V_n^* \subset X \sim U_{n+1}$, depending on whether U follows or precedes V in the ordering. In either case dist $[U_n^*,V_n^*] \geqq 2^{-n-1}$. It follows that if $U_n^\sim$ is defined to be the set of all points x such that the distance from x to U_n^* is less than 2^{-n-3}, then dist $[U_n^\sim,V_n^\sim] \geqq 2^{-n-2}$ and hence for each fixed n the family of all sets of the form $U_n^\sim$ is discrete. Let $\mathcal{V}$ be the family of $U_n^\sim$ for all n and all U in $\mathcal{U}$. Then $\mathcal{V}$ is an open cover of X, for if U is the first member of $\mathcal{U}$ to which x belongs, then surely $x \in U_n^\sim$ for some n. Evidently $U_n^\sim \subset U$, and consequently $\mathcal{V}$ is a σ-discrete open refinement of $\mathcal{U}$. ▮

22 *Notes* There are really two metrization problems. The topological problem has just been treated and the problem of uniform metrization will be considered in chapter 7 (statement and history are given there). Curiously enough a satisfactory solution of the latter was found much earlier than a solution of the former. Urysohn's theorem, although treating only a special case, was certainly the most satisfactory theorem of the topo-

logical sort until very recently. The key to the present reasonably satisfactory situation was furnished by two papers. Dieudonné [1] initiated a study of spaces with the property that each open cover has an open locally finite refinement (paracompact spaces; see chapter 5). A. H. Stone [1] showed that each metrizable space is paracompact (a special case of this theorem was earlier demonstrated by C. H. Dowker [1]). The σ-locally finite characterization was then discovered independently, by Nagata [1] and by Smirnov [1]. The σ-discrete characterization is due to Bing [1]. The proof of necessity (4.21) of the metrizability conditions is actually an initial fragment of Stone's proof of paracompactness.

Smirnov [2] has also showed that paracompactness together with local metrizability implies metrizability.

Finally a brief statement of the role of pseudo-metrizable spaces might be made. Most of the spaces which occur naturally in analysis are pseudo-metric rather than metric, and even in the metrization problem a construction via pseudo-metrics was convenient. Of course, one may always replace a pseudo-metric space by a related metric space (theorem 4.15), but the process of taking quotient spaces becomes a bit tedious and for most purposes the requirement $d(x,y) = 0$ iff $x = y$ is completely irrelevant. One is tempted to work exclusively with pseudo-metrics, but this has disadvantages, for example, when one seeks to construct a topological map. A possible way out is to redefine "topological map" to mean a relation which induces a one-to-one intersection and union-preserving map on the topologies.

PROBLEMS

A REGULAR SPACES

(a) Let X be a regular space and let $\mathfrak{D}$ be the family of all subsets of the form $\{x\}^-$ for x in X. Then $\mathfrak{D}$ is a decomposition of X, the projection of X onto the quotient space $\mathfrak{D}$ is both open and closed, and the quotient space is regular Hausdorff. (If A is a subset of X which is either open or closed, then $\{x\}^- \subset A$ whenever $x \in A$.)

(b) The product of regular spaces is again regular.

B CONTINUITY OF FUNCTIONS ON A METRIC SPACE

A function f on a pseudo-metric space (X,d) to a pseudo-metric space (Y,e) is continuous iff for each x in X and each $\epsilon > 0$ there is $\delta > 0$ such that $e(f(x),f(y)) < \epsilon$ if $d(x,y) < \delta$.

C PROBLEM ON METRICS

Let f be a continuous real-valued function defined on the set of all non-negative real numbers, such that $f(x) = 0$ iff $x = 0$, f is non-decreasing, and $f(x + y) \leq f(x) + f(y)$ for all non-negative numbers x and y. (A function satisfying this last condition is called *subadditive*.) If (X,d) is a metric space and $e(x,y) = f(d(x,y))$, then (X,e) is a metric space, and the metric topology of the space (X,e) is identical with that of (X,d). (A particular case of this result which occurs frequently in the literature: $f(x) = x/(1 + x)$.)

D HAUSDORFF METRIC FOR SUBSETS

Let (X,d) be a metric space of finite diameter, and let $\mathfrak{A}$ be the family of all closed subsets. For $r > 0$ and A in $\mathfrak{A}$ let $V_r(A) = \{x: \text{dist}\,(x,A) < r\}$, and define, for members A and B of $\mathfrak{A}$, $d'(A,B) = \inf\,\{r: A \subset V_r(B) \text{ and } B \subset V_r(A)\}$. d' is the Hausdorff metric; it is not the same as the distance between sets used in the text.

(a) $(\mathfrak{A},d')$ is a metric space, and the map which carries x in X into $\{x\}$ in $\mathfrak{A}$ is an isometry of X onto a subspace of $\mathfrak{A}$.

(b) The topology of the Hausdorff metric for $\mathfrak{A}$ is not determined by the metric topology for X. For example, let X be the set of all positive real numbers, let $d(x,y) = |x/(1 + x) - y/(1 + y)|$, and let $e(x,y) = \min\,[1,|x - y|]$. Then the metric topologies of (X,d) and (X,e) are identical, but those of $(\mathfrak{A},d')$ and $(\mathfrak{A},e')$ are different. (In $(\mathfrak{A},d')$ the set of all positive integers is an accumulation point of the family of all its finite subsets.)

Note For information and references on this topic see Michael [2].

E EXAMPLE (THE ORDINALS) ON THE PRODUCT OF NORMAL SPACES

The product of normal spaces is not generally normal.* Let Ω_0 be the set of all ordinal numbers less than the first uncountable ordinal Ω and let Ω' be $\Omega_0 \cup \{\Omega\}$, each with the order topology.

(a) *Interlacing lemma* Let $\{x_n, n \in \omega\}$ and $\{y_n, n \in \omega\}$ be two se-

* It is possible to do part of this problem a little more efficiently using methods from the following chapter. However, the facts given here will be useful later. I believe the example is due to J. Dieudonné and A. P. Morse, independently.

quences in Ω_0 such that $x_n \leqq y_n \leqq x_{n+1}$ for each n. Then both sequences converge, and to the same point of Ω_0.

(b) If A and B are closed disjoint subsets of Ω_0, then Ω is not an accumulation point of both A and B.

(c) Both Ω_0 and Ω' are normal. (If A and B are closed disjoint subsets and the first point of $A \cup B$ belongs to A, find a finite sequence $a_0, b_0, a_1, \cdots a_n$ (or b_n) such that $a_i \in A$, $b_i \in B$, no point of A is between a_i and b_i, and no point of B is between b_i and a_{i+1}, for each i. The intervals $(a_i,b_i]$ are both open and closed.)

(d) If f is a function on Ω_0 to Ω_0 such that $f(x) \geqq x$ for each x, then for some x in Ω_0, the point (x,x) is an accumulation point of the graph of f. (Define a sequence, inductively, such that $x_{n+1} = f(x_n)$, observe that $x_n \leqq f(x_n) \leqq x_{n+1}$, and use the interlacing lemma.)

(e) The product $\Omega_0 \times \Omega'$ is not normal. (Let A be the set of all points (x,x), and let $B = \Omega_0 \times \{\Omega\}$. If U is a neighborhood of A let $f(x)$ be the smallest ordinal larger than x such that $(x,f(x)) \notin U$. Then (d) applies.)

F EXAMPLE (THE TYCHONOFF PLANK) ON SUBSPACES OF NORMAL SPACES

A subspace of a normal space may fail to be normal. Let Ω' be the set of ordinal numbers not greater than the first uncountable ordinal Ω, and let ω' be the set of ordinals not greater than the first infinite ordinal, ω, each with the order topology. The product $\Omega' \times \omega'$ is called the *Tychonoff plank*. It is not difficult to prove directly that the plank is normal; however, this fact is an immediate consequence of a theorem of the next chapter. Let X be $(\Omega' \times \omega') \sim \{(\Omega,\omega)\}$, so that X is the plank with a corner point removed. Let A be the set of all points of X with first coordinate Ω and let B be the set of all points with second coordinate ω. Then there are no disjoint neighborhoods of A and B. (If U is a neighborhood of A, then for x in ω let $f(x)$ be the first ordinal such that if $y > f(x)$, then $(y,x) \in U$. The supremum of the values of f is less than Ω.)

G EXAMPLE: PRODUCTS OF QUOTIENTS AND NON-REGULAR HAUSDORFF SPACES

Let X be a regular Hausdorff space which is not normal, and let A and B be disjoint closed sets such that each neighborhood of A intersects each neighborhood of B. Let Δ be the set of all (x,x) for x in X (Δ is the identity relation on X).

(a) Let $R = \Delta \cup (A \times A)$. Then R is closed in $X \times X$ and the quotient space X/R is a Hausdorff space which is not regular. (The members of the quotient space are A, and $\{x\}$ for x in $X \sim A$.)

(b) Let $S = \Delta \cup (A \times A) \cup (B \times B)$. Then S is closed in $X \times X$, but X/S is not a Hausdorff space. (The members of X/S are A, B, and $\{x\}$ for each x in $X \sim (A \cup B)$.)

(c) There is a natural map of $X \times X$ onto $(X/S) \times (X/S)$ which carries (x,y) into $(S[x],S[y])$. It is natural to ask whether this map is open, provided X/S is given the quotient topology and $(X/S) \times (X/S)$ and $X \times X$ are given the product topologies. (This is equivalent to asking whether the product of quotients is topologically equivalent to the quotient of the product.) If S is the relation defined in (b), then the map is not open. (Consider the neighborhood $X \times X \sim (A \times A \cup B \times B \cup \Delta)$ of $A \times B$.)

H HEREDITARY, PRODUCTIVE, AND DIVISIBLE PROPERTIES

A property P of a space is *hereditary* iff each subspace of a space with P also has P; it is *productive* iff the product of spaces enjoying P has P; and it is *divisible* iff the quotient space of each space with P has P. Consider the properties: T_1, H = Hausdorff, R = regular, CR = completely regular, T = Tychonoff, N = normal, C = connected, S = separable, C_I = first axiom of countability, C_{II} = second countability axiom, M = metrizable, and L = Lindelöf. The following table is filled out + or −, depending on whether the property at the head of the column is or is not of the sort listed on the left. Show by example (most of the necessary examples have already been mentioned in the problems) or proof that the listing is correct.

	T_1	H	R	CR	T	N	C	S	C_I	C_{II}	M	L
Hereditary	+	+	+	+	+	−	−	−	+	+	+	−
Productive	+	+	+	+	+	−	+	−	−	−	−	−
Divisible	−	−	−	−	−	−	+	+	−	−	−	+

Quite different results are obtained if one varies the problem by considering only closed subspaces, or only open maps.

I HALF-OPEN INTERVAL SPACE

Let X be the set of all real numbers with the half-open interval topology (a base is the family of all half-open intervals $[a,b)$; see **1.K** and **1.L**). Then:

(a) X is regular.

(b) X is normal. (Recall that every open cover of X has a countable subcover.)

(c) The product space $X \times X$ is not normal. (Let $Y = \{(x,y): x + y = 1\}$, let A be the set of all members of Y with first coordinate irrational, and let $B = Y \sim A$. Assume that U and V are disjoint neighborhoods of A and B, and for x in A let $f(x) = \sup \{e: [x,e) \times [1 - x,e) \subset U\}$. Then f is a function on the set of all irrational numbers and f is never zero. The contradiction depends on the fact that for some positive integer n there is a rational number which is an accumulation point of $\{x: f(x) \geqq 1/n\}$. This fact is an immediate consequence of the theorem that the space of real numbers (with the usual topology) is of the second category (see chapter 7), but a direct proof seems awkward.)

Note This example is due to Sorgenfrey [1].

J THE SET OF ZEROS OF A REAL CONTINUOUS FUNCTION

A subset of a topological space is called a G_δ iff it is the intersection of the members of a countable family of open sets.

(a) If f is a continuous real valued function on X, then $f^{-1}[0]$ is a G_δ. (The set $\{0\}$ is a G_δ in the space of all real numbers.)

(b) If A is a closed G_δ in a normal topological space X, then there exists a continuous real-valued function f such that $A = f^{-1}[0]$.

K PERFECTLY NORMAL SPACES

A topological space is called *perfectly normal* iff it is normal and each closed subset is a G_δ.

(a) Each pseudo-metrizable space is perfectly normal.

(b) The product of an uncountable number of unit intervals is not perfectly normal. (A G_δ in such a space cannot consist of a single point.)

L CHARACTERIZATION OF COMPLETELY REGULAR SPACES

A topological space is completely regular iff it is homeomorphic to a subspace of a product of pseudo-metric spaces.

M UPPER SEMI-CONTINUOUS DECOMPOSITION OF A NORMAL SPACE

The image of a normal topological space under a closed continuous map is a normal space.

Chapter 5

COMPACT SPACES

The notion of a compact topological space is (like every concept studied in this book) an abstraction of certain important properties of the set of real numbers. The classic theorem of Heine-Borel-Lebesgue asserts that every open cover of a closed and bounded subset of the space of real numbers has a finite subcover. This theorem has extraordinarily profound consequences, and, like most good theorems, its conclusion has become a definition. A topological space is **compact (bicompact)** if and only if each open cover has a finite subcover.* A subset A of a topological space is compact iff it is, with the relative topology, compact; equivalently A is compact iff every cover of A by sets which are open in X has a finite subcover.

EQUIVALENCES

This section is devoted to characterizations of compactness in terms of closed sets, convergence, bases, and subbases.

A family $\mathcal{A}$ of sets has the **finite intersection property** iff the intersection of the members of each finite subfamily of $\mathcal{A}$ is non-void. The De Morgan formulae (0.2) on complements furnish the connection between this notion and the concept of compactness.

* The term "compact" has also been used to denote "sequentially compact" and "countably compact" (in the terminology of the problems at the end of this chapter). N. Bourbaki and his colleagues reserve the term "compact" for compact Hausdorff spaces.

1 Theorem *A topological space is compact if and only if each family of closed sets which has the finite intersection property has a non-void intersection.*

PROOF If $\mathcal{A}$ is a family of subsets of a topological space X, then, according to the De Morgan formulae, $X \sim \bigcup\{A: A \in \mathcal{A}\} = \bigcap\{X \sim A: A \in \mathcal{A}\}$ and hence $\mathcal{A}$ is a cover of X iff the intersection of the complements of the members of $\mathcal{A}$ is void. The space X is compact iff each family of open sets, such that no finite subfamily covers X, fails to be a cover, and this is true iff each family of closed sets which possesses the finite intersection property has a non-void intersection. ▮

2 Theorem *A topological space X is compact if and only if each net in X has a cluster point.*

Consequently, X is compact if and only if each net in X has a subnet which converges to some point of X.

PROOF Let $\{S_n, n \in D\}$ be a net in the compact topological space X and for each n in D let A_n be the set of all points S_m for $m \geqq n$. Then the family of all sets A_n has the finite intersection property because D is directed by $\geqq$, and consequently the family of all closures A_n^- also has the finite intersection property. Since X is compact there is a point s which belongs to each A_n^-, and according to theorem 2.7 such a point s is a cluster point of the net $\{S_n, n \in D\}$. To prove the converse proposition let X be a topological space in which every net has a cluster point and let $\mathcal{A}$ be a family of closed subsets of X such that $\mathcal{A}$ has the finite intersection property. Define $\mathcal{B}$ to be the family of all finite intersections of members of $\mathcal{A}$; then $\mathcal{B}$ has the finite intersection property and since $\mathcal{A} \subset \mathcal{B}$, it is sufficient to show $\bigcap\{B: B \in \mathcal{B}\}$ non-void. The intersection of two members of $\mathcal{B}$ is a member of $\mathcal{B}$ and therefore $\mathcal{B}$ is directed by $\subset$. If we choose a member S_B from each B in $\mathcal{B}$, then $\{S_B, B \in \mathcal{B}\}$ is a net in X and consequently has a cluster point s. If B and C are members of $\mathcal{B}$ such that $C \subset B$, then $S_C \in C \subset B$; therefore the net $\{S_B, B \in \mathcal{B}\}$ is eventually in the closed set B and hence the cluster point s belongs to B. Therefore s belongs to each member of $\mathcal{B}$ and the intersection of the members of $\mathcal{B}$ is non-void. Finally, the second statement of the theorem follows from the fact (2.6) that a point is a cluster point of a net iff some subnet converges to it. ∎

Under certain circumstances compactness can be characterized in terms of the existence of accumulation points of subsets. The following sequence of lemmas and the subsequent theorem indicate the situation. The problems at the end of the chapter show that the limitations imposed are necessary. It is convenient to use a variant of the notion of accumulation point in stating the results. A point x is an **ω-accumulation point** of a set A iff each neighborhood of x contains infinitely many points of A. Each ω-accumulation point of a set is also an accumulation point, and if the space is T_1 the converse holds.

3 LEMMA *Every sequence in a topological space has a cluster point if and only if every infinite set has an ω-accumulation point.*

PROOF Suppose that every sequence has a cluster point and that A is an infinite subset. Then there is a sequence of distinct points (a one-to-one sequence) in A, and each cluster point of such a sequence is clearly an ω-accumulation point of A. Conversely, if every infinite subset of a topological space has an accumulation point and $\{S_n, n \in \omega\}$ is a sequence in the space, then one of two situations must occur. Either the range of the sequence is infinite, in which case each ω-accumulation point of this infinite set is a cluster point of the sequence, or else the range of the sequence is finite. In the latter case, for some point x of the space, $S_n = x$ for infinitely many non-negative integers n, and x is a cluster point of the sequence. ∎

4 LEMMA *If X is a Lindelöf space and every sequence in X has a cluster point, then X is compact.*

PROOF It must be shown that each open cover of X has a finite subcover. Because of the hypothesis it may be assumed that the open cover consists of sets $A_0, A_1, \cdots, A_n \cdots$, for n in ω. Proceeding inductively, let $B_0 = A_0$ and for each p in ω let B_p be the first of the sequence of A's which is not covered by $B_0 \cup B_1 \cup \cdots \cup B_{p-1}$. If this choice is impossible at any stage, then the sets already selected are the required finite subcover. Otherwise it is possible to select a point b_p in B_p for each p in ω such that

$b_p \notin B_i$ for $i < p$. Let x be a cluster point of this sequence. Then $x \in B_p$ for some p, and since x is a cluster point, $b_q \in B_p$ for some $q > p$. But this is a contradiction. ∎

The following theorem summarizes information on sequences and subsequences, accumulation points and compactness.

5 THEOREM *If X is a topological space, then the conditions below are related as follows. For all spaces* (a) *is equivalent to* (b) *and* (d) *implies* (a). *If X satisfies the first axiom of countability, then* (a), (b), *and* (c) *are equivalent. If X satisfies the second axiom of countability, then all four conditions are equivalent. If X is pseudo-metric, then each of the four conditions implies that X satisfies the second countability axiom and all four are equivalent.*

(a) *Every infinite subset of X has an ω-accumulation point.*
(b) *Every sequence in X has a cluster point.*
(c) *For each sequence in X there is a subsequence converging to a point of X.*
(d) *The space X is compact.*

PROOF Lemma 5.3 states that (a) is equivalent to (b) and since a sequence is a net, 5.2 shows that (d) always implies (b). If X satisfies the first axiom of countability then (b) and (c) are equivalent by 2.8. If X satisfies the second axiom of countability, then every open cover has a countable subcover, lemma 5.4 applies, and hence all four statements are equivalent. If X is pseudo-metric, then X satisfies the first axiom of countability, the first three conditions are equivalent, each is implied by compactness, and the theorem will be proved if it is shown that a pseudo-metric space such that each infinite subset has an accumulation point is separable and hence satisfies the second axiom of countability. Suppose that X is such a pseudo-metric space. For r positive consider the family of all sets A such that the distance between any two distinct points of A is at least r. It is easily seen that this family has a maximal member A_r by 0.25. The set A_r must be finite, for the $r/2$ sphere about each point of X contains at most one member of A_r and therefore A_r has no accumulation point. Moreover, the r-sphere about each point x of X must intersect A_r because A_r is maximal and otherwise x could be adjoined to A_r. Finally the union A of sets A_r, for r

the reciprocal of a positive integer, is surely countable and A is clearly dense in X. ∎

If $\mathcal{B}$ is a base for the topology of a compact space X and $\mathcal{A}$ is a cover of X by members of $\mathcal{B}$, then there is a finite subcover of $\mathcal{A}$. Conversely, suppose that $\mathcal{B}$ is a base for the topology and that every cover by members of $\mathcal{B}$ has a finite subcover. If $\mathcal{C}$ is an arbitrary open cover of X define $\mathcal{A}$ to be the family of all members of $\mathcal{B}$ which are subsets of some member of $\mathcal{C}$. Because $\mathcal{B}$ is a base, the family $\mathcal{A}$ is a cover of X, and consequently there is a finite subcover $\mathcal{A}'$ of $\mathcal{A}$. For each member of $\mathcal{A}'$ we may select a member of $\mathcal{C}$ which contains it, and the result is a finite subcover of $\mathcal{C}$. This shows that, if "a base for a topology is compact," then the space is compact. This is a useful but not a very profound result. The corresponding theorem on subbases is both profound and useful.

6 THEOREM (ALEXANDER) *If $\mathcal{S}$ is a subbase for the topology of a space X such that every cover of X by members of $\mathcal{S}$ has a finite subcover, then X is compact.*

PROOF For brevity let us agree that a family of subsets of X is inadequate iff it fails to cover X, and is finitely inadequate iff no finite subfamily covers X. Then the definition of compactness of X can be stated: each finitely inadequate family of open sets is inadequate. Observe that the class of finitely inadequate families of open sets is of finite character and therefore each finitely inadequate family is contained in a maximal family by Tukey's lemma 0.25(c). Such a maximal finitely inadequate family $\mathcal{A}$ has a special property which is established as follows.* If $C \notin \mathcal{A}$ and C is open, then by maximality there is a finite subfamily $A_1, \cdots A_m$ of $\mathcal{A}$ such that $C \cup A_1 \cup \cdots A_m = X$. Hence no open set containing C belongs to $\mathcal{A}$. If D is another open set and $D \notin \mathcal{A}$, then there is $B_1, \cdots, B_n$ in $\mathcal{A}$ such that $D \cup B_1 \cup \cdots \cup B_n = X$ and $(C \cap D) \cup A_1 \cup \cdots \cup A_m \cup B_1 \cup \cdots \cup B_n = X$ by a simple set theoretic calculation. It follows that $C \cap D \notin \mathcal{A}$. Consequently, if no member of a finite family of open sets belongs to $\mathcal{A}$, then no open set containing the intersection belongs to $\mathcal{A}$; restated, if a member of $\mathcal{A}$ contains a finite intersection $C_1 \cap C_2 \cdots \cap C_p$ of open sets, then some $C_i \in \mathcal{A}$.

* Problem 2.I is precisely the result needed here.

The proof of the theorem is now straightforward. Suppose that $\mathcal{S}$ is a subbase such that each open cover by subbase elements has a finite subcover (that is, each finitely inadequate subfamily is inadequate) and suppose that $\mathcal{B}$ is a finitely inadequate family of open subsets of X. Then there is a maximal family $\mathcal{A}$ of this sort containing $\mathcal{B}$ and it is sufficient to show that $\mathcal{A}$ is inadequate. The family $\mathcal{S} \cap \mathcal{A}$ of all members of $\mathcal{A}$ which belong to $\mathcal{S}$ is finitely inadequate and hence $\mathcal{S} \cap \mathcal{A}$ does not cover X. Consequently the theorem will be proved if it is shown that each point in $\bigcup \{A: A \in \mathcal{A}\}$ belongs to $\bigcup \{A: A \in \mathcal{S} \cap \mathcal{A}\}$. Because $\mathcal{S}$ is a subbase each point x of a member A of $\mathcal{A}$ belongs to some finite intersection of members of $\mathcal{S}$ which is contained in A. The paragraph above shows that some one of this finite family belongs to $\mathcal{A}$, hence $\bigcup \{A: A \in \mathcal{A}\} = \bigcup \{A: A \in \mathcal{S} \cap \mathcal{A}\}$, and the theorem is proved. ▮

COMPACTNESS AND SEPARATION PROPERTIES

In this section the consequences of compactness in conjunction with the so-called separation axioms will be examined. In each case the theorem proved is the assumed separation axiom (Hausdorff, regular, completely regular) with the word "point" replaced by "compact set." A simple but important corollary on continuous mappings of compact spaces into Hausdorff spaces is derived, and finally we prove a separation theorem of A. D. Wallace which includes most of the earlier theorems.

It is always true that a closed subset A of a compact space X is compact, for each net in A has a subnet which converges to a point which belongs to A because A is closed. (A proof based directly on the definition of compactness is almost as simple.) The converse theorem is false, for if A is a proper non-void subset of an indiscrete space X (only X and the void set are open), then A is surely compact but not closed. This cannot happen if X is a Hausdorff space.

7 THEOREM *If A is a compact subset of a Hausdorff space X and x is a point of $X \sim A$, then there are disjoint neighborhoods of x and of A.*

Consequently each compact subset of a Hausdorff space is closed.

PROOF Since X is Hausdorff there is a neighborhood U of each point y of A such that x does not belong to the closure U^-. Because A is compact there is a finite family $U_0, U_1, \cdots, U_n$ of open sets covering A such that $x \notin U_i^-$ for $i = 0, 1, \cdots, n$. If $V = \bigcup \{U_i: i = 0, 1, \cdots, n\}$, then $A \subset V$ and $x \notin V^-$. Consequently $X \sim V^-$ and V are disjoint neighborhoods of x and A. ▮

8 THEOREM *Let f be a continuous function carrying the compact topological space X onto the topological space Y. Then Y is compact, and if Y is Hausdorff and f is one to one then f is a homeomorphism.*

PROOF If $\mathfrak{A}$ is an open cover of Y, then the family of all sets of the form $f^{-1}[A]$, for A in $\mathfrak{A}$, is an open cover of X which has a finite subcover. The family of images of members of the subcover is a finite subfamily of $\mathfrak{A}$ which covers Y and consequently Y is compact. Suppose that Y is Hausdorff and f is one to one. If A is a closed subset of X, then A is compact and hence its image $f[A]$ is compact and therefore closed. Then $(f^{-1})^{-1}[A]$ is closed for each closed set A and f^{-1} is continuous. ▮

9 THEOREM *If A and B are disjoint compact subsets of a Hausdorff space X, then there are disjoint neighborhoods of A and B.*
Consequently each compact Hausdorff space is normal.

PROOF For each x in A there is by 5.7 a neighborhood of x and a neighborhood of B which are disjoint. Consequently there is a neighborhood U of x whose closure is disjoint from B, and since A is compact there is a finite family $U_0, U_1, \cdots, U_n$ such that U_i^- is disjoint from B for $i = 0, 1, \cdots, n$ and $A \subset V = \bigcup \{U_i: i = 0, 1, \cdots, n\}$. Then V is a neighborhood of A and $X \sim V^-$ is a neighborhood of B which is disjoint from V. ▮

10 THEOREM *If X is a regular topological space, A a compact subset, and U a neighborhood of A, then there is a closed neighborhood V of A such that $V \subset U$.*
Consequently each compact regular space is normal.

PROOF Because X is regular, for each x in A there is an open neighborhood W of x such that $W^- \subset U$, and by compactness there is a finite open cover $W_0, W_1, \cdots, W_n$ of A such that $W_i^- \subset U$ for each i. Then $V = \bigcup \{W_i^-: i = 0, 1, \cdots, n\}$ is the required neighborhood of A. ▌

11 THEOREM *If X is a completely regular space, A is a compact subset and U is a neighborhood of A, then there is a continuous function f on X to the closed interval* [0,1] *such that f is one on A and zero on $X \sim U$.*

PROOF For each x in A there is a continuous function g which is one at x and zero on $X \sim U$. The set $\{y: g(y) > \frac{1}{2}\}$ is open in X and hence if h is defined by $h(y) = \min [2g(y),1]$, then h is continuous, has values in [0,1], is zero on $X \sim U$, and is one on a neighborhood of x. Because A is compact there is a finite family $h_0, h_1, \cdots h_n$ of continuous functions on X to [0,1] such that $A \subset \bigcup \{h_i^{-1}[1]: i = 0, 1, \cdots, n\}$ and each h_i is zero on $X \sim U$. The function f whose value at x is $\max \{h_i(x): i = 0, 1, \cdots, n\}$ is the required function. ▌

Each of the last two theorems has a formulation which is superficially different; the statement "A is compact and U a neighborhood of A" can be replaced by "if A is compact and B is a disjoint closed set," and the conclusion changed in the obvious way.

Most of the results of this section are easy consequences of the following theorem.

12 THEOREM (WALLACE) *If X and Y are topological spaces, A and B are compact subsets of X and Y respectively, and W is a neighborhood of $A \times B$ in the product space $X \times Y$, then there are neighborhoods U of A and V of B such that $U \times V \subset W$.*

PROOF For each member (x,y) of $A \times B$ there are open neighborhoods R of x and S of y such that $R \times S \subset W$. Since B is compact, for a fixed x in A there are neighborhoods R_i of x and corresponding open sets S_i, for $i = 0, 1, \cdots n$, such that $B \subset Q = \bigcup \{S_i: i = 0, 1, \cdots, n\}$. If $P = \bigcap \{R_i: i = 0, 1, \cdots n\}$, then P is a neighborhood of x and Q is a neighborhood of B such that $P \times Q \subset W$. Since A is compact there are open sets P_i

in X and Q_i in Y, for $i = 0, 1, \cdots m$, such that each Q_i is a neighborhood of B, $P_i \times Q_i \subset W$, and $A \subset \bigcup \{P_i: i = 0, 1, \cdots m\} = U$. Then U and $V = \bigcap \{Q_i: i = 0, 1, \cdots, m\}$ are neighborhoods of A and B respectively, $U \times V$ is a subset of W, and the theorem follows. ▌

PRODUCTS OF COMPACT SPACES

The classical theorem of Tychonoff on the product of compact spaces is unquestionably the most useful theorem on compactness. It is probably the most important single theorem of general topology. This section is devoted to the Tychonoff theorem and a few of its consequences.

13 THEOREM (TYCHONOFF) *The cartesian product of a collection of compact topological spaces is compact relative to the product topology.*

PROOF Let $Q = \times\{X_a: a \in A\}$ where each X_a is a compact topological space and Q has the product topology. Let $\mathcal{S}$ be the subbase for the product topology consisting of all sets of the form $P_a^{-1}[U]$ where P_a is the projection into the a-th coordinate space and U is open in X_a. In view of theorem 5.6 the space Q will be compact if each subfamily $\mathcal{A}$ of $\mathcal{S}$, such that no finite subfamily of $\mathcal{A}$ covers Q, fails to cover Q. For each index a let $\mathcal{B}_a$ be the family of all open sets U in X_a such that $P_a^{-1}[U] \in \mathcal{A}$. Then no finite subfamily of $\mathcal{B}_a$ covers X_a and hence by compactness there is a point x_a such that $x_a \in X_a \sim U$ for each U in $\mathcal{B}_a$. The point x whose a-th coordinate is x_a then belongs to no member of $\mathcal{A}$ and consequently $\mathcal{A}$ is not a cover. ▌

We give an alternate proof of Tychonoff's theorem which does not depend on the Alexander theorem 5.6.

ALTERNATE PROOF (BOURBAKI) It will be proved that if $\mathcal{B}$ is a family of subsets of the product and $\mathcal{B}$ has the finite intersection property, then $\bigcap \{B^-: B \in \mathcal{B}\}$ is not void. The class of all families which possess the finite intersection property is of finite character and consequently we may assume that $\mathcal{B}$ is maximal with respect to this property by Tukey's lemma 0.25(c). Because $\mathcal{B}$

is maximal each set which contains a member of $\mathcal{B}$ belongs to $\mathcal{B}$ and the intersection of two members of $\mathcal{B}$ belongs to $\mathcal{B}$. Moreover, if C intersects each member of $\mathcal{B}$, then $C \in \mathcal{B}$ by maximality.* Finally, the family of projections of members of $\mathcal{B}$ into a coordinate space X_a has the finite intersection property and it is therefore possible to choose a point x_a in $\bigcap\{P_a[B]^-: B \in \mathcal{B}\}$. The point x whose a-th coordinate is x_a then has the property: each neighborhood U of x_a intersects $P_a[B]$ for every B in $\mathcal{B}$, or equivalently $P_a^{-1}[U] \in \mathcal{B}$, for each neighborhood U of x_a in X_a. Therefore finite intersections of sets of this form belong to $\mathcal{B}$. Then each neighborhood of x which belongs to the defining base for the product topology belongs to $\mathcal{B}$ and hence intersects each member of $\mathcal{B}$. Therefore x belongs to B^- for each B in $\mathcal{B}$, and the theorem is proved. ∎

Several important applications of Tychonoff's theorem occur in the chapter on function spaces; for the moment we consider a very simple consequence. A subset of a pseudo-metric space is **bounded** iff it is of finite diameter. Thus a subset of the space of real numbers is bounded iff it has both an upper and lower bound. The following is the classical theorem of Heine-Borel-Lebesgue.

14 THEOREM *A subset of Euclidean n-space is compact if and only if it is closed and bounded.*

PROOF Let A be a compact subset of E_n. Then A is closed because E_n is a Hausdorff space. Because of compactness A can be covered by a finite family of open spheres of radius one, and because each of these is bounded A is bounded. To prove the converse suppose that A is a closed and bounded subset of E_n. Let B_i be the image of A under the projection into the i-th coordinate space, and notice that each B_i is bounded because the projection decreases distances. Then $A \subset \times\{B_i: i = 0, 1, \cdots, n-1\}$, and this set is a subset of a product of closed bounded intervals of real numbers. Since A is a closed subset of the product, and the product of compact spaces is compact, the proof reduces to showing that a closed interval $[a,b]$ is compact relative to the usual topology. Let $\mathcal{C}$ be an open cover of $[a,b]$ and

* We are evidently reproving part of proposition 2.I.

let c be the supremum of all members x of $[a,b]$ such that some finite subfamily of $\mathfrak{C}$ covers $[a,x]$. (The set is not void because a is a member.) Choose U in $\mathfrak{C}$ such that $c \in U$, and choose a member d of the open interval (a,c) such that $[d,c] \subset U$. There is a finite subfamily of $\mathfrak{C}$ which covers $[a,d]$, and this family with U adjoined covers $[a,c]$. Unless $c = b$ the same finite subfamily covers an interval to the right of c, which contradicts the choice of c. The theorem follows. ▮

The closed unit interval is compact and consequently each cube (the product of closed unit intervals) is compact. The following characterization of Tychonoff spaces (completely regular T_1-spaces) is then almost self-evident.

15 THEOREM *A topological space is a Tychonoff space if and only if it is homeomorphic to a subspace of a compact Hausdorff space.*

PROOF By 4.6, each Tychonoff space is homeomorphic to a subset of a cube, which is a compact Hausdorff space. Conversely, each compact Hausdorff space is normal and consequently (Urysohn's lemma 4.4) is a Tychonoff space, and each subspace is therefore a Tychonoff space. ▮

The product of more than a finite number of non-compact spaces fails to be compact in a rather spectacular way. A set in a topological space is **nowhere-dense** in the space iff its closure has a void interior.

16 THEOREM *If an infinite number of the coordinate spaces are non-compact, then each compact subset of the product is nowhere dense.*

PROOF Suppose that $\times\{X_a: a \in A\}$ has a compact subset B with an interior point x. Then B contains a neighborhood U of x which is a member of the defining base and is therefore of the form $\bigcap\{P_a^{-1}[V_a]: a \in F\}$, where F is a finite subset of A and V_a is open in X_a. If b is a member of $A \sim F$, then $P_b[B] = X_b$ and X_b is therefore compact because it is the continuous image of a compact space. Hence all but a finite number of the coordinate spaces are compact. ▮

LOCALLY COMPACT SPACES

A topological space is **locally compact** iff each point has at least one compact neighborhood. A compact space is automatically locally compact, every discrete space is locally compact, and each closed subspace of a locally compact space is itself locally compact (the intersection of a closed set and a compact set is a closed subset of the latter, and hence compact). Many of the pleasant properties of compact spaces are shared by locally compact spaces. The following proposition is a convenient tool for the study of such spaces.

17 THEOREM *If X is a locally compact topological space which is either Hausdorff or regular, then the family of closed compact neighborhoods of each point is a base for its neighborhood system.*

PROOF Let x be a point of X, C a compact neighborhood of x, and U an arbitrary neighborhood of x. If X is regular, then there is a closed neighborhood V of x which is a subset of the intersection of U and the interior of C, and evidently V is closed and compact. If X is Hausdorff and W is the interior of $U \cap C$, then, since W^- is a compact Hausdorff space, W contains a closed compact set V which is a neighborhood of x in W^- by 5.9; but V is also a neighborhood of x in W (that is, with respect to the relativized topology for W) and is therefore a neighborhood of x in X. ∎

In particular it follows that every locally compact Hausdorff space is regular; actually a stronger statement is true.

18 THEOREM *If U is a neighborhood of a closed compact subset A of a regular locally compact topological space X, then there is a closed compact neighborhood V of A such that $A \subset V \subset U$.*

Moreover, there is a continuous function f on X to the closed unit interval such that f is zero on A and one on $X \sim V$.

PROOF For each point x of A there is a neighborhood W which is a closed compact subset of U. Since A is compact it may be covered by a finite family of such neighborhoods and their union is a closed compact neighborhood V of A. Then V with the relative topology is a regular compact space which is therefore normal (5.10). Hence there is a continuous function g on V to the closed unit interval such that g is zero on A and one on $V \sim V^0$ (V^0 is the interior of V). Let f equal g on V and one on $X \sim V$. Then f is continuous because V^0 and $X \sim V$ are separated and f is continuous on V and $X \sim V^0$. (Problem 3.B.) ∎

It follows that each locally compact, regular, topological space is completely regular and each locally compact Hausdorff space is a Tychonoff space.

It is not true that the continuous image of a locally compact space is locally compact, for every discrete space is locally compact and each topological space is the continuous one-to-one image of a discrete space (using the same set, the discrete topology, and the identity function). If a function is both open and continuous, then the image of a compact neighborhood of a point is a compact neighborhood of the image point, and consequently the image of a locally compact space is locally compact. This simple fact and an earlier result give a precise description of those product spaces which are locally compact.

19 THEOREM *If a product is locally compact, then each coordinate space is locally compact and all except a finite number of coordinate spaces are compact.*

PROOF If a product is locally compact, then each coordinate space is locally compact because projection into a coordinate space is open. If infinitely many coordinate spaces are non-compact, then each compact subset of the product is nowhere dense, according to 5.16, and no point has a compact neighborhood. ∎

QUOTIENT SPACES

In this section the investigation of quotient spaces which was begun in chapter 3 is continued. We are interested in the consequences of compactness and the single theorem of the section summarizes some of the pleasant properties which result from the additional assumption. It has already been observed that the continuous image of a compact space is compact, but without additional hypotheses the image space may still be quite unattractive. For example, if X is the closed unit interval with the usual topology and $\mathfrak{D}$ is the decomposition consisting of all subsets of the form $\{x: x - a \textit{ is rational}\}$, then the quotient space is compact and the projection onto the quotient space is open, but the quotient topology is indiscrete (only the space and the void set open). It turns out that, if the members of $\mathfrak{D}$ are compact and the decomposition is upper semi-continuous, then the quotient space inherits many of the properties of X.

20 THEOREM *Let X be a topological space, let $\mathfrak{D}$ be an upper semi-continuous decomposition of X whose members are compact, and let $\mathfrak{D}$ have the quotient topology. Then $\mathfrak{D}$ is, respectively, Hausdorff, regular, locally compact, or satisfies the second axiom of countability, provided X has the corresponding property.*

PROOF For convenience let us agree that a subset of X is admissible iff it is the union of members of $\mathfrak{D}$. In view of the definition of upper semi-continuity each neighborhood in X of a member A of $\mathfrak{D}$ contains an admissible neighborhood, and hence the image under projection of a neighborhood of A in X is a neighborhood of A in $\mathfrak{D}$. Moreover, projection carries closed sets into closed sets (3.12). Suppose that X is a Hausdorff space and that A and B are distinct members of $\mathfrak{D}$. Then by 5.9 there are disjoint neighborhoods (in X) of A and B, these contain disjoint admissible neighborhoods, and the projections of the latter are the required disjoint neighborhoods of A and B in $\mathfrak{D}$. If X is regular, $A \in \mathfrak{D}$, and $\mathfrak{U}$ is a neighborhood of A in $\mathfrak{D}$, then the union U of the members of $\mathfrak{U}$ is a neighborhood of A in X. In view of 5.10 there is a closed neighborhood of A contained in U, and the image under projection of this neighborhood is the required neighborhood of A in $\mathfrak{D}$. If X is locally compact, then evidently there is a compact neighborhood of each member of $\mathfrak{D}$, and the image under projection is a compact neighborhood in $\mathfrak{D}$.

Finally, suppose there is a countable base $\mathfrak{B}$ for the topology of X. The family $\mathfrak{U}$ of unions of finite subfamilies of $\mathfrak{B}$ is countable. For each member U of $\mathfrak{U}$ let U' be the union of all members of $\mathfrak{D}$ which are subsets of U, and let $\mathfrak{J}$ be the family of all sets U' for U in $\mathfrak{U}$. Then the images of the members of $\mathfrak{J}$ are open and it will be shown that the collection of images is a base for the quotient topology. This will follow if for each A in $\mathfrak{D}$ and each neighborhood V of A there is U in $\mathfrak{I}$ such that $A \subset U \subset V$. But A may be covered by a finite number of the members of $\mathfrak{B}$ such that the union W of these members, which is a member of $\mathfrak{U}$, is contained in V. If $U = W'$, then $U \varepsilon \mathfrak{I}$ and $A \subset U \subset V$, and the theorem follows. ∎

There is an interesting corollary to this theorem. If X is separable metric and the members of an upper semi-continuous decomposition are compact, then the quotient space is Hausdorff, normal, and satisfies the second axiom of countability, and is consequently metrizable.

COMPACTIFICATION

In studying a non-compact topological space X it is often convenient to construct a space which contains X as a subspace and is itself compact. For example, it is frequently useful to adjoin two points, $+\infty$ and $-\infty$, to the space of real numbers. The resulting space is sometimes called the *extended* real numbers; it is linearly ordered by agreeing that $+\infty$ is the largest member and $-\infty$ is the smallest. With this ordering (an extension of the usual ordering) it turns out that every non-void subset of the extended real numbers has both an infimum and a supremum and the space is compact relative to its order topology (5.C). The extended reals are a *compactification* of the space of real numbers, in a sense which will presently be made precise. Of course, this device is primarily a convenience. It does not add to our knowledge of the real numbers. However, it does permit the use of the standard compactness arguments and it simplifies many proofs materially.

The simplest sort of compactification of a topological space is made by adjoining a single point. This procedure is familiar in analysis, for in function theory the complex sphere is constructed by adjoining a single point, ∞, to the Euclidean plane and specifying that the neighborhoods of ∞ are the complements of bounded subsets of the plane. This construction can be duplicated for an arbitrary topological space; the clue to the topology to be introduced in the enlarged space is the fact that the complement of an open neighborhood of ∞ in the complex sphere is compact.

The **one point compactification** * of a topological space X is the set $X^* = X \cup \{\infty\}$ with the topology whose members are the open subsets of X and all subsets U of X^* such that $X^* \sim U$ is a closed compact subset of X. Of course, it must be verified that this specification gives a topology for X^*. This verification is made in the proof of the following proposition.

21 THEOREM (ALEXANDROFF) *The one point compactification X^* of a topological space X is compact and X is a subspace. The space X^* is Hausdorff if and only if X is locally compact and Hausdorff.*

PROOF A set U is open in X^* iff (a) $U \cap X$ is open in X and (b) whenever $\infty \in U$, then $X \sim U$ is compact. Consequently finite intersections and arbitrary unions of sets open in X^* intersect X in open sets. If ∞ is a member of the intersection of two open subsets of X^*, then the complement of the intersection is the union of two closed compact subsets of X and is therefore closed and compact. If ∞ belongs to the union of the members of a family of open subsets of X^*, then ∞ belongs to some member U of the family, and the complement of the union is a closed subset of the compact set $X \sim U$ and is therefore closed and compact. Consequently X^* is a topological space and X is a subspace. If $\mathfrak{U}$ is an open cover of X^*, then ∞ is a member of some U in $\mathfrak{U}$ and $X \sim U$ is compact, and hence there is a finite subcover of $\mathfrak{U}$. Therefore X^* is compact. If X^* is a Hausdorff space, then its open subspace X is a locally compact Hausdorff space. Finally it must be shown that X^* is a Hausdorff space if X is a locally compact Hausdorff space. It is only necessary to show that, if $x \in X$, then there are disjoint neighborhoods of x and ∞. But since X is locally compact and Hausdorff there is a closed compact neighborhood U of x in X and $X^* \sim U$ is the required neighborhood of ∞. ∎

If X is a compact topological space, then ∞ is an isolated point of the one point compactification (that is, $\{\infty\}$ is both open and closed). Conversely, if ∞ is an isolated point of X^*, then X is closed in X^* and is therefore compact.

The one point compactification is of a very special sort, and

* This definition is actually incomplete until ∞ is defined. Any element which is not a member of X, for example X, will do.

we wish to consider other methods of embedding a topological space in a compact space. It is convenient to allow a topological embedding rather than insist that the original be actually a subspace of the constructed compact space. With this in mind, a **compactification** of a topological space X is defined to be a pair (f,Y), where Y is a compact topological space and f is a homeomorphism of X onto a dense subspace of Y. (To be consistent, the one point compactification of X should be the pair (i,X^*), where i is the identity function.) A compactification (f,Y) is called Hausdorff iff Y is a Hausdorff space. A relation is defined on the collection of all compactifications of a space X by agreeing that $(f,Y) \geqq (g,Z)$ iff there is a continuous map h of Y onto Z such that $h \circ f = g$. Equivalently $(f,Y) \geqq (g,Z)$ iff the function $g \circ f^{-1}$ on $f[X]$ to Z has a continuous extension h which carries Y onto Z. If the function h can be taken to be a homeomorphism, then (f,Y) and (g,Z) are said to be **topologically equivalent.** In this case both of the relations $(f,Y) \geqq (g,Z)$ and $(g,Z) \geqq (f,Y)$ hold, for h^{-1} is a continuous map of Z onto Y such that $f = h^{-1} \circ g$.

22 THEOREM *The collection of all compactifications of a topological space is partially ordered by* $\geqq$. *If* (f,Y) *and* (g,Z) *are Hausdorff compactifications of a space and* $(f,Y) \geqq (g,Z) \geqq (f,Y)$, *then* (f,Y) *and* (g,Z) *are topologically equivalent.*

PROOF If $(f,Y) \geqq (g,Z) \geqq (h,U)$, where these are compactifications of a space X, then there are continuous functions j on Y to Z and k on Z to U such that $g = j \circ f$ and $h = k \circ g$ and hence $h = k \circ j \circ f$ and $(f,Y) \geqq (h,U)$. Consequently $\geqq$ partially orders the collection of all compactifications of X. If (f,Y) and (g,Z) are Hausdorff compactifications each of which follows the other relative to the ordering $\geqq$, then both $f \circ g^{-1}$ and $g \circ f^{-1}$ have continuous extensions j and k to all of Z and Y respectively. Since $k \circ j$ is the identity map on the dense subset $g[X]$ of Z and Z is Hausdorff, $k \circ j$ is the identity map of Z onto itself and similarly $j \circ k$ is the identity map of Y onto Y. Consequently (f,Y) and (g,Z) are topologically equivalent. ∎

The smallest compactification of a compact Hausdorff space X is X itself (more precisely (i,X) where i is the identity map on

X). One would expect that the one point compactification of a non-compact space would be the smallest relative to the ordering $\geqq$. If we restrict our attention to Hausdorff compactifications this is actually the case (a corollary to 5.G), although it is easy to see that there is generally no compactification which is smaller than every other. On the other hand, if X is a space which has a Hausdorff compactification (by 5.15 such a space is a Tychonoff space), then there is a largest compactification which we now construct.

For each topological space X let $F(X)$ be the family of all continuous functions on X to the closed unit interval Q. The cube $Q^{F(X)}$ (the product of the unit interval Q taken $F(X)$ times) is compact by the Tychonoff theorem. The evaluation map e carries a member x of X into the member $e(x)$ of $Q^{F(X)}$ whose f-th coordinate is $f(x)$ for each f in $F(X)$. Evaluation is a continuous map of X into the cube $Q^{F(X)}$, and if X is a Tychonoff space, then e is a homeomorphism of X onto a subspace of $Q^{F(X)}$. (The embedding lemma 4.5 states these facts explicitly.) The **Stone-Čech compactification** is the pair $(e, \beta(X))$ where $\beta(X)$ is the closure of $e[X]$ in the cube $Q^{F(X)}$. We take time out for a lemma before showing the crucial property of this compactification.

23 LEMMA *If f is a function on a set A to a set B and f^* is the map of Q^B into Q^A defined by $f^*(y) = y \circ f$ for all y in Q^B, then f^* is continuous.*

PROOF A map into a product space is continuous iff the map followed by each projection is continuous (3.3). If a is a member of A, then $P_a \circ f^*(y) = P_a(y \circ f) = y(f(a))$. But $y(f(a))$ is simply the projection of y into the $f(a)$-coordinate space of Q^B and this is a continuous map. ▌

The construction outlined in this lemma is worthy of notice, for it is used systematically in dealing with function spaces. Observe that the function f^* induced by f goes in the direction opposite to that of f, in the sense that f carries A into B while f^* carries Q^B into Q^A.

With the aid of this lemma the principal theorem on the Stone-Čech compactification becomes a routine though mildly intricate calculation.

24 THEOREM (STONE-ČECH) *If X is a Tychonoff space and f is a continuous function on X to a compact Hausdorff space Y, then there is a continuous extension of f which carries the compactification $\beta(X)$ into Y. (More precisely, if $(e,\beta(X))$ is the Stone-Čech compactification, then $f \circ e^{-1}$ can be extended to a continuous function on $\beta(X)$ to Y.)*

PROOF Given f define f^* on $F(Y)$ to $F(X)$ by letting $f^*(a) = a \circ f$ for each a in $F(Y)$. Continuing, define f^{**} on $Q^{F(X)}$ to $Q^{F(Y)}$ by letting $f^{**}(q) = q \circ f^*$ for each q in $Q^{F(X)}$. Let e be the evaluation map of X into $Q^{F(X)}$ and let g be the evaluation map of Y into $Q^{F(Y)}$. The following diagram shows the situation.

$$\begin{array}{ccc} \beta(X) \subset Q^{F(X)} & \xrightarrow{\quad f^{**} \quad} & Q^{F(Y)} \supset \beta(Y) \\ \uparrow e & & \uparrow g \\ X & \xrightarrow{\quad f \quad} & Y \end{array}$$

The map e is a homeomorphism, and the map g is a homeomorphism of Y onto $\beta(Y)$ because Y is compact Hausdorff. The map f^{**} is continuous by lemma 5.23 and, if it is shown that $f^{**} \circ e = g \circ f$, then it will follow that $g^{-1} \circ f^{**}$ is the required continuous extension of $f \circ e^{-1}$. If x is a member of X and h a member of $F(Y)$, then $(f^{**} \circ e)(x)(h) = (e(x) \circ f^*)(h) = e(x)(h \circ f) = h \circ f(x) = g(f(x))(h) = (g \circ f)(x)(h)$ because of the definitions of f^{**}, f^*, e, and g respectively. The theorem follows. ∎

The extension property of the foregoing theorem shows that the Stone-Čech compactification $(e,\beta(X))$ follows every other Hausdorff compactification in the ordering $\geq$ and is therefore the largest such compactification. If (f,Y) has this extension property, then $(f,Y) \geq (e,\beta(X))$ and consequently is topologically equivalent to $(e,\beta(X))$ by 5.22. Hence the compactification $(e,\beta(X))$ is characterized (to a topological equivalence) by the extension property of theorem 5.24.

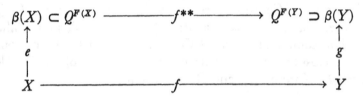

25 *Note* The results above (M. H. Stone [6] and Čech, [1]) furnish a maximal compactification. Many other smaller compactifications have been constructed for various purposes. There is a very large literature on the subject and it is only possible to cite a few sample contributions. For a recent contribution to one of the oldest compactification theories (Carathéodory's prime end theory) see Ursell and Young [1]. Freudenthal [1] examines a compactification which is maximal in a class much more restricted than that majorized by $\beta(X)$. A general discussion of compactification is given by Myškis ([1], [2], and [3]). He distinguishes between "external" descriptions of a compactification (such as that of $\beta(X)$, and the almost periodic compactification of a group as sketched in 7.**T**) and "internal" descriptions (for example the Alexandroff one point compactification and the Wallman (5.**R**)). The relation between internal and external description of a compactification is frequently the key to the usefulness of the notion. Certain parts of the internal structure of $\beta(X)$ have been discussed (see Nagata [2], Smirnov [3], and Wallace [2]). The compactification $\beta(X)$ is also related to the notion of absolute closure; see, for example, M. H. Stone [6], A. D. Alexandroff [1], Katětov [1], and Ramanathan [1].

LEBESGUE'S COVERING LEMMA

There is an extremely useful lemma of Lebesgue which states that, if $\mathfrak{U}$ is an open cover of a closed interval of real numbers, then there is a positive number r such that, if $|x - y| < r$, then both x and y belong to some member of the cover. In a certain sense each open cover covers the points of the interval "uniformly." In this section we prove this lemma and a topological variant which will apply to arbitrary compact spaces. The latter result may be considered to be an introduction to the ideas of the next section on paracompactness.

26 THEOREM *If $\mathfrak{U}$ is an open cover of a compact subset A of a pseudo-metric space (X,d), then there is a positive number r such that the open r-sphere about each point of A is contained in some member of $\mathfrak{U}$.*

PROOF Let $U_1, \cdots, U_n$ be a finite subcover of the open cover $\mathfrak{U}$ of A, let $f_i(x) = \text{dist}\,[x, X \sim U_i]$, and let $f(x) = \max\,[f_i(x)\colon i = 1, \cdots, n]$. Then each f_i is continuous and consequently f is continuous. Each point of A belongs to some U_i and hence

$f(x) \geqq f_i(x) > 0$ for each x in A. The set $f[A]$ is a compact subset of the positive real numbers and consequently there is a positive real number r such that $f(x) > r$ for all x in A. Hence for each x in A there is i such that $f_i(x) > r$ and it follows that the open r-sphere about x is contained in U_i. ▮

There is a useful corollary of the foregoing theorem. If A is a compact subset of a pseudo-metric space and U is a neighborhood of A, then there is a positive number r such that U contains the open r-sphere about every point of A; that is, the distance from A to $X \sim U$ is positive.

Theorem 5.26 may be rephrased in a suggestive way. If V is the set of all pairs of points of X such that $d(x,y) < r$, then $V[x] = \{y: (x,y) \in V\}$ is simply the open r-sphere about x. The set V is an open subset of $X \times X$ and contains the diagonal Δ (the set of all pairs (x,x) for x in X). The foregoing theorem then implies the following topological result: If $\mathfrak{U}$ is an open cover of a compact pseudo-metric space, then there is a neighborhood V of the diagonal in $X \times X$ such that for each point x the set $V[x]$ is contained in some member of $\mathfrak{U}$. This variant of the Lebesgue lemma turns out to be correct for arbitrary compact regular spaces.

A cover $\mathfrak{U}$ of a topological space is called an **even** cover iff there is a neighborhood V of the diagonal in $X \times X$ such that for each x the set $V[x]$ is contained in some member of $\mathfrak{U}$. In other words, the family of all sets of the form $V[x]$ refines $\mathfrak{U}$. Recall that a cover $\mathfrak{A}$ is a refinement of $\mathfrak{U}$ iff each member of $\mathfrak{A}$ is a subset of some member of $\mathfrak{U}$, and that a family $\mathfrak{B}$ of sets is locally finite iff there is a neighborhood of each point of the space which intersects only finitely many members of $\mathfrak{B}$. A family of sets is **closed** iff each member is closed.

27 THEOREM *If an open cover of a space has a closed locally finite refinement then it is an even cover.*

Consequently each open cover of a compact regular space is even.

PROOF Let $\mathfrak{U}$ be an open cover of a topological space X and let $\mathfrak{A}$ be a closed locally finite refinement. For each A in $\mathfrak{A}$ choose a member U_A of $\mathfrak{U}$ such that $A \subset U_A$, and let $V_A = (U_A \times U_A) \cup ((X \sim A) \times (X \sim A))$. Evidently V_A is an open neighborhood of the diagonal in $X \times X$, and, if $x \in A$, then $V_A[x] = U_A$. Letting $V = \bigcap \{V_A : A \in \mathfrak{A}\}$, it follows that for each point x the set $V[x] \subset V_A[x] = U_A$ and consequently the family of sets of the form $V[x]$ is a refinement of $\mathfrak{U}$. It remains to be proved that V is a neighborhood of the diagonal. For each point (x,x) of the diagonal choose a neighborhood W of x such that W intersects only finitely many members of $\mathfrak{A}$. If $W \cap A$ is void, then $W \subset X \sim A$ and $W \times W \subset V_A$. It follows that V contains the intersection of $W \times W$ with a finite number of the sets V_A and is therefore a neighborhood of (x,x).

Finally, if X is compact and regular, then each open cover $\mathfrak{U}$ has a closed finite refinement (cover X by open subsets whose closures refine $\mathfrak{U}$) and hence each open cover is even. ▮

* *PARACOMPACTNESS*

A topological space is **paracompact** iff it is regular * and each open cover has an open locally finite refinement. The purpose of this section is to prove the equivalence of paracompactness and a number of other conditions. The methods used are closely related to those of chapter 6.

Recall that a family $\mathfrak{A}$ of subsets of a topological space is discrete iff there is a neighborhood of each point of the space which intersects at most one member of the family. The family $\mathfrak{A}$ is σ-discrete (σ-locally finite) iff it is the union of countably many discrete (respectively locally finite) subfamilies. The principal theorem of the section can now be stated; its proof is given in the sequence of lemmas which follows the statement.

28 Theorem *If X is a regular topological space, then the following statements are equivalent.*

(a) *The space X is paracompact.*
(b) *Each open cover of X has a locally finite refinement.*
(c) *Each open cover of X has a closed locally finite refinement.*
(d) *Each open cover of X is even.*
(e) *Each open cover of X has an open σ-discrete refinement.*
(f) *Each open cover of X has an open σ-locally finite refinement.*

* The usual definition of paracompact specifies "Hausdorff" instead of "regular." It is not hard to show that a Hausdorff space is regular if each open cover has an open locally finite refinement.

The pattern of proof is (a) → (b) → (c) → (d) → (e) → (f) → (b) → (a). The first of these implications is clear, and the following lemma demonstrates the second.

29 LEMMA *If X is regular and each open cover has a locally finite refinement, then each open cover has a closed locally finite refinement.*

PROOF If $\mathfrak{U}$ is an open cover of X, then there is an open cover $\mathfrak{V}$ such that the family of closures of members of $\mathfrak{V}$ refines $\mathfrak{U}$, because X is regular. (For each x, if $x \in U$ there is an open neighborhood V of x such that $V^- \subset U$.) Let $\mathfrak{A}$ be a locally finite refinement of $\mathfrak{V}$. Then the family $\mathfrak{B}$ of all closures of members of $\mathfrak{A}$ is locally finite, and each member of $\mathfrak{B}$ is a subset of V^- for some V in $\mathfrak{V}$. Hence $\mathfrak{B}$ is the required closed locally finite refinement of $\mathfrak{U}$. ∎

For any topological space an open cover which has a closed locally finite refinement is even, according to 5.27. Hence statement (c) of the theorem implies (d). Before proving the next implication we prove two lemmas which are of some interest in themselves. For convenience we review some of the facts which will be needed (see the section on relations in chapter 0). If U is a subset of $X \times X$ and $x \in X$, then $U[x]$ is the set of all points y such that $(x,y) \in U$. If A is a subset of X, then $U[A] = \{y: (x,y) \in U \text{ for some } x \text{ in } A\}$; clearly $U[A]$ is the union of the sets $U[x]$ for x in A. The set $\{(x,y): (y,x) \in U\}$ is denoted by U^{-1}, and U is called symmetric if $U = U^{-1}$. The set $U \cap U^{-1}$ is always symmetric. If U and V are subsets of $X \times X$, then $U \circ V$ is the set of all pairs (x,z) such that for some y in X it is true that $(x,y) \in V$ and $(y,z) \in U$. In other words $(x,z) \in U \circ V$ iff $(x,z) \in V^{-1}[y] \times U[y]$ for some y, and consequently $U \circ V$ is the union of the sets $V^{-1}[y] \times U[y]$ for y in X. In particular if V is symmetric, then $V \circ V = \bigcup \{V[y] \times V[y]: y \in X\}$. Finally, for each subset A of X it is true that $(U \circ V)[A] = U[V[A]]$.

30 LEMMA *Let X be a topological space such that each open cover is even. If U is a neighborhood of the diagonal in $X \times X$ then there is a symmetric neighborhood V of the diagonal such that $V \circ V \subset U$.*

PROOF For each point x of X there is a neighborhood $W(x)$ such that $W(x) \times W(x) \subset U$, because U is a neighborhood of the diagonal. The family $\mathcal{W}$ of all sets of the form $W(x)$ is an open cover of X and there is therefore a neighborhood R of the diagonal such that the family of all sets $R[x]$ refines $\mathcal{W}$, and hence $R[x] \times R[x] \subset U$ for each x. Finally, let $V = R \cap R^{-1}$. Then V is a symmetric neighborhood of the diagonal and $V[x] \times V[x] \subset U$ for all x. Since $V \circ V$ is the union of the sets $V[x] \times V[x]$ it follows that $V \circ V \subset U$. ▮

The preceding lemma has the following intuitive content. Let us say two points x and y are at most U-distance apart if $(x,y) \in U$. Then there is V such that, if x and y, and y and z, are at most V-distance apart, then x and z are at most U-distance apart.

The following lemma shows that paracompact spaces satisfy a very strong normality condition.

31 LEMMA *Let X be a topological space such that each open cover is even and let $\mathcal{A}$ be a locally finite (or a discrete) family of subsets of X. Then there is a neighborhood V of the diagonal in $X \times X$ such that the family of all sets $V[A]$ for A in $\mathcal{A}$ is locally finite (respectively discrete).*

PROOF If $\mathcal{A}$ is a locally finite family of subsets there is an open cover $\mathcal{U}$ of X such that each member of $\mathcal{U}$ intersects only finitely many of the members of the family $\mathcal{A}$. Let U be a neighborhood of the diagonal such that the sets $U[x]$ refine $\mathcal{U}$. By the preceding lemma there is a neighborhood V of the diagonal such that $V \circ V \subset U$, and it may be supposed that $V = V^{-1}$. If $V \circ V[x] \cap A$ is void, then $V[x]$ must be disjoint from $V[A]$ because: if $y \in V[x] \cap V[A]$, then $(y,x) \in V^{-1} = V$, $(z,y) \in V$ for some z in A, and hence $(z,x) \in V \circ V$. Then $z \in V \circ V[x]$ and this is a contradiction. Consequently if $V[x]$ intersects $V[A]$, then $V \circ V[x]$ intersects A, and it follows that the family of all sets $V[A]$ for A in $\mathcal{A}$ is locally finite. If "finitely many" is replaced by "at most one," then a proof of the corresponding proposition for discrete families is obtained. ▮

If V is an open subset of $X \times X$, then $V[x]$ is open for every point x of X, because $V[x]$ is the inverse image of V under the continuous map which carries each point y of X into (x,y). If

A is a subset of X, then $V[A]$ is open because it is the union of the sets $V[x]$ for x in A. Consequently the preceding lemma permits us to enlarge each member of a locally finite or discrete family to an open set and still preserve the character of the family. In particular, if each open cover $\mathfrak{U}$ of a regular space has a locally finite refinement $\mathfrak{A}$, then the lemma applies (we have shown that (b) → (c) → (d) in 5.28) and there is an open neighborhood V of the diagonal such that the family of all sets $V[A]$ for A in $\mathfrak{A}$ is locally finite. The latter family may fail to be a refinement of $\mathfrak{U}$, but this is easily remedied by choosing U_A in $\mathfrak{U}$ such that $A \subset U_A$ and then letting $W_A = U_A \cap V[A]$. The family which is constructed in this fashion is clearly an open locally finite refinement of $\mathfrak{U}$ and it follows that the space is paracompact; that is, (b) → (a) in 5.28.

There is an obvious corollary to 5.31. A family consisting of two closed disjoint subsets is evidently discrete and hence:

32 Corollary *A paracompact space is normal.*

The proof of 5.28 will be complete if we establish two facts: If X is regular and each open cover is even, then each open cover has an open σ-discrete refinement, and if each open cover of X has an open σ-locally finite refinement, then each open cover has a locally finite refinement. (Evidently (e) → (f) in 5.28.)

33 Lemma *If X is a space such that each open cover is even, then every open cover of X has an open σ-discrete refinement.*

PROOF The proof, like that of 4.21, is an application of A. H. Stone's trick. (This lemma can be deduced from 4.21 and the results of chapter 6.) Because of lemma 5.31 it is sufficient to find a σ-discrete refinement of an open cover $\mathfrak{U}$, since such a σ-discrete refinement can then be "expanded" to an open σ-discrete refinement. Let V be an open neighborhood of the diagonal such that the family of all sets $V[x]$ for x in X refines $\mathfrak{U}$. Let $V_0 = V$ and select, inductively, V_n to be an open symmetric neighborhood of the diagonal such that $V_n \circ V_n \subset V_{n-1}$ for each positive integer n. Let $U_1 = V_1$ and, inductively, let $U_{n+1} = V_{n+1} \circ U_n$. It is easy to see that $U_n \subset V_0$ for each n, and it follows that for each n the family of all $U_n[x]$ for x in X refines

$\mathfrak{U}$. Choose a relation $<$ which well-orders X (see 0.25) and for each n and each x let $U_n{}^*(x) = U_n[x] \sim \bigcup \{U_{n+1}[y]: y < x\}$. For each fixed n the family $\mathfrak{U}_n$ of all sets $U_n{}^*(x)$ is discrete, as may be demonstrated as follows. Clearly $U_n{}^*(x)$ is disjoint from $V_{n+1}[U_n{}^*(y)]$ if $x \neq y$ because of the construction. If for some z in X the neighborhood $V_{n+1}[z]$ intersects $U_n{}^*(y)$, then $z \in V_{n+1}[U_n{}^*(y)]$ and $V_{n+1}[U_n{}^*(y)]$ is a neighborhood of z which intersects no set $U_n{}^*(x)$ for $x \neq y$. It follows that the family $\mathfrak{U}_n$ is discrete and it remains to be proved that each point of X belongs to some member of some $\mathfrak{U}_n$. For x in X choose y to be the first point of X such that x belongs to $U_n[y]$ for some n. Then surely $x \in U_n{}^*(y)$ for some n. ∎

34 Lemma *If each open cover of a space has an open σ-locally finite refinement, then each open cover has a locally finite refinement.*

PROOF Let $\mathfrak{U}$ be an open cover and let $\mathfrak{V}$ be an open σ-locally finite refinement. Suppose that $\mathfrak{V} = \bigcup \{\mathfrak{V}_n: n \in \omega\}$ where each $\mathfrak{V}_n$ is an open locally finite family. For each n and each member V of $\mathfrak{V}_n$ let $V^* = V \sim \bigcup \{U: U \in \mathfrak{V}_k \textit{ for some } k < n\}$, and let $\mathfrak{W}$ be the family of all sets of the form V^*. Then $\mathfrak{W}$ is a cover of X and a refinement of $\mathfrak{U}$. Finally, for x in X let n be the first integer such that x belongs to some member V of $\mathfrak{V}_n$. Then V is a neighborhood of x which is disjoint from every member of $\mathfrak{W}$ save those which were constructed from the families $\mathfrak{V}_k$ for $k \leqq n$. It follows that $\mathfrak{W}$ is locally finite. ∎

Theorem 4.21 states that each open cover of a pseudo-metrizable space has an open σ-discrete refinement. This fact and theorem 5.28 of this section then give the corollary:

35 Corollary *Each pseudo-metrizable space is paracompact.*

In conclusion it should be remarked that subspaces, quotients, and products of paracompact spaces are usually not paracompact. Moreover, a space may be locally metrizable, locally compact, Hausdorff, normal, and satisfy the first axiom of countability and still fail to be paracompact. The requisite examples are given in the problems at the end of this chapter.

36 *Notes* There is another characterization of paracompactness which might be added to the list given in 5.28. A regular space

is paracompact iff it is fully normal (see problem 5.V). This characterization is due to A. H. Stone [1]. The equivalences (b), (c), (e), and (f) of theorem 5.28 are due to E. Michael [1]. As far as I know, equivalence (d) was first noticed by J. S. Griffin and myself.

The σ-discrete characterization of paracompactness might well be taken as a definition of countable dimension (see Hurewicz and Wallman [1; 32] and Eilenberg [1]). There is an F_σ-theorem (Michael *loc. cit.*) which is also suggestive of dimension theory.

PROBLEMS

A EXERCISE ON REAL FUNCTIONS ON A COMPACT SPACE

(a) If A is a non-void compact subset of the space of real numbers, then both the supremum and the infimum of A belong to A.

(b) Each continuous real valued function f on a compact space X assumes a maximum and a minimum value. That is, there are points x and y of the space such that $f(x)$ and $f(y)$ are respectively the supremum and infimum of f on X.

(c) Let f be a continuous real valued function f on a compact space X. If f is always positive, then f is bounded away from zero, in the sense that there is $e > 0$ such that $f(x) > e$ for x in X.

B COMPACT SUBSETS

(a) The intersection of two compact subsets of a topological space may fail to be compact. The intersection of the members of an arbitrary family of closed and compact subsets is closed and compact. (Clearly two compact subsets with non-compact intersection must be subsets of a space which is not Hausdorff. Let X be the product of the space of real numbers and an indiscrete space which has two members.)

(b) The closure of a compact subset of a topological space may fail to be compact. However, the closure of a compact subset of a regular space is compact.

(c) If A and B are disjoint closed subsets of a pseudo-metric space and A is compact, then there is a member x of A such that dist $(A,B) =$ dist $(x,B) > 0$. (The function dist (x,B) is continuous in x and is positive for x in A.)

(d) If A and B are disjoint closed and compact subsets of a pseudo-metric space, then there are members x of A and y of B such that $d(x,y) = \text{dist}(A,B)$.

C COMPACTNESS RELATIVE TO THE ORDER TOPOLOGY

Let X be a set which is linearly ordered by a relation $<$ and let X have the order topology (see 1.I). Then every closed, order-bounded subset of X is compact iff X is order-complete relative to $<$. (The family of all subsets of X of the form $\{x: a < x\}$ or $\{x: x < a\}$ is a sub-base for the order topology for X and Alexander's subbase theorem 5.6 applies. A proof which is independent of 5.6 can be made via the argument which was used in 5.14.)

D ISOMETRIES OF COMPACT METRIC SPACES

Let X and Y be metric spaces, let X be compact, let f be an isometry of X onto a subspace of Y, and let g be an isometry of Y onto a subspace of X. Then f maps X onto Y. (If h is an isometry of X onto a proper subset of itself and $x \in X \sim h[X]$ let $a = \text{dist}(x,h[X])$. Define a sequence inductively by letting $x_0 = x$ and $x_{n+1} = h(x_n)$ and prove that, if $m \neq n$, then $d(x_m,x_n) \geqq a$.)

E COUNTABLY COMPACT AND SEQUENTIALLY COMPACT SPACES

A topological space is *countably compact* iff every countable open cover has a finite subcover. A space is *sequentially compact* iff every sequence has a convergent subsequence.

(a) A space is countably compact iff each sequence has a cluster point.

(b) A T_1-space is countably compact iff every infinite set has an accumulation point. (See 5.3.)

(c) A T_1-space is countably compact iff every infinite open cover has a proper subcover. (If A is an infinite set with no accumulation point, then each subset of A is closed. One may construct an open cover $\mathfrak{U}$ by choosing an open neighborhood of each point of A which contains no other point of A and then adjoining, if necessary, $X \sim A$. Then $\mathfrak{U}$ has no proper subcover. On the other hand, if $\mathfrak{V}$ is an open cover with no proper subcover then each member V of $\mathfrak{V}$ contains a point belonging to no other member of $\mathfrak{V}$.)

(d) A space satisfying the first countability axiom is countably compact iff it is sequentially compact (5.5).

(e) With the order topology, the set Ω_0 of all ordinal numbers less than the first uncountable ordinal Ω is locally compact, Hausdorff, satisfies the first axiom of countability, and is sequentially compact, but is not compact.

Note Proposition (c) is due to Arens and Dugundji [1].

F COMPACTNESS; THE INTERSECTION OF COMPACT CONNECTED SETS

(a) Let $\mathcal{A}$ be a family of closed compact sets such that $\bigcap\{A: A \in \mathcal{A}\}$ is a subset of an open set U. Then there is a finite subfamily $\mathcal{F}$ of $\mathcal{A}$ such that $\bigcap\{A: A \in \mathcal{F}\} \subset U$.

(b) If $\mathcal{A}$ is a family of compact subsets of a Hausdorff space X such that finite intersections of members of $\mathcal{A}$ are connected, then $\bigcap\{A: A \in \mathcal{A}\}$ is connected.

G PROBLEM ON LOCAL COMPACTNESS

If X is a Hausdorff space and Y is a dense locally compact subspace, then Y is open.

H NEST CHARACTERIZATION OF COMPACTNESS

A topological space X is compact iff each nest of closed non-void sets has a non-void intersection. (Recall that a nest is a family of sets which is linearly ordered by inclusion. If each nest of closed non-void sets has a non-void intersection and $\mathcal{A}$ is a family of closed sets with the finite intersection property, let $\mathcal{B}$ be a maximal family of closed sets which contains $\mathcal{A}$ and has the finite intersection property, and let $\mathcal{N}$ be a maximal nest in $\mathcal{B}$. Examination of the properties of $\mathcal{B}$ and of $\mathcal{N}$ leads to a proof. An entirely different proof can be based on well ordering, using part of the procedure outlined in the next problem.)

I COMPLETE ACCUMULATION POINTS

A point x is a *complete accumulation point* of a subset A of a topological space iff for each neighborhood U of x the sets A and $A \cap U$ have the same cardinal number. A topological space is compact iff each infinite subset has a complete accumulation point. (If X is not compact choose an open cover $\mathcal{A}$ with no finite subcover such that the cardinal number c of $\mathcal{A}$ is as small as possible. Let C be a well-ordered set of cardinal c such that the set of predecessors of each member has a cardinal less than c. (It is shown in the appendix that c is such a set.) Let f be a one-to-one map of C onto $\mathcal{A}$. Then for each member b of C the union $\bigcup\{f(a): a < b\}$ does not cover X and, in fact, the complement of this union must have cardinal number at least as great as c

It is therefore possible to choose x_b from the complement such that $x_a \neq x_b$ for $a < b$. Consider the set of all x_b.)

J EXAMPLE: UNIT SQUARE WITH DICTIONARY ORDER

Let X be the cartesian product of the closed unit interval Q with itself ordered by dictionary (lexicographic) order. (That is, $(a,b) < (c,d)$ iff $a < c$ or $a = c$ and $b < d$.) With the order topology X is compact, connected, and Hausdorff. It satisfies the first countability axiom but is not separable and is hence not metrizable.

K EXAMPLE (THE ORDINALS) ON NORMALITY AND PRODUCTS

The product of a locally compact, normal Hausdorff space and a compact Hausdorff space may fail to be normal. (The difficult part has already been established in 4.**E** and it is only necessary to show that Ω' and Ω_0 are compact and locally compact Hausdorff respectively. Ω' is the space of ordinals less than or equal to Ω and Ω_0 is the set of ordinals less than Ω, each with the order topology.)

L THE TRANSFINITE LINE

Let A be a well-ordered set, let the half-open interval $[0,1)$ have the usual order, let $A \times [0,1)$ have the dictionary (lexicographic) order, and let $A \times [0,1)$ have the order topology. Discuss the properties of this space.

M EXAMPLE: THE HELLY SPACE

The *Helly space* is the family H of all non-decreasing functions on the closed unit interval Q with values in Q. It is a subset of the product space Q^Q, and its topology is the relative product topology. The space H has the following properties:

(a) H is compact Hausdorff. (It is a closed subspace of Q^Q.)

(b) H satisfies the first axiom of countability and is hence sequentially compact. (The set of points of discontinuity of each member of H is countable. This fact, and the fact that Q is separable, must be used in constructing a countable base for the neighborhood system of a point h of H.)

(c) H is separable. (A countable dense set can be constructed using the rationals.)

(d) H is not metric. (For t in Q let $f_t(x)$ be 0 for $x < t$, 1 for $x > t$, and let $f_t(t) = \frac{1}{2}$. The family A of all functions of the form f_t is uncountable and no member of A is an accumulation point of A. But each subspace of a compact metric space is separable.)

N EXAMPLES ON CLOSED MAPS AND LOCAL COMPACTNESS

(a) Let X be the space of real numbers with the usual topology, let I be the set of integers, and let $\mathfrak{D}$ be the decomposition whose members are I and all sets $\{x\}$ for x in $X \sim I$. Then the projection of X onto the quotient space is closed and continuous, but the quotient space is not locally compact nor does it satisfy the first axiom of countability.

(b) Let Ω_0 be the set of all ordinal numbers less than Ω, with the order topology, let A be a closed uncountable set whose complement is also uncountable, and let $\mathfrak{D}$ be the decomposition whose members are A and all sets $\{x\}$ for x in $\Omega_0 \sim A$. Then the projection of Ω_0 onto the quotient space is continuous and closed and the quotient space is compact, but it fails to satisfy the first axiom of countability. (Use the interlacing lemma 4.E.)

O CANTOR SPACES

The *Cantor discontinuum* (*middle third* set) is the set of all members of the closed unit interval which have a triadic expansion in which the digit one does not occur. (It will be convenient throughout this problem to use only irrational triadic expansions, that is, expansions which are not identically zero from some point on. Each real number has a unique irrational expansion, as noted in 0.14.) The discontinuum is called the middle third set because: The (open) middle third of the interval [0,1] is precisely the set of numbers whose triadic expansions have ones in the first place after the "decimal" point. The middle third of each of the remaining intervals consists of points whose expansions have ones in the second but not the first place. Continuing, it is clear that the discontinuum can be obtained by successive deletion of middle thirds.

A product space 2^A (that is, all functions on a set A to the discrete space whose only members are 0 and 1, with the product topology) is called a *Cantor space*.

(a) The Cantor discontinuum is homeomorphic to 2^ω. For x in 2^ω let $f(x)$ be the member of [0,1] whose triadic expansion has the digit $2x(p)$ in the p-th place.)

(b) Each point of the discontinuum is an accumulation point and the complement of the discontinuum is an open dense subset of the real numbers.

(c) If A is a closed non-void subset of 2^ω, then there is a continuous function r on 2^ω to A such that $r(x) = x$ for x in A. (It is a little easier to see the proof if one looks at the Cantor discontinuum, which is the homeomorphic image of 2^ω.)

(d) Each compact Hausdorff space is the continuous image of a closed subset of some Cantor space. (Let F be the family of all functions f on 2 such that $f(0)$ and $f(1)$ are closed subsets of the compact Hausdorff space X and $f(0) \cup f(1) = X$. If x is a member of 2^F and $f \in F$, then $f(x_f)$ is a closed subset of X. The intersection $\bigcap \{f(x_f): f \in F\}$ is void or consists of a single point, and in the latter case this point is defined to be $\phi(x)$. One can prove that the domain of ϕ is a closed subset of 2^F; if U is a subset of X, then $\phi^{-1}[U] = \{x: x$ *is a member of domain* ϕ *and* $\bigcap \{f(x_f): f \in F\} \subset U\}$.)

(e) Each compact metric space X is the continuous image of 2^ω. (Instead of the family F of the previous proof one may construct a smaller family which will play the same role. If $U_0, \cdots, U_n, \cdots$ is a base for the topology of X let $f_n(0) = U_n^-$ and $f_n(1) = X \sim U_n$.)

(f) Each Cantor space 2^A satisfies the countable chain condition; that is, each disjoint family of open sets is countable. (If $\mathfrak{U}$ is a disjoint family of open subsets of 2^A, then one may suppose that the members of $\mathfrak{U}$ belong to the defining base for the product topology; each member is, in a natural sense, the intersection of a finite number of half-spaces. For some integer n there is then an infinite (in fact, uncountable) disjoint family, each member of which is the intersection of precisely n half-spaces. A simple argument on disjointness completes the proof.

There is a shorter, more sophisticated proof. A Cantor space with coordinatewise addition, modulo 2, is a compact topological group and hence there is a Haar measure (see Halmos [1; 254]). Since this measure is finite and is positive for open sets the countable chain condition is clear.)

(g) Not every compact Hausdorff space is the continuous image of the Cantor set. (The one point compactification of an uncountable discrete space does not satisfy the countable chain condition.)

Notes Proposition (b) is due to Cantor, (e) to P. Alexandroff and Urysohn, and (f) and (g) to J. W. Tukey. Proposition (g) is also a corollary of some results of Szpilrajn [1].

P CHARACTERIZATION OF THE STONE-ČECH COMPACTIFICATION

Let (f,Y) be a Hausdorff compactification of the topological space X such that for each bounded continuous real-valued function g on X the function $g \circ f^{-1}$ has a continuous extension. Then (f,Y) is topologically equivalent to the Stone-Čech compactification $(e,\beta(X))$. (Consider the definition of $\beta(X)$.)

Q EXAMPLE (THE ORDINALS) ON COMPACTIFICATION

Let Ω' be the set of all ordinal numbers less than or equal to Ω, and let $\Omega_0 = \Omega' \sim \{\Omega\}$. Assign each the order topology. Then the Stone-Čech compactification $\beta(\Omega_0)$ is homeomorphic to Ω'. (This will follow from the preceding problem if it is shown that every bounded real-valued continuous function f on Ω_0 is eventually constant,* in the sense that for some x in Ω_0, if $y > x$, then $f(y) = f(x)$. If f is a bounded continuous real-valued function and r and s are real numbers such that $r > s$, then the interlacing lemma 4.E shows that one of the sets $\{x: f(x) \geqq r\}$ and $\{x: f(x) \leqq s\}$ is countable. Using this fact it is not hard to see that f is eventually constant. The hypothesis that f be bounded is actually not essential.)

Note This result is due to Tong [1].

R THE WALLMAN COMPACTIFICATION

Let X be a T_1-space, let $\mathfrak{F}$ be the family of all closed subsets of X, and let $w(X)$ be the collection of all subfamilies $\mathfrak{A}$ of $\mathfrak{F}$ which possess the finite intersection property and are maximal in $\mathfrak{F}$ relative to this property.

(a) If $\mathfrak{A} \in w(X)$, then the intersection of two members of $\mathfrak{A}$ is a member of $\mathfrak{A}$; dually, if A and B are members of $\mathfrak{F} \sim \mathfrak{A}$, then $A \cup B$ is a member of $\mathfrak{F} \sim \mathfrak{A}$. (See 2.I.)

(b) For each point x of X let $\phi(x) = \{A\colon A \in \mathfrak{F} \text{ and } x \in A\}$. Then ϕ is a one-to-one map of X into $w(X)$.

(c) For each open subset U of X let $U^* = \{\mathfrak{A}\colon \mathfrak{A} \in w(X) \text{ and } A \subset U \text{ for some } A \text{ in } \mathfrak{A}\}$. Then $w(X) \sim U^* = \{\mathfrak{A}\colon X \sim U \in \mathfrak{A}\}$. If U and V are open subsets of X, then $(U \cap V)^* = U^* \cap V^*$ and $(U \cup V)^* = U^* \cup V^*$.

(d) Let $w(X)$ have the topology with a base the family of all sets of the form U^* for U open in X. Then $w(X)$ is compact, the map ϕ is continuous, and $\phi(X)$ is dense in $w(X)$. (Show compactness via the finite intersection property argument for complements of members of the base.)

(e) If X is normal, then $w(X)$ is Hausdorff.

(f) If f is a bounded continuous real-valued function on X, then $f \circ \phi^{-1}$ may be extended continuously to all of $w(X)$. (If a continuous extension is impossible, then by a little argument it can be shown that there are closed disjoint subsets R and S of the reals such that $f^{-1}[R]$

* This curious property of Ω_0 has been used by E. Hewitt [1] in constructing a regular Hausdorff space X such that every continuous real-valued function on X is constant.

and $f^{-1}[S]$ are disjoint but the closures of the images under ϕ of these sets intersect. But if A and B are closed disjoint subsets of X, then $\{\mathfrak{A}: A \in \mathfrak{A}\}$ and $\{\mathfrak{A}: B \in \mathfrak{A}\}$ are disjoint and closed in $w(X)$.)

(g) If $w(X)$ is Hausdorff, then the Wallman compactification is topologically equivalent to the Stone-Čech compactification. (See 5.P.)

Notes The principal virtue of the Wallman compactification (Wallman [1]) lies in the fact that the correspondence carrying U into U^* preserves finite intersections and unions. Moreover, the topology for X is carried onto a base for the topology for $w(X)$ by the correspondence, and from this fact it follows that the dimension of X (in the covering sense) and the dimension of $w(X)$ are identical, and X and $w(X)$ have isomorphic Čech homology groups. See Samuel [1] for a related construction.

S BOOLEAN RINGS: STONE REPRESENTATION THEOREM

Let $(R,+,\cdot)$ be a Boolean ring (see 2.K), let S' be the set of all ring homomorphisms of R into I_2 (= the integers mod 2), and let $S = S' \sim \{0\}$, where 0 is the homomorphism which is identically zero. Then S' is a subset of the product $I_2{}^R$. The *Stone space* of the ring R is S with the relative product topology (I_2 is assigned the discrete topology).

A *Boolean space* is a Hausdorff space such that the family of all sets which are both compact and open is a base for the topology. A Boolean space is automatically locally compact. The *characteristic ring* of a Boolean space is the ring of all continuous functions f into I_2 such that $f^{-1}[1]$ is compact (that is, all functions to I_2 which vanish outside a compact set; sometimes called functions with a compact support).

(a) The Stone space of a Boolean ring R is a Boolean space and is compact whenever R has a unit. (In this case $S = \{h: h \in S' \text{ and } h(1) = 1\}$.)

(b) *Stone-Weierstrass mod* 2 Let $\mathfrak{F}$ be the characteristic ring of a Boolean space X and let $\mathfrak{G}$ be a subring of $\mathfrak{F}$ which has the two point property (that is, for distinct points x and y of X and for a and b in I_2 there is g in $\mathfrak{G}$ such that $g(x) = a$ and $g(y) = b$). Then $\mathfrak{F} = \mathfrak{G}$.

(If X is compact, then $\mathfrak{G}$ has the two point property whenever $1 \in \mathfrak{G}$ and $\mathfrak{G}$ distinguishes points, in the sense that for distinct points x and y of X there is g in $\mathfrak{G}$ such that $g(x) \neq g(y)$. A routine but instructive compactness argument serves to establish (b). One might begin by showing that for a compact subset Y of X and a point x of $X \sim Y$ there is g in $\mathfrak{G}$ such that $g(x) = 0$ and g on Y is one.)

(c) *Representation theorem* Each Boolean ring is isomorphic (under the evaluation map) to the characteristic ring of its Stone space. (For

r in R the evaluation at r, $e(r)$, is the function on S whose value at a member s of S is $s(r)$. This theorem depends on the existence of enough homomorphisms 2.**K** and the foregoing proposition (b).)

(d) If X is a Boolean space, $\mathfrak{F}$ its characteristic ring, and $\mathfrak{I}$ a maximal proper ideal in $\mathfrak{F}$, then $\mathfrak{I} = \{f: f(x) = 0\}$ for some x in X. (Show first that unless there is a point at which all members of $\mathfrak{I}$ vanish, then $\mathfrak{I} = \mathfrak{F}$.)

(e) *Dual representation theorem* If X is a Boolean space, then X is homeomorphic (under the evaluation map) to the Stone space of its characteristic ring. (A maximal ideal is the set of zeros of a unique homomorphism into I_2 and every such set of zeros is a maximal ideal. The preceding proposition (d) shows essentially that the evaluation map carries X onto the Stone space.)

Notes The results above are due to M. H. Stone [3].

There is an interesting variation of the process of representing a Boolean space. If X is a Boolean space let $\mathfrak{F}$ be the ring of all continuous functions on X to I_2. (The requirement that $f^{-1}[1]$ be compact is omitted.) The evaluation map of X into the Stone space S of $\mathfrak{F}$ turns out to be a homeomorphism again, but S is compact and it is, in fact, homeomorphic to the Stone-Čech compactification $\beta(X)$. We omit the proof of this fact as well as the characterizations of ideals and subrings of a Boolean ring in terms of the Stone space.

Finally, this problem is so arranged that the pattern can be transferred to the algebra of all continuous real-valued functions f on a locally compact Hausdorff space X such that, for $e > 0$, $\{x: |f(x)| \geqq e\}$ is compact. The most difficult step in reproducing the pattern is the Stone-Weierstrass theorem, 7.**R**, of which (b) above is a miniature. It also turns out that, if X is a Tychonoff space, then the space of all real homomorphisms of the algebra of bounded continuous functions on X is homeomorphic to $\beta(X)$, very much like the situation sketched in the previous paragraph.

T COMPACT CONNECTED SPACES (THE CHAIN ARGUMENT)

Let (X,d) be a compact pseudo-metric space. For each positive number e, define an *e-chain* from a point x of X to a point y to be a finite sequence of points, the first of which is x, the last y, such that the distance between successive points is less than e. For each subset A of X, $C_e(A)$ is defined to be the set of all points which can be joined to points of A by an e-chain and $C(A)$ is defined to be $\bigcap \{C_e(A): e > 0\}$. An equivalent definition: Let $V_0(A) = A$, $V_1(A) = \{x: \text{dist}\,(x,A) < e\}$

and inductively $V_{n+1}(A) = V_1(V_n(A))$. Set $C_e(A) = \bigcup\{V_n(A): n \in \omega\}$.

(a) For each $e > 0$ and each set A the set $C_e(A)$ is open and closed.

(b) If A is a connected subset of X, then $C(A)$ is connected. Hence $C(\{x\})$ is the component C_x of X about x for each point x. (If $C(A)$ is the union of disjoint closed sets B and D let f = [dist (B,D)]/3 and show by 5.G that $C_e(A) \subset \{x\colon \text{dist}(x, B \cup D) < f\}$ for some positive e.)

(c) If A is a subset of X, then $C(A) = \bigcup\{C_x: x \in A^-\}$. (If $x \notin C(A)$, then $x \notin C_e(A)$ for some positive e.)

(d) The decomposition of X into components is upper semi-continuous.

(e) If X is connected and U is an open neighborhood of a point x, then the closure of some component of U intersects $X \sim U$. (If not, there is a compact neighborhood V of the closure of the component which is contained in U. The component about x of V is contained in the interior V^0 of V and using (c) one can show that there are open and closed subsets of V containing $V \sim V^0$ and x respectively.)

(f) No closed connected subset of X which contains more than one point is the union of a countable disjoint family of closed subsets. (Proposition (e) plays a critical role in this proof. If the set $\bigcup\{A_n: n \in \omega\}$ is closed and connected and the sets A_n are closed and disjoint it is possible to find a closed connected set which is disjoint from A_1 and intersects more than one of the sets A_n.)

(g) Let X be the subset $\{(x,y): x^2y^2 = 1\}$ of the Euclidean plane with the usual metric. Then X is locally compact and any two points can be joined by an e-chain for each $e > 0$, but X is not connected.

Notes The results of this problem generalize very naturally to compact Hausdorff (or compact regular) spaces. The even covering theorem 5.27 gives the necessary mechanism.

Lest proposition (e) make one over-optimistic on the properties of connected sets, the classic example of Knaster and Kuratowski [1] should be mentioned. There is a connected subspace X of the Euclidean plane and a point x of X such that $X \sim \{x\}$ contains no connected set.

U FULLY NORMAL SPACES

If $\mathcal{U}$ is a family of subsets of a set X and x is a point of X, then the ***star*** at x of $\mathcal{U}$ is the union of the members of $\mathcal{U}$ to which x belongs. A cover $\mathcal{V}$ is a ***star-refinement*** of $\mathcal{U}$ iff the family of stars of $\mathcal{V}$ at points of X is a refinement of $\mathcal{U}$. A topological space is ***fully normal*** iff each open cover has an open star-refinement. Then: A regular topological space is fully normal iff it is paracompact. (If X is paracompact the even covering property together with 5.30 yields an easy proof of full normality. On the other hand, if X is fully normal, $\mathfrak{U}$ is an open cover and $\mathfrak{V}$ is an open star-refinement of $\mathfrak{U}$, then $\bigcup\{V \times V: V \in \mathfrak{V}\}$ is a neighborhood of the diagonal.)

Note The definition of full normality is due to J. W. Tukey [1], who proved many useful properties. The equivalence with paracompactness was proved by A. H. Stone [1].

V POINT FINITE COVERS AND METACOMPACT SPACES

A family of subsets of X is *point finite* iff no point of X belongs to more than a finite number of members of the family. A topological space is *metacompact* iff each open cover has a point finite refinement.

(a) Let $\mathfrak{U}$ be a point finite open cover of a normal space X. Then it is possible to select an open set $G(U)$ for each U in $\mathfrak{U}$ in such a way that $G(U)^- \subset U$ and the family of all sets $G(U)$ is a cover of X. (Choose a maximal member of the class of all functions F satisfying the conditions: the domain of F is a subfamily of $\mathfrak{U}$, $F(U)$ is an open set whose closure is contained in U for each U in the domain of F and $\bigcup\{F(U): U \in \text{domain } F\} \cup \bigcup\{V: V \in \mathfrak{U} \textit{ and } V \notin \text{domain } F\} = X$. Point finiteness of $\mathfrak{U}$ implies the existence of a maximal F.)

(b) A point finite cover of a set has a minimal subcover (that is, a subcover no proper subfamily of which is a cover).

(c) A metacompact T_1-space is countably compact (see 5.E) iff it is compact.

Note Propositions (b) and (c) are taken directly from Arens and Dugundji [1].

W PARTITION OF UNITY

A *partition of unity* on a topological space X is a family F of continuous functions on X to the set of non-negative real numbers such that $\sum\{f(x): f \in F\} = 1$ for each x in X, and all but a finite number of members of F vanish on some neighborhood of each point of X. A partition F of unity is *subordinate* to a cover $\mathfrak{U}$ of X iff each member of F vanishes outside some member of $\mathfrak{U}$. Then: For each locally finite open cover $\mathfrak{U}$ of a normal space there is a partition of unity which is subordinate to $\mathfrak{U}$. A slightly stronger result may be proved: If $\mathfrak{U}$ is a locally finite open cover of a normal space, then it is possible to select a non-negative continuous function f_U for each U in $\mathfrak{U}$ such that f_U is 0 outside U and is everywhere less than or equal to one, and $\sum\{f_U(x): U \in \mathfrak{U}\} = 1$ for all x. (See 5.V(a) above.)

Note As far as I know, this result (approximately) is due independently to Hurewicz, Bochner, and Dieudonné.

X THE BETWEEN THEOREM FOR SEMI-CONTINUOUS FUNCTIONS

Let g and h be, respectively, lower and upper semi-continuous real-valued functions on a paracompact space X, and suppose that $h(x) < g(x)$ for all x in X. Then there is a continuous real-valued function p on X such that $h(x) < p(x) < g(x)$ for each x. (Let $\mathfrak{U}$ be the family of all open subsets U of X such that the supremum of h on U is less than the infimum of g on U, and let F be a partition of unity which is subordinate to $\mathfrak{U}$. For each f in F choose k_f such that, if $f(x) \neq 0$, then $h(x) < k_f < g(x)$, and let $p(x) = \sum\{k_f f(x): f \in F\}$. The value of p at a point x is then an average of numbers, all of which lie between $h(x)$ and $g(x)$.)

Notes The result above can be improved by first finding a countable refinement for the family $\mathfrak{U}$. The proposition then holds for countably paracompact spaces (that is, spaces such that each countable open cover has a locally finite refinement). The converse of the sharpened form of the theorem is true. Dowker [2] has proved the equivalence of: (1) X is countably paracompact and normal, (2) the product of X and the closed unit interval is normal, and (3) the proposition above. Dowker also shows that a perfectly normal space (normal and each closed subset is a G_δ) is countably paracompact. It is not known whether a normal Hausdorff space must be countably paracompact.

Y PARACOMPACT SPACES

(a) Each regular Lindelöf space is paracompact.

(b) A topological space is defined to be *σ-compact* iff it is the union of a countable family of compact subsets. Each σ-compact space is a Lindelöf space.

(c) If a regular space is the union of the members of an open discrete family of Lindelöf subspaces, then it is paracompact. Consequently each locally compact group is paracompact. (Consider the family of cosets modulo the smallest subgroup containing a fixed compact neighborhood of the identity.)

(d) The half-open interval space of problems 1.K and 4.I is regular and Lindelöf and hence paracompact. The cartesian product of this space with itself is not normal and is therefore not paracompact.

(e) With the order topology the set of all ordinals which are less than the first uncountable ordinal is not paracompact. (Consider the cover consisting of all sets of the form $\{x: x < a\}$. The supremum of each member of an arbitrary refinement of this cover is less than Ω.)

Notes Proposition (a) above is due to Morita [1]. For further information on paracompactness (an F_σ-theorem, products, etc.) see Michael [1]. Bing [1] has studied a normality condition which is intermediate to normality and paracompactness. In this connection it might be emphasized that lemma 5.31 states a noteworthy normality property of paracompact spaces.

Chapter 6

UNIFORM SPACES

There are several properties of metric spaces which are not topological but are closely connected with topological properties. We give examples of the sort of connections contemplated, postponing the definitions and proofs. The property of being a Cauchy sequence is not a topological invariant, for the map f such that $f(x) = 1/x$ is a homeomorphism of the space of positive real numbers onto itself which carries the Cauchy sequence $\{1/(n+1) \colon n \in \omega\}$ into the non-Cauchy sequence $\{n + 1, n \in \omega\}$. However, it is possible to derive topological results from statements about Cauchy sequences; for example, a subset A of the space of all real numbers is closed if and only if each Cauchy sequence in A converges to some point of A. The reverse sort of implication may also occur; thus, each continuous function on a compact metric space is uniformly continuous. In this case we deduce from a topological premise (that the space is compact) a non-topological conclusion (that a function is uniformly continuous). This chapter is devoted to a study of quasi-topological results of this sort.

The mathematical construct employed in studying uniformity properties is called a uniform space. A brief discussion will indicate how this notion, which is due to A. Weil [1], applies.

A sequence $\{x_n, n \in \omega\}$ in a pseudo-metric space (X,d) is called a Cauchy sequence iff $d(x_m,x_n)$ converges to zero as m and n become large. This notion is not meaningful in an arbitrary topological space; in order to define a Cauchy sequence it is necessary to know, in some sense, for what pairs the distance $d(x,y)$ is small. This statement may be made precise in the following way. If $V_{d,r} = \{(x,y): d(x,y) < r\}$, then $\{x_n, n \in \omega\}$ is a Cauchy sequence iff for each positive r it is true that (x_m,x_n) is a member of $V_{d,r}$ for m and n large. The notion of uniform continuity can also be formulated in terms of the family of all sets of the form $V_{d,r}$. This suggests consideration of a set X and a special family of subsets of $X \times X$.

If X is a topological group, then a sequence $\{x_n, n \in \omega\}$ may be called a Cauchy sequence iff $x_m x_n^{-1}$ is near the identity e of the group when m and n are large. Again, the information needed to make this definition is information about pairs of points. We need to know which pairs of points (x,y) are such that xy^{-1} is near the identity e. For each neighborhood U of e let $V_U = \{(x,y): xy^{-1} \in U\}$. Then clearly the family of all sets of the form V_U determines which sequences are Cauchy.

A uniform space is defined to be a set X together with a family of subsets of $X \times X$ which satisfies certain natural conditions. This follows the pattern suggested by both of the preceding examples. However, it should be emphasized that this is by no means the only framework in which uniformity can be studied. It is possible to study a set X together with a distinguished family of pseudo-metrics for X, or to distinguish a collection of covers of X which are to be uniform covers (roughly in the sense of the Lebesgue covering lemma 5.26). One may also consider "metrics" with values in a structure less restricted than that of the real numbers. All of these notions are essentially equivalent, as indicated in the problems at the end of the chapter.

Finally, it must be said that there are uniformity properties of metric spaces which apparently do not generalize to less restricted situations. The last section is devoted to a study of some of these.

UNIFORMITIES AND THE UNIFORM TOPOLOGY

We will be concerned with subsets of a cartesian product $X \times X$ of a set with itself. These subsets are relations in the sense of chapter 0, and for convenience we review some of the earlier definitions and results about them. A relation is a set of ordered pairs, and if U is a relation the inverse relation U^{-1} is the set of all pairs (x,y) such that $(y,x) \in U$. The operation of taking inverses is involutory in the sense that $(U^{-1})^{-1}$ is always U. If $U = U^{-1}$, then U is called symmetric. If U and V are relations, then the composition $U \circ V$ is the set of all pairs (x,z) such that for some y it is true that $(x,y) \in V$ and $(y,z) \in U$. Composition is associative, that is, $U \circ (V \circ W) = (U \circ V) \circ W$, and it is always true that $(U \circ V)^{-1} = V^{-1} \circ U^{-1}$. The set of all pairs (x,x) for x in X is called the identity relation, or the diagonal, and is denoted by $\Delta(X)$ or simply Δ. For each subset A of X the set $U[A]$ is defined to be $\{y\colon (x,y) \in U \textit{ for some } x \textit{ in } A\}$, and if x is a point of X, then $U[x]$ is $U[\{x\}]$. For each U and V and each A it is true that $(U \circ V)[A] = U[V[A]]$. Finally a simple lemma will be needed.

1 LEMMA *If V is symmetric, then $V \circ U \circ V = \bigcup \{V[x] \times V[y]\colon (x,y) \in U\}$.*

PROOF By definition $V \circ U \circ V$ is the set of all pairs (u,v) such that $(u,x) \in V$, $(x,y) \in U$ and $(y,v) \in V$ for some x and some y. Since V is symmetric this is the set of all (u,v) such that $u \in V[x]$ and $v \in V[y]$ for some (x,y) in U. But $u \in V[x]$ and $v \in V[y]$ iff $(u,v) \in V[x] \times V[y]$, and hence $V \circ U \circ V = \{(u,v)\colon (u,v) \in V[x] \times V[y] \textit{ for some } (x,y) \textit{ in } U\} = \bigcup \{V[x] \times V[y]\colon (x,y) \in U\}$. ∎

A **uniformity** for a set X is a non-void family $\mathcal{U}$ of subsets of $X \times X$ such that

(a) each member of $\mathcal{U}$ contains the diagonal Δ;
(b) if $U \in \mathcal{U}$, then $U^{-1} \in \mathcal{U}$;
(c) if $U \in \mathcal{U}$, then $V \circ V \subset U$ for some V in $\mathcal{U}$;
(d) if U and V are members of $\mathcal{U}$, then $U \cap V \in \mathcal{U}$; and
(e) if $U \in \mathcal{U}$ and $U \subset V \subset X \times X$, then $V \in \mathcal{U}$.

The pair $(X,\mathcal{U})$ is a **uniform space**.

The metric antecedents of the conditions above are not hard to discern. The first is derived from the condition that $d(x,x) = 0$ and the second derives from the symmetry condition $d(x,y) = d(y,x)$. The third is a vestigal form of the triangle inequality

—it says roughly that for r-spheres there are $(r/2)$-spheres. The fourth and fifth resemble axioms for the neighborhood system of a point and they will be used to derive the corresponding properties for a neighborhood system relative to a topology which will presently be defined.

There may be many different uniformities for a set X. The largest of these is the family of all those subsets of $X \times X$ which contain Δ and the smallest is the family whose only member is $X \times X$. If X is the set of real numbers the **usual uniformity** for X is the family $\mathfrak{U}$ of all subsets U of $X \times X$ such that $\{(x,y): |x - y| < r\} \subset U$ for some positive number r. Each member of $\mathfrak{U}$ is a neighborhood of the diagonal Δ (the line with equation $y = x$), but it is to be emphasized that not every neighborhood of the diagonal is a member of $\mathfrak{U}$. For example, the set $\{(x,y): |x - y| < 1/(1 + |y|)\}$ is a neighborhood of Δ but not a member of $\mathfrak{U}$.

It is not generally true that the union or the intersection of two uniformities for X is a uniformity. However, the union of a collection of uniformities generates a uniformity in a rather natural sense. A subfamily $\mathfrak{B}$ of a uniformity $\mathfrak{U}$ is a **base** for $\mathfrak{U}$ iff each member of $\mathfrak{U}$ contains a member of $\mathfrak{B}$. If $\mathfrak{B}$ is a base for $\mathfrak{U}$, then $\mathfrak{B}$ determines $\mathfrak{U}$ entirely, for a subset U of $X \times X$ belongs to $\mathfrak{U}$ iff U contains a member of $\mathfrak{B}$. A subfamily $\mathfrak{S}$ is a **subbase** for $\mathfrak{U}$ iff the family of finite intersections of members of $\mathfrak{S}$ is a base for $\mathfrak{U}$. These definitions are entirely analogous to the definitions of base and subbase for a topology.

2 THEOREM *A non-void family $\mathfrak{B}$ of subsets of $X \times X$ is a base for some uniformity for X if and only if*

(a) *each member of $\mathfrak{B}$ contains the diagonal Δ;*
(b) *if $U \in \mathfrak{B}$, then U^{-1} contains a member of $\mathfrak{B}$;*
(c) *if $U \in \mathfrak{B}$, then $V \circ V \subset U$ for some V in $\mathfrak{B}$; and*
(d) *the intersection of two members of $\mathfrak{B}$ contains a member.*

The straightforward proof of this proposition is omitted.

The property of being a subbase for some uniformity is less easy to characterize. However, the following simple result is adequate for our needs.

3 Theorem *A family $\mathcal{S}$ of subsets of $X \times X$ is a subbase for some uniformity for X if*

(a) *each member of $\mathcal{S}$ contains the diagonal Δ,*
(b) *for each U in $\mathcal{S}$ the set U^{-1} contains a member of $\mathcal{S}$, and*
(c) *for each U in $\mathcal{S}$ there is V in $\mathcal{S}$ such that $V \circ V \subset U$.*

In particular, the union of any collection of uniformities for X is the subbase for a uniformity for X.

PROOF It must be shown that the family $\mathcal{B}$ of finite intersections of members of $\mathcal{S}$ satisfies the conditions of 6.2. This follows easily from the observation: If $U_1, \cdots U_n$ and $V_1, \cdots, V_n$ are subsets of $X \times X$, if $U = \bigcap\{U_i: i = 1, \cdots, n\}$ and $V = \bigcap\{V_i: i = 1, \cdots, n\}$, then $V \subset U^{-1}$ (or $V \circ V \subset U$) whenever $V_i \subset U_i^{-1}$ (respectively, $V_i \circ V_i \subset U_i$) for each i. ∎

If $(X,\mathcal{U})$ is a uniform space the **topology $\mathcal{T}$ of the uniformity** $\mathcal{U}$, or the **uniform topology,** is the family of all subsets T of X such that for each x in T there is U in $\mathcal{U}$ such that $U[x] \subset T$. (This is precisely the generalization of the metric topology, which is the family of all sets which contain a sphere about each point.) It must be verified that $\mathcal{T}$ is indeed a topology, but this offers no difficulty: In view of the definition, the union of members of $\mathcal{T}$ is surely a member of $\mathcal{T}$. If T and S are members of $\mathcal{T}$ and $x \in T \cap S$, then there are U and V in $\mathcal{U}$ such that $U[x] \subset T$ and $V[x] \subset S$, and hence $(U \cap V)[x] \subset T \cap S$; consequently $T \cap S \in \mathcal{T}$ and $\mathcal{T}$ is a topology.

The relation between a uniformity and the uniform topology will now be examined.

4 Theorem *The interior of a subset A of X relative to the uniform topology is the set of all points x such that $U[x] \subset A$ for some U in $\mathcal{U}$.*

PROOF It must be shown that the set $B = \{x: U[x] \subset A$ *for some U in* $\mathcal{U}\}$ is open relative to the uniform topology, for B surely contains every open subset of A and, if B is open, then it must necessarily be the interior of A. If $x \in B$, then there is a member U of $\mathcal{U}$ such that $U[x] \subset A$ and there is V in $\mathcal{U}$ such that $V \circ V \subset U$. If $y \in V[x]$, then $V[y] \subset V \circ V[x] \subset U[x] \subset A$, and hence $y \in B$. Hence $V[x] \subset B$ and B is open. ∎

It follows immediately that $U[x]$ is a neighborhood of x for each U in the uniformity $\mathfrak{U}$, and consequently the family of all sets $U[x]$ for U in $\mathfrak{U}$ is a base for the neighborhood system of x (the family is actually identical with the neighborhood system but this is of no great importance). The following proposition is then clear.

5 THEOREM *If $\mathfrak{B}$ is a base (or subbase) for the uniformity $\mathfrak{U}$, then for each x the family of sets $U[x]$ for U in $\mathfrak{B}$ is a base (subbase respectively) for the neighborhood system of x.*

The uniform topology for X may be used to construct a product topology for $X \times X$. As might be expected, members of the uniformity have a special structure relative to this topology.

6 THEOREM *If U is a member of the uniformity $\mathfrak{U}$, then the interior of U is also a member; consequently the family of all open symmetric members of $\mathfrak{U}$ is a base for $\mathfrak{U}$.*

PROOF The interior of a subset M of $X \times X$ is the set of all (x,y) such that, for some U and some V in $\mathfrak{U}$, $U[x] \times V[y] \subset M$. Since $U \cap V \in \mathfrak{U}$ the interior of M is $\{(x,y): V[x] \times V[y] \subset M$ *for some V in* $\mathfrak{U}\}$. If $U \in \mathfrak{U}$ there is a symmetric member V of $\mathfrak{U}$ such that $V \circ V \circ V \subset U$ and, according to lemma 6.1, $V \circ V \circ V = \bigcup \{V[x] \times V[y]: (x,y) \in V\}$. Hence every point of V is an interior point of U and, since the interior of U contains V, it is a member of $\mathfrak{U}$. ∎

In view of the foregoing theorem every member of a uniformity is a neighborhood of the diagonal. It is to be emphasized that the converse of this proposition is false. There may be many very different uniformities for X, all having the same topology and hence the same family of neighborhoods of the diagonal.

7 THEOREM *The closure, relative to the uniform topology, of a subset A of X is $\bigcap \{U[A]: U \in \mathfrak{U}\}$. The closure of a subset M of $X \times X$ is $\bigcap \{U \circ M \circ U: U \in \mathfrak{U}\}$.*

PROOF A point x belongs to the closure of a subset A of X iff $U[x]$ intersects A for each U in $\mathfrak{U}$. But $U[x]$ intersects A iff $x \in U^{-1}[A]$, and since each member of $\mathfrak{U}$ contains a symmetric member, $x \in A^-$ iff $x \in U[A]$ for each U in $\mathfrak{U}$. The first statement is then proved. Similarly, if U is a symmetric member of $\mathfrak{U}$, then $U[x] \times U[y]$ intersects a subset M of $X \times X$ iff $(x,y) \in U[u] \times U[v]$ for some (u,v) in M, that is, iff $(x,y) \in \bigcup\{U[u] \times U[v]: (u,v) \in M\}$. Since by lemma 6.1 this last set is $U \circ M \circ U$ it follows that $(x,y) \in M^-$ iff $(x,y) \in \bigcap\{U \circ M \circ U: U \in \mathfrak{U}\}$. ∎

8 THEOREM *The family of closed symmetric members of a uniformity $\mathfrak{U}$ is a base for $\mathfrak{U}$.*

PROOF If $U \in \mathfrak{U}$ and V is a member of $\mathfrak{U}$ such that $V \circ V \circ V \subset U$, then $V \circ V \circ V$ contains the closure of V in view of the preceding theorem; hence U contains a closed member W of $\mathfrak{U}$ and $W \cap W^{-1}$ is a closed symmetric member. ∎

It will be shown presently that a uniform space (more precisely a space with a uniform topology) is always completely regular. At the moment it is easy to see that such a space is regular, for each neighborhood of a point x contains a neighborhood $V[x]$ such that V is a closed member of $\mathfrak{U}$, and $V[x]$ is consequently closed. Therefore a space with a uniform topology is a Hausdorff space iff each set consisting of a single point is closed. Since the closure of the set $\{x\}$ is $\bigcap\{U[x]: U \in \mathfrak{U}\}$, the space is Hausdorff iff $\bigcap\{U: U \in \mathfrak{U}\}$ is the diagonal Δ. In this case $(X,\mathfrak{U})$ is said to be **Hausdorff** or **separated.**

UNIFORM CONTINUITY; PRODUCT UNIFORMITIES

If f is a function on a uniform space $(X,\mathfrak{U})$ with values in a uniform space $(Y,\mathfrak{V})$, then f is **uniformly continuous** relative to $\mathfrak{U}$ and $\mathfrak{V}$ iff for each V in $\mathfrak{V}$ the set $\{(x,y): (f(x), f(y)) \in V\}$ is a member of $\mathfrak{U}$. This condition may be rephrased in several ways. For each function f on X to Y let f_2 be the induced function on $X \times X$ to $Y \times Y$ which is defined by $f_2(x,y) = (f(x),f(y))$. Then f is uniformly continuous iff for each V in $\mathfrak{V}$ there is U in $\mathfrak{U}$ such that $f_2[U] \subset V$. We also have: if $\mathfrak{S}$ is a subbase for $\mathfrak{V}$, then f is uniformly continuous iff $f_2^{-1}[V] \in \mathfrak{U}$ for each V in $\mathfrak{S}$, because f_2^{-1} preserves unions and intersections. If Y is the set of real numbers and $\mathfrak{V}$ is the usual uniformity, then it follows that f is uniformly continuous iff for each positive number r there is U in $\mathfrak{U}$ such that $|f(x) - f(y)| < r$ whenever $(x,y) \in U$. If X is also the space of real numbers with the usual uniformity, then f is uniformly continuous iff for each positive number r there is a positive number s such that $|f(x) - f(y)| < r$ whenever $|x - y| < s$.

It is evident that, if f is on X to Y and g is a function on Y, then $(g \circ f)_2 = g_2 \circ f_2$, and from this it follows that the composition of two uniformly continuous functions is again uniformly continuous. If f is one-to-one map of X onto Y and both f and f^{-1} are uniformly continuous, then f is a **uniform isomorphism,** and the spaces X and Y (more precisely $(X,\mathfrak{U})$ and $(Y,\mathfrak{V})$) are said to be **uniformly equivalent.** The composition of two uniform isomorphisms, the inverse of a uniform isomorphism, and the identity map of a space onto itself are all uniform isomorphisms, and consequently the collection of all uniform spaces is divided into equivalence classes, consisting of uniformly equivalent spaces. A property which when possessed by one uniform space is also possessed by every uniformly isomorphic space is a **uniform invariant.** With a few exceptions the properties studied in this chapter are uniform invariants.

As might be expected, uniform continuity implies continuity relative to the uniform topology.

9 THEOREM *Each uniformly continuous function is continuous relative to the uniform topology, and hence each uniform isomorphism is a homeomorphism.*

PROOF Let f be a uniformly continuous function on $(X,\mathfrak{U})$ to $(Y,\mathfrak{V})$ and let U be a neighborhood of $f(x)$. Then there is V in $\mathfrak{V}$ such that $V[f(x)] \subset U$, and $f^{-1}[V[f(x)]] = \{y: f(y) \in V[f(x)]\} = \{y: (f(x), f(y)) \in V\} = f_2^{-1}[V][x]$, and this is a neighborhood of x. Hence $f^{-1}[U]$ is a neighborhood of x and continuity is proved. ∎

If f is a function on a set X to a uniform space $(Y,\mathfrak{V})$, then it is not generally true that the family of all sets $f_2^{-1}[V]$ for V in $\mathfrak{V}$ is a uniformity for X. The difficulty is that there may be a subset of $X \times X$ which contains some set $f_2^{-1}[V]$, but is not the inverse of any subset of $Y \times Y$. However, this difficulty is not profound; the family of all $f_2^{-1}[V]$ is the base for a uniformity $\mathfrak{U}$ for X, as we now verify. It is clear that f_2^{-1} preserves inclusions, intersections, and inverses (that is, $f_2^{-1}[V^{-1}] = [f_2^{-1}[V]]^{-1}$), and consequently it is only necessary to show that for each member U of $\mathcal{V}$ there is V in $\mathcal{V}$ such that $f_2^{-1}[V] \circ f_2^{-1}[V] \subset f_2^{-1}[U]$. But if $V \circ V \subset U$ and (x,y) and (y,z) belong to $f_2^{-1}[V]$, then both $(f(x),f(y))$ and $(f(y),f(z))$ belong to V, and hence $(f(x),f(z)) \in V \circ V$. It follows that the family of inverses of members of $\mathcal{V}$ is indeed a base for a uniformity $\mathcal{U}$ for X. It is clear that f is uniformly continuous relative to $\mathcal{U}$ and $\mathcal{V}$, and in fact $\mathcal{U}$ is smaller than every other uniformity for which f is uniformly continuous.

If $(X,\mathcal{U})$ is a uniform space and Y is a subset of X, then in view of the preceding discussion there is a smallest uniformity $\mathcal{V}$ such that the identity map of Y into X is uniformly continuous. It is clear that the members of $\mathcal{V}$ are simply the intersections of the members of $\mathcal{U}$ with $Y \times Y$ (sometimes called the trace of $\mathcal{U}$ on $Y \times Y$). The uniformity $\mathcal{V}$ is called the **relativization** of $\mathcal{U}$ to Y, or the **relative uniformity** for Y, and $(Y, \mathcal{V})$ is called a uniform subspace of the space $(X,\mathcal{U})$. We omit the simple verification of the fact that the topology of the relative uniformity $\mathcal{V}$ is the relativized topology of $\mathcal{U}$.

We have seen that there is always a unique smallest uniformity which makes a map of a set X into a uniform space uniformly continuous. This proposition may be extended to a family F of functions such that each member f of F maps X into a uniform space $(Y_f,\mathcal{U}_f)$. The family of all sets of the form $f_2^{-1}[U] = \{(x,y): (f(x),f(y)) \in U\}$, for f in F and U in $\mathcal{U}_f$, is a subbase for a uniformity $\mathcal{U}$ for X, and $\mathcal{U}$ is the smallest uniformity such that each map f is uniformly continuous. (Theorem 6.3 shows that the family of sets of the form $f_2^{-1}[U]$ is a subbase for a uniformity, and evidently $\mathcal{U}$ makes each f uniformly continuous and is smaller than every other uniformity with this property.) It is in precisely this way that the product uniformity is defined. If $(X_a,\mathcal{U}_a)$ is a uniform space for each member a of an index set A, then the **product uniformity** for $\times\{X_a: a \in A\}$ is the smallest uniformity such that projection into each coordinate space is uniformly continuous. The family of all sets of the form $\{(x,y): (x_a,y_a) \in U\}$, for a in A and U in $\mathcal{U}_a$, is a subbase for the product uniformity. If x is a member of the product, then a subbase for the neighborhood system of x (relative to the uniform topology) may be constructed from the subbase for the product uniformity. Hence the family of all sets of the form $\{y: (x_a, y_a) \in U\}$ is a subbase for the neighborhood system of x. It follows that a base for the neighborhood system of x relative to the topology of the product uniformity is the family of finite intersections of sets of the form $\{y: y_a \in U[x_a]\}$ for a in A and U in $\mathfrak{U}_a$. But the same family is also a base for the neighborhood system of x relative to the product topology, and consequently the product topology is the topology of the product uniformity. This statement is the first half of the following theorem.

10 THEOREM *The topology of the product uniformity is the product topology.*

A function f on a uniform space to a product of uniform spaces is uniformly continuous if and only if the composition of f with each projection into a coordinate space is uniformly continuous.

PROOF If f is uniformly continuous with values in the product $\times\{X_a: a \in A\}$, then each projection P_a is uniformly continuous and the composition $P_a \circ f$ is uniformly continuous. If $P_a \circ f$ is uniformly continuous for each a in A and U is a member of the uniformity of X_a, then $\{(u,v): (P_a \circ f(u), P_a \circ f(v)) \in U\}$ is a member of the uniformity $\mathcal{V}$ of the domain of f. But this set can be written in the form $f_2^{-1}[\{(x,y): (x_a,y_a) \in U\}]$. Hence the inverse under f_2 of each member of a subbase for the product uniformity belongs to $\mathcal{V}$ and f is therefore uniformly continuous. ∎

The next proposition begins the development of the relation between uniformities and pseudo-metrics for X.

11 THEOREM *Let $(X,\mathfrak{U})$ be a uniform space and let d be a pseudo-metric for X. Then d is uniformly continuous on $X \times X$ relative to the product uniformity if and only if the set $\{(x,y): d(x,y) < r\}$ is a member of $\mathfrak{U}$ for each positive number r.*

PROOF Let $V_{d,r} = \{(x,y): d(x,y) < r\}$. It must be shown that $V_{d,r} \in \mathfrak{U}$ for each positive r iff d is uniformly continuous with respect to the product uniformity for $X \times X$. If U is a member of $\mathfrak{U}$, then the sets $\{((x,y),(u,v)): (x,u) \in U\}$ and $\{((x,y),(u,v)): (y,v) \in U\}$ belong to the product uniformity, and it is easy to see that the family of all sets of the form $\{((x,y),(u,v)): (x,u) \in U$ *and*

$(y,v) \in U\}$ is a base for the product uniformity. Hence if d is uniformly continuous, then for each positive r there is U in $\mathfrak{U}$ such that, if (x,u) and (y,v) belong to U, then $|d(x,y) - d(u,v)| < r$. In particular, letting $(u,v) = (y,y)$, it follows that, if $(x,y) \in U$, then $d(x,y) < r$. Then $U \subset V_{d,r}$ and consequently $V_{d,r} \in \mathfrak{U}$. To prove the converse observe that, if both (x,u) and (y,v) belong to $V_{d,r}$, then $|d(x,y) - d(u,v)| < 2r$ because $d(x,y) \leq d(x,u) + d(u,v) + d(y,v)$ and $d(u,v) \leq d(x,u) + d(x,y) + d(y,v)$. It follows that, if $V_{d,r} \in \mathfrak{U}$ for each positive r, then d is uniformly continuous. ∎

METRIZATION

The purpose of this section is to compare uniform spaces and pseudo-metrizable spaces. The comparison is an example of the standard procedure for testing the effectiveness of a generalization. The generalization is compared with the mathematical object which it purports to generalize in order to discover the extent to which the basic concepts have been isolated. In this case (as in many other instances) the comparison yields a representation of the generalized object in terms of its progenitor. A uniformity will be assigned to each family of pseudo-metrics for a set X, and the principal result of the section states that every uniformity is derived in this fashion from the family of its uniformly continuous pseudo-metrics. It will also be shown that a uniformity can be derived from a single pseudo-metric if and only if the uniformity has a countable base.

Each pseudo-metric d for a set X generates a uniformity in the following way. For each positive number r let $V_{d,r} = \{(x,y): d(x,y) < r\}$. Clearly $(V_{d,r})^{-1} = V_{d,r}$, $V_{d,r} \cap V_{d,s} = V_{d,t}$ where $t = \min\,[r,s]$, and $V_{d,r} \circ V_{d,r} \subset V_{d,2r}$. It follows that the family of all sets of the form $V_{d,r}$ is a base for a uniformity for X. This uniformity is called the **pseudo-metric uniformity,** or the **uniformity generated by d.** A uniform space $(X,\mathfrak{U})$ is said to be **pseudo-metrizable** (or **metrizable**) if and only if there is a pseudo-metric (metric, respectively) d such that $\mathfrak{U}$ is the uniformity generated by d. The uniformity generated by a pseudo-metric d can be described in another way. According to 6.11 a pseudo-metric d is uniformly continuous relative to a uniformity $\mathfrak{V}$ (more precisely, relative to the product uniformity constructed from $\mathcal{V}$) if and only if $V_{d,r} \in \mathcal{V}$ for each positive r. The uniformity $\mathcal{U}$ derived from d can then be characterized as the smallest uniformity which makes d uniformly continuous on $X \times X$. It should be noticed that the pseudo-metric topology is identical with the uniform topology of $\mathcal{U}$, because $V_{d,r}[x]$ is the open r-sphere about x and the family of sets of this form is a base for the neighborhood system of x relative to both topologies.

The crucial step in the metrization theorem for uniform spaces is provided by the following lemma.

12 METRIZATION LEMMA *Let* $\{U_n, n \in \omega\}$ *be a sequence of subsets of* $X \times X$ *such that* $U_0 = X \times X$, *each* U_n *contains the diagonal, and* $U_{n+1} \circ U_{n+1} \circ U_{n+1} \subset U_n$ *for each* n. *Then there is a non-negative real-valued function* d *on* $X \times X$ *such that*

(a) $d(x,y) + d(y,z) \geq d(x,z)$ *for all* x, y, *and* z; *and*
(b) $U_n \subset \{(x,y): d(x,y) < 2^{-n}\} \subset U_{n-1}$ *for each positive integer* n.

If each U_n *is symmetric, then there is a pseudo-metric* d *satisfying condition* (b).

PROOF Define a real-valued function f on $X \times X$ by letting $f(x,y) = 2^{-n}$ iff $(x,y) \in U_{n-1} \sim U_n$ and $f(x,y) = 0$ iff (x,y) belongs to each U_n. The desired function d is constructed from its "first approximation" f by a chaining argument. For each x and each y in X let $d(x,y)$ be the infimum of $\sum\{f(x_i,x_{i+1}): i = 0, \cdots, n\}$ over all finite sequences $x_0, x_1, \cdots, x_{n+1}$ such that $x = x_0$ and $y = x_{n+1}$. It is evident that d satisfies the triangle inequality and since $d(x,y) \leq f(x,y)$ it follows that $U_n \subset \{(x,y): d(x,y) < 2^{-n}\}$. If each U_n is symmetric, then $f(x,y) = f(y,x)$ for each pair (x,y) and consequently d is a pseudo-metric in this case. The proof is completed by showing that $f(x_0,x_{n+1}) \leq 2\sum\{f(x_i,x_{i+1}): i = 0, \cdots, n\}$, from which it will follow that, if $d(x,y) < 2^{-n}$, then $f(x,y) < 2^{-n+1}$, hence $(x,y) \in U_{n-1}$, and $\{(x,y): d(x,y) < 2^{-n}\} \subset U_{n-1}$. The proof is by induction on n, and the inequality is clearly valid for $n = 0$. For convenience, call the number $\sum\{f(x_i,x_{i+1}): i = r, \cdots, s\}$ the length of the chain from r to $s + 1$, and let a be the length of the chain from

0 to $n+1$. Let k be the largest integer such that the chain from 0 to k is of length at most $a/2$, and notice that the chain from $k+1$ to $n+1$ has length at most $a/2$. By the induction hypothesis, each of $f(x_0,x_k)$ and $f(x_{k+1},x_{n+1})$ is at most $2(a/2) = a$, and surely $f(x_k,x_{k+1})$ is at most a. If m is the smallest integer such that $2^{-m} \leqq a$, then (x_0,x_k), (x_k,x_{k+1}) and (x_{k+1},x_{n+1}) all belong to U_m and therefore $(x_0,x_{n+1}) \in U_{m-1}$. Hence $f(x_0,x_{n+1}) \leqq 2^{-m+1} \leqq 2a$ and the lemma is proved. ∎

If a uniformity $\mathfrak{U}$ for X has a countable base $V_0, V_1, \cdots, V_n \cdots$, then it is possible to construct by induction a family $U_0, U_1, \cdots, U_n \cdots$ such that each U_n is symmetric, $U_n \circ U_n \circ U_n \subset U_{n-1}$ and $U_n \subset V_n$ for each positive integer n. The family of sets U_n is then a base for $\mathfrak{U}$, and upon applying the metrization lemma it follows that the uniform space $(X,\mathfrak{U})$ is pseudo-metrizable. Hence:

13 METRIZATION THEOREM *A uniform space is pseudo-metrizable if and only if its uniformity has a countable base.*

This theorem clearly implies that a uniform space is metrizable iff it is Hausdorff and its uniformity has a countable base.

14 *Notes* To the best of my knowledge this theorem first appears in Alexandroff and Urysohn [2]. These authors were seeking a solution to the topological metrization problem (see 4.18), and the result they state is (approximately): a topological Hausdorff space $(X,\mathfrak{I})$ is metrizable iff there is a uniformity with a countable base such that $\mathfrak{I}$ is the uniform topology. This is a rather unsatisfactory solution to the topological metrization problem but (with a slightly strengthened conclusion) is precisely the metrization theorem for uniform spaces. Chittenden [1] first proved a "uniform" form of 6.13 and his proof was later drastically simplified by A. H. Frink [1] and by Aronszajn [1]. The preceding proof is Bourbaki's arrangement of Frink's. The first appearance of 6.13 in the form just given occurs in André Weil's classic monograph [1] in which he introduces the notion of uniform space. ∎

A uniformity for a set X may be derived from a family P of pseudo-metrics in the following fashion. Letting $V_{p,r} = \{(x,y): p(x,y) < r\}$, the family of all sets of the form $V_{p,r}$ for p in P and r positive is the subbase for a uniformity $\mathfrak{U}$ for X. This uniformity $\mathfrak{U}$ is defined to be the **uniformity generated by *P*.** The uniformity may be described in several instructive ways. According to 6.11 a pseudo-metric p is uniformly continuous on $X \times X$ relative to the product uniformity derived from $\mathfrak{V}$ iff $V_{p,r} \in \mathfrak{V}$ for each positive r. Consequently the uniformity generated by P is the smallest uniformity which makes each member p of P uniformly continuous on $X \times X$. Another description: For a fixed member p of P the family of all sets $V_{p,r}$ for r positive is a base for the uniformity of the pseudo-metric space (X,p). If $\mathfrak{V}$ is a uniformity for X, then the identity map of $(X,\mathfrak{V})$ into (X,p) is uniformly continuous iff $V_{p,r} \in \mathfrak{V}$ for each positive r. It follows that the uniformity $\mathfrak{U}$ is the smallest such that for each p in P the identity map of X into (X,p) is uniformly continuous. This fact yields yet another description. Let Z be the product $\times\{X: p \in P\}$ (that is, the product of X with itself as many times as there are members of P) and let f be the map of X into Z defined by $f(x)_p = x$ for each x in X and each p in P. Let the p-th coordinate space of this product be assigned the uniformity of the pseudo-metric p, and let Z have the product uniformity. The projection of Z into the p-th coordinate space is the identity map of X onto the pseudo-metric space (X,p), and it therefore follows from 6.10 that the uniformity generated by P is the smallest having the property that the map of X into Z is uniformly continuous. But f is one to one and is consequently a uniform isomorphism of X onto a subspace of the product of pseudo-metric spaces.

It is clearly of some importance to know which uniformities are generated by families of pseudo-metrics—this might be called the generalized metrization problem for uniform spaces. The solution to the problem is a direct application of the preceding results. Let $(X,\mathfrak{U})$ be a uniform space and let P be the family of all pseudo-metrics for X which are uniformly continuous on $X \times X$. The uniformity generated by P is smaller than $\mathfrak{U}$ in view of 6.11. But the metrization lemma 6.12 shows that for each member U of $\mathfrak{U}$ there is a member p of P such that $\{(x,y):$

$p(x,y) < 1/4\}$ is contained in U, and hence $\mathfrak{U}$ is smaller than the uniformity generated by P. Thus:

15 THEOREM *Each uniformity for X is generated by the family of all pseudo-metrics which are uniformly continuous on $X \times X$.*

There is an interesting corollary to the foregoing theorem. It has already been observed that, if a uniformity $\mathfrak{U}$ for X is generated by a family P of pseudo-metrics, then the space is uniformly isomorphic to a subspace of a product of pseudo-metric spaces, and it is possible to sharpen this result if $(X,\mathfrak{U})$ is Hausdorff. The uniformity $\mathfrak{U}$ is the smallest which makes the identity map of X into the pseudo-metric space (X,p) uniformly continuous for each p in P. The space (X,p) is isometric under a map h_p to a metric space (X_p,p^*), by theorem 4.15, and it follows that $\mathfrak{U}$ is the smallest uniformity making each of the maps h_p uniformly continuous. If a map h of X into $\times\{X_p: p \in P\}$ is defined by letting $h(x)_p = h_p(x)$, then by 6.10 the uniformity $\mathfrak{U}$ is the smallest such that h is uniformly continuous. If $(X,\mathfrak{U})$ is Hausdorff, then h must be one to one, and in this case h is a uniform isomorphism. The preceding theorem then implies the following result (Weil [1]).

16 THEOREM *Each uniform space is uniformly isomorphic to a subspace of the product of pseudo-metric spaces and each uniform Hausdorff space is uniformly isomorphic to a subspace of the product of metric spaces.*

The preceding theorem yields a characterization of those topologies which can be the uniform topology for some uniformity, for a topological space is completely regular if and only if it is homeomorphic to a subspace of a product of pseudo-metrizable spaces (4.**L**).

17 COROLLARY *A topology $\mathfrak{J}$ for a set X is the uniform topology for some uniformity for X if and only if the topological space $(X,\mathfrak{J})$ is completely regular.*

The remainder of this section is devoted to a clarification of the relationship between uniformities and pseudo-metrics. A family P of pseudo-metrics for a set X is said to be a **gage** iff there is a uniformity $\mathfrak{U}$ for X such that P is the family of all pseudo-metrics which are uniformly continuous on $X \times X$ relative to the product uniformity derived from $\mathfrak{U}$. The family P is called the gage of the uniformity $\mathfrak{U}$ and $\mathfrak{U}$ is the uniformity of P ($\mathfrak{U}$ is generated by P according to 6.15). Every family of pseudo-metrics generates a uniformity; it will also be said to generate the gage of this uniformity. A direct description of the gage generated by a family P of pseudo-metrics is possible. The family of all sets of the form $V_{p,r}$ for p in P and r positive is a subbase for the uniformity of the gage, and hence a pseudo-metric q is uniformly continuous on the product iff for each positive number s the set $V_{q,s}$ contains some finite intersection of sets $V_{p,r}$ for p in P. This remark establishes the following proposition.

18 THEOREM *Let P be a family of pseudo-metrics for a set X and let Q be the gage generated by P. Then a pseudo-metric q belongs to Q if and only if for each positive number s there is a positive number r and a finite subfamily $p_1, \cdots, p_n$ of P such that $\bigcap \{V_{p_i,r}: i = 1, \cdots, n\} \subset V_{q,s}$.*

Each concept which is based on the notion of a uniformity can be described in terms of a gage because each uniformity is completely determined by its gage. The following theorem is a dictionary of such descriptions. Recall that p-dist $(x,A) = \inf\{p(x,y): y \in A\}$ is the p-distance from a point x to a set A.

19 THEOREM *Let $(X,\mathfrak{U})$ be a uniform space and let P be the gage of $\mathfrak{U}$. Then:*

(a) *The family of all sets $V_{p,r}$ for p in P and r positive is a base for the uniformity $\mathfrak{U}$.*
(b) *The closure relative to the uniform topology of a subset A of X is the set of all x such that p-dist $(x,A) = 0$ for each p in P.*
(c) *The interior of a set A is the set of all points such that for some p in P and some positive number r the sphere $V_{p,r}[x] \subset A$.*
(d) *Suppose P' is a subfamily of P which generates P. A net $\{S_n, n \in D\}$ in X converges to a point s if and only if $\{p(S_n,s), n \in D\}$ converges to zero for each p in P'.*

(e) *A function f on X to a uniform space* $(Y,\mathcal{V})$ *is uniformly continuous if and only if for each member q of the gage Q of* $\mathcal{V}$ *it is true that* $q \circ f_2 \in P$. *(Recall* $f_2(x,y) = (f(x),f(y))$.*)*

Equivalently, f is uniformly continuous if and only if for each q in Q and each positive number s there is p in P and r positive such that, if $p(x,y) < r$, *then* $q(f(x),f(y)) < s$.

(f) *If* $(X_a,\mathcal{U}_a)$ *is a uniform space for each member a of an index set A and* P_a *is the gage of* $\mathcal{U}_a$ *then the gage of the product uniformity for* $\times\{X_a: a \in A\}$ *is generated by all pseudo-metrics of the form* $q(x,y) = p_a(x_a,y_a)$ *for a in A and* p_a *in* P_a.

The proof is omitted. It is a straightforward application of earlier results.

COMPLETENESS

This section is devoted to a number of elementary theorems based on the concept of a Cauchy net. A uniform space will be called complete iff each Cauchy net in the space converges to some point. The two most useful results of the section state that the product of complete spaces is complete, and that a uniformly continuous function f to a complete Hausdorff space has a uniformly continuous extension whose domain is the closure of the domain of f.

It will be supposed throughout that X is a set, $\mathcal{U}$ is a uniformity for X, and P is the gage of $\mathcal{U}$ (that is, P is the family of all pseudo-metrics for X which are uniformly continuous on $X \times X$). The definitions will be given in terms of both $\mathcal{U}$ and P, and the proofs use the formulation which is most convenient for the problem under consideration. The set $\{(x,y): p(x,y) < r\}$ will be denoted by $V_{p,r}$.

A net $\{S_n, n \in D\}$ in the uniform space $(X,\mathcal{U})$ is a **Cauchy net** iff for each member U of $\mathcal{U}$ there is N in D such that $(S_m,S_n) \in U$ whenever both m and n follow N in the ordering of D. This definition may be rephrased in terms of a net in $X \times X$. In this form it is stated: the net $\{S_n, n \in D\}$ is a Cauchy net iff the net $\{(S_m,S_n), (m,n) \in D \times D\}$ is eventually in each member of $\mathcal{U}$. (It is understood that $D \times D$ is given the product ordering.)

The family of all sets of the form $V_{p,r}$ for p in the gage P and r positive is a base for the uniformity $\mathfrak{U}$, and it follows that $\{S_n, n \in D\}$ is a Cauchy net iff $\{(S_m,S_n), (m,n) \in D \times D\}$ is eventually in each set of the form $V_{p,r}$. In other words, $\{S_n, n \in D\}$ is a Cauchy net if and only if $\{p(S_m,S_n), (m,n) \in D \times D\}$ converges to zero for each pseudo-metric p belonging to the gage P.

There is a simple lemma about Cauchy nets which is used often enough to deserve a formal statement.

20 Lemma *A net $\{S_n, n \in D\}$ in a uniform space $(X,\mathfrak{U})$ is a Cauchy net if and only if either of the following statements is true.*

(a) *The net $\{(S_m,S_n), (m,n) \in D \times D\}$ is eventually in each member of some subbase for the uniformity $\mathfrak{U}$.*

(b) *The net $\{p(S_m,S_n), (m,n) \in D \times D\}$ converges to zero for each p in some family of pseudo-metrics which generates the gage P.*

PROOF If a family Q of pseudo-metrics generates P, then the family of all $V_{p,r}$ for p in Q and r positive is a subbase for the uniformity, so that the proof of (b) reduces to that of (a). To prove (a) notice that, if a net (for example $\{(S_m,S_n), (m,n) \in D \times D\}$) is eventually in each of a finite number of sets, it is then eventually in their intersection. ∎

The following proposition relates Cauchy nets to convergence relative to the uniform topology.

21 Theorem *Each net which converges to a point relative to the uniform topology is a Cauchy net. A Cauchy net converges to each of its cluster points.*

PROOF If $\{S_n, n \in D\}$ converges to a point s, then $\{d(S_n,s), n \in D\}$ converges to zero for each member d of the gage P. Since $d(S_m,S_n) \leqq d(S_m,s) + d(S_n,s)$, it follows that $\{d(S_m,S_n), (m,n) \in D \times D\}$ converges to zero and the net is therefore a Cauchy net. Suppose that $\{S_n, n \in D\}$ is a Cauchy net and s is a cluster point. Then for d in P and r positive there is N in D such that, if $m \geqq N$ and $n \geqq N$, then $d(S_m,S_n) < r/2$. Since s is a cluster point, there is p in D such that $d(S_p,s) \leqq r/2$ and $p \geqq N$. Then $d(S_n,s) \leqq d(S_n,S_p) + d(S_p,s) < r$ if $n \geqq N$, and it follows that the net converges to s. ∎

A uniform space is **complete** iff every Cauchy net in the space converges to a point of the space. Evidently each closed subspace of a complete space $(X,\mathfrak{U})$ is complete. If $(X,\mathfrak{U})$ is Hausdorff and $(Y,\mathfrak{V})$ is a complete subspace, then Y is closed in X, for a net in Y which converges to a point x of X is necessarily a Cauchy net, and x is the unique limit point. This obvious result is one of the most useful facts about completeness.

22 THEOREM *A closed subspace of a complete space is complete, and a complete subspace of a Hausdorff uniform space is closed.*

Before proceeding it may be worth while to mention several examples of complete spaces. If the uniformity $\mathfrak{U}$ is the largest possible uniformity for X (that is, consists of all subsets of $X \times X$ which contain the diagonal), then $(X,\mathfrak{U})$ is complete. The smallest uniformity for X also yields a complete space. If a uniform space $(X,\mathfrak{U})$ is compact relative to the uniform topology, then it is complete, for every net has a cluster point and consequently by theorem 6.21 each Cauchy net converges to some point. The space of real numbers is complete relative to the usual uniformity. This may be seen by verifying that each Cauchy net is eventually in some bounded subset A of the space of real numbers and is therefore eventually in the compact set A^-.

There is a characterization of completeness which is suggestive of compactness. Recall that a family of sets has the finite intersection property iff no finite intersection of members of the family is void, and a topological space is compact iff the intersection of the members of each family of closed sets with the finite intersection property is non-void. To describe completeness another qualification is put on the family. A family $\mathfrak{A}$ of subsets of a uniform space $(X,\mathfrak{U})$ contains **small sets** iff for each U in $\mathfrak{U}$ there is a member A of $\mathfrak{A}$ such that A is a subset of $U[x]$ for some point x. Another formulation is: for each U in $\mathfrak{U}$ there is A in $\mathfrak{A}$ such that $A \times A \subset U$. In terms of the gage P of the uniform space, a family $\mathfrak{A}$ contains small sets iff for each positive r and each d in P there is A in $\mathfrak{A}$ such that the d-diameter of A is less than r. We omit the proof that these three statements are equivalent.

23 THEOREM * *A uniform space is complete if and only if each family of closed sets which has the finite intersection property and contains small sets has a non-void intersection.*

PROOF Let $(X,\mathfrak{U})$ be a complete uniform space and $\mathfrak{A}$ a family of closed sets which has the finite intersection property and contains small sets. If $\mathfrak{F}$ is the family of all finite intersections of members of $\mathfrak{A}$, then $\mathfrak{F}$ is directed by $\subset$, and for each F in $\mathfrak{F}$ we may choose a point x_F in F. The net $\{x_F, F \in \mathfrak{F}\}$ is a Cauchy net because, if A and B follow a member F of $\mathfrak{F}$ in the ordering $\subset$ (that is, $A \subset F$ and $B \subset F$), then x_A and x_B belong to F, and $\mathfrak{F}$ contains small sets. Consequently, $\{x_F: F \in \mathfrak{F}\}$ converges to a point and since the net is eventually in each member of $\mathfrak{F}$ the point must belong to every member of $\mathfrak{F}$. Hence the intersection $\bigcap \{A: A \in \mathfrak{A}\}$ is non-void. To prove the converse let $\{x_n, n \in D\}$ be a Cauchy net, and for each n in D let A_n be the set of all points x_m for $m \geqq n$. Then the family $\mathfrak{A}$ of all sets of the form A_n has the finite intersection property, and since the net is Cauchy the family $\mathfrak{A}$ contains small sets. There is hence a point y which belongs to the intersections of the closures, $\bigcap \{A_n^-: n \in D\}$, and, according to 2.7, the point y is a cluster point of the net $\{x_n, n \in D\}$. Since $\{x_n, n \in D\}$ is a Cauchy net it converges to y. ∎

One might suspect that a uniform space satisfying the first axiom of countability would be complete if every Cauchy sequence in the space converged to a point of the space. Unfortunately this suspicion is unfounded, but the following feeble result is correct.

24 THEOREM *A pseudo-metrizable uniform space is complete if and only if every Cauchy sequence in the space converges to a point.*

PROOF If a uniform space is complete, then each Cauchy net in X, and in particular each Cauchy sequence in X, converges to a point. On the other hand, suppose that (X,d) is a pseudo-metric space such that every Cauchy sequence converges to a point, and that $\mathfrak{A}$ is a family of closed subsets of X which has the finite intersection property and contains small sets. For

* A filter is a **Cauchy filter** if it contains small sets. Then the theorem can be stated: a space is complete iff each Cauchy filter converges to some point.

each non-negative integer n select a member A_n of $\mathcal{A}$ which is of diameter less than 2^{-n} and select a point x_n belonging to A_n. If m and n are large, then $d(x_m,x_n)$ is small because x_m and x_n belong to A_m and A_n respectively, these two sets intersect, and each has small diameter. Hence $\{x_n, n \in \omega\}$ is a Cauchy sequence and therefore converges to a point y of X. If B is an arbitrary member of $\mathcal{A}$, then dist $(x_n,B) < 2^{-n}$ because B intersects A_n, and it follows that y belongs to the closure of B. Since $\mathcal{A}$ is a family of closed sets y belongs to every member of $\mathcal{A}$. ∎

The usual method of proving completeness consists in showing the space in question is uniformly isomorphic to a closed subspace of a product of complete spaces and then appealing to the following theorem. The proof of this theorem requires the fact that the image of a Cauchy net under a uniformly continuous map is a Cauchy net—a fact which is evident from the definition.

25 THEOREM *The product of uniform spaces is complete if and only if each coordinate space is complete.*

A net in the product is a Cauchy net if and only if its projection into each coordinate space is a Cauchy net.

PROOF Suppose that $(Y_a,\mathcal{U}_a)$ is a complete uniform space for each member a of an index set A. For each a the projection of a Cauchy net into Y_a is a Cauchy net and hence converges to a point, say, y_a. Then the net in the product converges to the point y with a-th coordinate y_a and consequently the product is complete. The simple proof of the converse is omitted.

If $\{x_n, n \in D\}$ is a net in the product which projects into a Cauchy net in each coordinate space, then for each member U of $\mathcal{U}_a$ the net $\{(x_m,x_n), (m,n) \in (D \times D)\}$ is eventually in the inverse under projection of U. That is, $\{(x_m,x_n), (m,n) \in (D \times D)\}$ is eventually in $\{(x,z)\colon (x_a,z_a) \in U\}$. Since the family of sets of this form is a subbase for the product uniformity it follows (6.20) that $\{x_n, n \in D\}$ is a Cauchy net. ∎

A function f is **uniformly continuous on a subset** A of a uniform space $(X,\mathcal{U})$ iff its restriction to A, $f \mid A$, is uniformly continuous with respect to the relativized uniformity. If the range space is complete and Hausdorff * and f is uniformly continuous

* This requirement is not necessary for the existence of an extension, but is necessary for the uniqueness.

on its domain A, then there is a unique uniformly continuous extension whose domain is the closure of A.

26 THEOREM *Let f be a function whose domain is a subset A of a uniform space $(X,\mathcal{U})$ and whose values lie in a complete Hausdorff uniform space $(Y,\mathcal{V})$. If f is uniformly continuous on A, then there is a unique uniformly continuous extension f^- of f whose domain is the closure of A.*

PROOF The function f is a subset of $X \times Y$ (we do not distinguish between a function and its graph) and the desired extension is the closure f^- of f in $X \times Y$. (A pair (x,y) belongs to f^- iff there is a net in A converging to x such that the image net converges to y.) The domain of f^- is evidently the closure of A. We will show that, if W is a member of $\mathcal{V}$, then there is U in $\mathcal{U}$ such that, if (x,y) and (u,v) are members of f^- and $x \in U[u]$, then $y \in W[v]$. Since Y is Hausdorff this will show that f^- is a function and that f^- is uniformly continuous. Choose a member V of $\mathcal{V}$ which is closed and symmetric and such that $V \circ V \subset W$ and choose a member U of $\mathcal{U}$ which is open and symmetric and such that $f[U[x]] \subset V[f(x)]$ for each x in A; suppose (x,y) and (u,v) belong to f^- and $x \in U[u]$. Then the intersection of $U[x]$ and $U[u]$ is open and there is consequently z in A such that both x and u belong to $U[z]$. Both y and v belong to the closure of $f[U[z]]$, by the definition of f^-, and hence both y and v belong to $V[f(z)]$. Hence $(y,v) \in V \circ V \subset W$ and $y \in W[v]$. ▮

COMPLETION

It is the purpose of this section to show that each uniform space is uniformly isomorphic to a dense subspace of a complete uniform space. It is therefore possible to adjoin "ideal elements" to a uniform space in such a way as to obtain a complete uniform space. The procedure is suggestive of the compactification process of chapter 5, but there is one significant difference: the completion of a uniform space is (essentially) unique.

For a metric space X it is possible to find a complete metric space X^* such that X is *isometric* to a dense subspace of X^* (not just uniformly isomorphic). We base the general construction of a completion on this preliminary result.

27 THEOREM *Each metric (or pseudo-metric) space can be mapped by a one-to-one isometry onto a dense subset of a complete metric (respectively pseudo-metric) space.*

PROOF It is only necessary to prove the theorem for a pseudo-metric space (X,d), since the corresponding result for metric spaces then follows from 4.15. Let X^* be the class of all Cauchy sequences in X, and for members S and T of X^* let $d^*(S,T)$ be the limit of $d(S_m,T_m)$ as m becomes large (formally, the limit of $\{d(S_m,T_m), m \in \omega\}$). It is easy to verify that d^* is a pseudo-metric for X^*. Let F be the map which carries each point x of X into the sequence which is constantly equal to x; that is, $F(x)_n = x$ for all n. Evidently F is a one-to-one isometry and it remains to prove that $F[X]$ is dense in X^* and X^* is complete. The first of these statements is almost self-evident; if $S \in X^*$ and n is large, then $F(S_n)$ is near S. To show X^* complete, first observe that it is sufficient to show that each Cauchy sequence in $F[X]$ converges to a point of X^* because $F[X]$ is dense in X^*. Finally, each Cauchy sequence in $F[X]$ is of the form $F \circ S = \{F(S_n), n \in \omega\}$, where S is a Cauchy sequence in X, and $F \circ S$ converges in X^* to the member S of X^*. ▌

Each uniform space is uniformly isomorphic to a subspace of a product of pseudo-metric spaces, and each Hausdorff uniform space is uniformly isomorphic to a product of metric spaces, by 6.16. The preceding theorem implies that a metric or pseudo-metric space is uniformly isomorphic to a subspace of a complete space of the same sort. It follows without difficulty that:

28 THEOREM *Each uniform space is uniformly isomorphic to a dense subspace of a complete uniform space. Each Hausdorff uniform space is uniformly isomorphic to a dense subspace of a complete Hausdorff uniform space.*

A **completion** of a uniform space $(X,\mathfrak{U})$ is a pair, $(f,(X^*,\mathfrak{U}^*))$ where $(X^*,\mathfrak{U}^*)$ is a complete uniform space and f is a uniform isomorphism of X into a dense subspace of X^*. The completion is Hausdorff iff $(X^*,\mathfrak{U}^*)$ is a Hausdorff uniform space. The foregoing theorem can then be stated: *Each (Hausdorff) uniform space has a (Hausdorff) completion.*

There is a uniqueness property for Hausdorff completions. If f and g are uniform isomorphisms of X onto dense subspaces of complete Hausdorff uniform spaces X^* and X^{**}, then both $g \circ f^{-1}$ and $f \circ g^{-1}$ have uniformly continuous extensions to all of X^* and X^{**} respectively, by 6.26. It follows that the extension of $g \circ f^{-1}$ is a uniform isomorphism of X^* onto X^{**}. Stated roughly: *the Hausdorff completion of a Hausdorff uniform space is unique to a uniform isomorphism.*

COMPACT SPACES

Each completely regular topology $\mathfrak{I}$ for a set X is the uniform topology for some uniformity $\mathfrak{U}$, but the uniformity is usually not unique. If $(X,\mathfrak{I})$ is compact and regular, then it turns out that there is precisely one uniformity whose topology is $\mathfrak{I}$. In this case the topology determines the uniformity, topological invariants are uniform invariants, and the theory takes a particularly simple form. This section is devoted to a proof of the uniqueness theorem just quoted and to two other propositions. As before, we use either the uniformity of a space or the corresponding gage of uniformly continuous pseudo-metrics as convenience dictates.

29 THEOREM *If $(X,\mathfrak{U})$ is a compact uniform space, then every neighborhood of the diagonal Δ in $X \times X$ is a member of $\mathfrak{U}$ and every pseudo-metric which is continuous on $X \times X$ is a member of the gage of $\mathfrak{U}$.*

PROOF Let $\mathfrak{B}$ be the family of closed members of $\mathfrak{U}$ and let V be an arbitrary open neighborhood of Δ. If $(x,y) \in \bigcap \{U: U \in \mathfrak{B}\}$, then, since $\mathfrak{B}$ is a base for $\mathfrak{U}$, y belongs to every neighborhood of x and hence (x,y) belongs to every neighborhood of Δ. It follows that $\bigcap \{U: U \in \mathfrak{B}\}$ is a subset of V. Since each member U of $\mathfrak{B}$ is compact and V is open the intersection of some finite subfamily of $\mathfrak{B}$ is also a subset of V and hence $V \in \mathfrak{U}$.

If a pseudo-metric d for X is continuous on $X \times X$, then for each positive r the set $\{(x,y): d(x,y) < r\}$ is a neighborhood of the diagonal. Hence d is uniformly continuous and therefore belongs to the gage of $\mathfrak{U}$. ∎

Each compact regular topological space is completely regular and its topology is therefore the uniform topology for some uniformity. This uniformity has just been identified.

30 Corollary *If $(X,\mathfrak{I})$ is a compact regular topological space, then the family of all neighborhoods of the diagonal Δ is a uniformity for X and $\mathfrak{I}$ is the uniform topology.*

There is another corollary.

31 Theorem *Each continuous function on a compact uniform space to a uniform space is uniformly continuous.*

PROOF If f is a continuous function on X to Y, then f_2, where $f_2(x,y) = (f(x),f(x))$, is a continuous function on $X \times X$ to $Y \times Y$. Consequently if d belongs to the gage of Y the composition $d \circ f_2$ is continuous on $X \times X$. It follows from theorem 6.29 that $d \circ f_2$ belongs to the gage of X, and hence the function f is uniformly continuous. ∎

Each compact uniform space $(X,\mathfrak{U})$ can be written as the union of a finite number of small sets, in the sense that for each pseudo-metric d belonging to the gage of $\mathfrak{U}$ and each positive r there is a finite cover of X by sets of d-diameter less than r. This is a direct consequence of compactness, since X can be covered by a finite number of $r/3$ spheres about points and each of these is of diameter less than r. A uniform space $(X,\mathfrak{U})$ is **totally bounded** (or **precompact**) iff X is the union of a finite number of sets of d-diameter less than r for each pseudo-metric d of the gage of $\mathfrak{U}$ and each positive r. In terms of $\mathfrak{U}$ this can be stated: for each U in $\mathfrak{U}$ the set X is the union of a finite number of sets B such that $B \times B \subset U$, or, equivalently, for each U in $\mathfrak{U}$ there is a finite subset F of X such that $U[F] = X$. A subset Y of a uniform space is called totally bounded iff Y, with the relativized uniformity, is totally bounded.

There is a simple but very useful relation between compactness and total boundedness.

32 Theorem *A uniform space $(X,\mathfrak{U})$ is totally bounded if and only if each net in X has a Cauchy subnet.*

Consequently a uniform space is compact if and only if it is totally bounded and complete.

PROOF Suppose S is a net in a totally bounded uniform space $(X,\mathfrak{U})$. The existence of a Cauchy subnet is an obvious consequence of problem 2.J, but we sketch the proof without using the earlier result. Let $\mathfrak{A}$ be the family of all subsets A of X such that S is frequently in A. Then $\{X\} \subset \mathfrak{A}$ and by the maximal principle 0.25 there is a maximal subfamily $\mathfrak{B}$ of $\mathfrak{A}$ which contains $\{X\}$ and has the finite intersection property. Because of maximality it is true that, if a finite union $B_1 \cup \cdots \cup B_n$ of members of $\mathfrak{A}$ belongs to $\mathfrak{B}$, then $B_i \in \mathfrak{B}$ for some i (see 2.I for details). Since X is totally bounded it may be covered by a finite number of small sets, and it follows that $\mathfrak{B}$ contains small sets. Finally, it follows from 2.5 that there is a subnet of S which is eventually in each member of $\mathfrak{B}$, and evidently this subnet is Cauchy.

If $(X,\mathfrak{U})$ is not totally bounded, then for some U in $\mathfrak{U}$ and for every finite subset F of X it is true that $U[F] \neq X$. It follows that one may find by induction a sequence $\{x_n, n \in \omega\}$ such that $x_n \notin U[x_p]$ if $p < n$. Clearly the sequence $\{x_n, n \in \omega\}$ has no Cauchy subnet.

Finally, if $(X,\mathfrak{U})$ is complete and totally bounded, then each net has a subnet which converges to a point of X and hence the space is compact. It has already been observed that a compact space is complete. ∎

There is one other very useful lemma concerning compact spaces. The proposition is an extension of the Lebesgue covering lemma 5.26. A cover of a subset A of a uniform space $(X,\mathfrak{U})$ is a **uniform cover** iff there is a member U of $\mathfrak{U}$ such that the set $U[x]$ is a subset of some member of the cover for every x in A (that is, the family of $U[x]$ for x in A refines the cover). In terms of the gage of the uniformity $\mathfrak{U}$, a cover of A is uniform iff there is a member d of the gage and a positive number r such that the open sphere of d-radius r about each point of A is contained in some member of the cover.

33 THEOREM *Each open cover of a compact subset of a uniform space is a uniform cover.*

In particular, each neighborhood of a compact subset A contains a neighborhood of the form $U[A]$ where U is a member of the uniformity.

PROOF Let $\mathfrak{A}$ be an open cover of the compact subset A of the uniform space $(X,\mathfrak{U})$. Then for each x in A there is U in $\mathfrak{U}$ such that $U[x]$ is a subset of some member of $\mathfrak{A}$, and hence there is V in $\mathfrak{U}$ such that $V \circ V[x]$ is a subset of some member of $\mathfrak{A}$. Choose a finite number of members $x_1, \cdots, x_n$ of A and $V_1, \cdots, V_n$ of $\mathfrak{U}$ such that the sets $V_i[x_i]$ cover A and for each i it is true that $V_i \circ V_i[x_i]$ is a subset of some member of $\mathfrak{A}$. Finally, let $W = \bigcap \{V_i: i = 1, \cdots, n\}$. Then for each point y of A for some i the point y belongs to $V_i[x_i]$ and hence $W[y] \subset W \circ V_i[x_i] \subset V_i \circ V_i[x_i]$. Consequently $W[y]$ is a subset of some member of $\mathfrak{A}$. ∎

FOR METRIC SPACES ONLY

This section is devoted to two propositions concerning complete metric spaces. The results are among the most useful consequences of completeness, and it is unfortunate that no generalization to complete uniform spaces seems possible. The first proposition is the classic theorem of Baire on category; this theorem and one or two related results occupy most of the section. The last theorem of the section states that the image under a continuous uniformly open map of a complete metric space is again complete, provided the range space is Hausdorff. The proof relies on a lemma which we state in considerably more general form than is necessary for this proposition. The lemma (essentially a formalization of an argument of Banach) also yields directly the closed graph and open mapping theorems of normed linear space theory. (See problem 6.**R**.)

34 THEOREM (BAIRE) *Let X be either a complete pseudo-metric space or a locally compact regular space. Then the intersection of a countable family of open dense subsets of X is itself dense in X.*

PROOF We prove the theorem for locally compact regular spaces, adding in parentheses the modifications necessary to establish it for a complete pseudo-metric space. Suppose that $\{G_n, n \in \omega\}$ is a sequence of dense open subsets of X and that U is an arbitrary open non-void subset of X. It must be shown that $U \cap \bigcap \{G_n: n \in \omega\}$ is non-void. To this end choose inductively an open set V_0 such that V_0^- is a compact subset of $U \cap G_0$ (such that V_0^- is a subset of $U \cap G_0$ and has diameter less than one), and then for each positive integer n choose V_n such that V_n^- is a subset of $V_{n-1} \cap G_n$ (and the diameter of V_n is less than $1/n$). This choice is possible because G_n is dense and open. The family of all sets V_n^- for non-negative integers n has the finite intersection property, consists of closed sets and V_0^- is compact (the family contains small sets). Hence $\bigcap\{V_n^-: n \in \omega\}$ is non-void, and since $V_{n+1}^- \subset U \cap G_n$ it follows that $U \cap \bigcap\{G_n: n \in \omega\}$ is non-void. ∎

It should be remarked that the Baire theorem is a hybrid in that a topological conclusion (the intersection of a countable number of dense open sets is dense) is deduced from a non-topological premise (that the space is complete pseudo-metric). There is a purely topological statement which is equivalent. If $(X,\mathfrak{J})$ is a topological space such that for some pseudo-metric d for X the space (X,d) is complete and $\mathfrak{J}$ is the pseudo-metric topology, then the same conclusion holds. (Topological spaces for which there exists such a complete metric have been characterized in a different way, as noted in 6.**K**.)

A terminology has been devised which is very convenient in discussing questions related to the Baire theorem. A subset A of a topological space is **nowhere dense** in X iff the interior of the closure of A is void; otherwise stated, A is nowhere dense in X iff the open set $X \sim A^-$ is dense in X. It is evident that the finite union of nowhere dense sets is nowhere dense. A subset A of X is **meager** in X or of the **first category** in X iff A is the union of a countable family of nowhere dense sets. The Baire theorem can then be stated: the complement of a meager subset of a complete metric space is dense. (The complement of a meager set is sometimes called **co-meager** or **residual** in X.)

A set A is **non-meager** or of the **second category** in X iff it is not meager in X. The following result is a sort of a localization theorem. From the fact that a set A is non-meager we deduce the existence of points x such that A intersects each neighborhood of x in a non-meager set. It is sometimes said that A is of the second category at such points.

35 THEOREM *Let A be a subset of a topological space X and let $M(A)$ be the union of all open sets V such that $V \cap A$ is meager in X. Then $A \cap M(A)^-$ is meager in X.*

PROOF Let $\mathfrak{U}$ be a disjoint family of open sets which is maximal with respect to the property: if $U \in \mathfrak{U}$, then $U \cap A$ is meager. Such a family $\mathfrak{U}$ exists because of the maximal principle 0.25. Let $W = \bigcup\{U: U \in \mathfrak{U}\}$. The proof reduces to showing that $W \cap A$ is meager, for if this is known then $A \cap W^-$ is meager because $W^- \sim W$ is nowhere dense, and from the maximality of $\mathfrak{U}$ it follows that W^- contains every open set V such that $V \cap A$ is meager. To show that $W \cap A$ is meager, for each U in $\mathfrak{U}$ write $U \cap A$ in the form $\bigcup\{U_n: n \in \omega\}$ where U_n is nowhere dense. Then, because the family $\mathfrak{U}$ is disjoint, the set $\bigcup\{U_n: U \in \mathfrak{U}\}$ is nowhere dense for each non-negative integer n. Hence $W \cap A$ is meager. ▌

An important consequence of the preceding theorem is that if a subset A of a topological space is non-meager then there is a non-void open set V such that the intersection of A with every neighborhood of each point of V^- is non-meager.

The concluding theorem of this chapter shows that completeness is preserved by certain mappings. A map of a uniform space $(X,\mathfrak{U})$ into a uniform space $(Y,\mathfrak{V})$ is **uniformly open** iff for each U in $\mathfrak{U}$ there is V in $\mathfrak{V}$ such that $f[U[x]] \supset V[f(x)]$ for each x in X. It is not true that uniformly open maps preserve completeness for arbitrary uniform spaces; Köthe [1] has given an example of a complete linear topological space and a closed subspace such that the quotient space is not complete. The theorem, like the Baire theorem, is peculiar to pseudo-metric spaces.

The proof of the theorem which is given here depends on a lemma which has other profound consequences (see 6.**R**). The lemma concerns a relation R between points of a pseudo-metric space (X,d) and a uniform space $(Y,\mathfrak{V})$; that is, R is a subset of $X \times Y$. Let $U_r = \{(x,y): d(x,y) < r\}$, so that $U_r[x]$ is simply the r-sphere about x.

36 LEMMA *Let R be a closed subset of the product of a complete pseudo-metric space (X,d) with the uniform space $(Y,\mathfrak{V})$ and suppose that for each positive r there is V in $\mathfrak{V}$ such that $R[U_r[x]]^-$ contains $V[y]$ for each (x,y) in R. Then for each r and each positive e it is true that $R[U_{r+e}[x]] \supset R[U_r[x]]^- \supset V[y]$.*

PROOF The critical fact needed for the proof is: if A is a subset of X and $v \in R[A]^-$, then there is a set B of arbitrarily small diameter such that $v \in R[B]^-$ and $A \cap B$ is not void. This is true because: if r is arbitrary, if V is a symmetric member of $\mathcal{V}$ such that $R[U_r[x]]^- \supset V[y]$ for each member (x,y) of R, if v' is a point of $R[A]$ such that $v' \in V[v]$, and if u is a point of A such that $(u,v') \in R$, then $v \in V[v'] \subset R[U_r[u]]^-$, and the diameter of $U_r[u]$ is at most $2r$.

The lemma is now established as follows. Suppose that $v \in R[U_r[x]]^-$. It will be shown that $v \in R[U_{r+e}[x]]$, which will complete the proof. Let $A_0 = U_r[x]$, and select inductively, for each positive integer n, a subset A_n of X such that $v \in R[A_n]^-$, $A_n \cap A_{n-1}$ is not void, and the diameter of A_n is less than $e2^{-n}$. Since X is complete there is evidently a point u such that each neighborhood W of u contains some A_n (hence $v \in R[W]^-$). Clearly $d(x,u) < r + e$. For each neighborhood W of u and each neighborhood Z of v it is true that $R[W]$ intersects Z, and hence there is (u',v') in R with u' in W and v' in Z; that is, $R \cap (W \times Z)$ is non-void. Since R is closed $(u,v) \in R$ and the proof is complete. ∎

Suppose now that f is uniformly open and continuous, that X is complete and pseudo-metrizable, that Y is Hausdorff, and that Y^* is a Hausdorff completion of Y. Then (the graph of) f is a subset of $X \times Y^*$ which is closed because f is continuous, and satisfies the condition of the preceding lemma because the map of X into Y is uniformly open. Then the lemma implies that f is a uniformly open map of X into Y^*. Finally, since $f[X]$ contains $V[f[X]]$ for some V in $\mathcal{V}$, it must be true that $f[X]$ is closed (and open) in Y^*; hence $f[X]$ is complete.

37 COROLLARY *Let f be a continuous uniformly open map of a complete pseudo-metrizable space into a Hausdorff uniform space. Then the range of the map f is complete.*

PROBLEMS

A EXERCISE ON CLOSED RELATIONS

Let X and Y be topological spaces and let R be a closed subset of $X \times Y$. If A is a compact subset of X, then $R[A]$ is a closed subset of

Y. (If $y \notin R[A]$, then $A \times \{y\}$ is contained in the open set $(X \times Y) \sim R$, and theorem 5.12 may be applied.)

B EXERCISE ON THE PRODUCT OF TWO UNIFORM SPACES

Let $(X,\mathfrak{U})$ and $(Y,\mathfrak{V})$ be uniform spaces and for each U in $\mathfrak{U}$ and each V in $\mathfrak{V}$ let $W(U,V) = \{((x,y),(u,v)): (x,u) \in U \text{ and } (y,v) \in V\}$.

(a) The family of sets of the form $W(U,V)$ is a base for the product uniformity for $X \times Y$.

(b) If R is a subset of $X \times Y$, then $W(U,V)[R] = V \circ R \circ U^{-1} = \bigcup\{U[x] \times V[y]: (x,y) \in R\}$.

(c) The closure of a subset R of $X \times Y$ is $\bigcap\{V \circ R \circ U^{-1}: U \in \mathfrak{U} \text{ and } V \in \mathfrak{V}\}$.

C A DISCRETE NON-METRIZABLE UNIFORM SPACE

It should be observed that a uniform space $(X,\mathfrak{U})$ may fail to be metrizable even though the topology of $\mathfrak{U}$ is metrizable. Let Ω_0 be the set of all ordinals which are less than the first uncountable ordinal Ω, and for each member a of Ω_0 let $U_a = \{(x,y): x = y \text{ or } x \geq a \text{ and } y \geq a\}$. Then the family of all sets of the form U_a is a base for a uniformity $\mathfrak{U}$ for Ω_0 (observe that $U_a = U_a \circ U_a = U_a^{-1}$). The topology of this uniformity is the discrete topology and hence metrizable, but the uniform space $(\Omega_0,\mathfrak{U})$ is not metrizable.

D EXERCISE: UNIFORM SPACES WITH A NESTED BASE

Let $(X,\mathfrak{U})$ be a Hausdorff uniform space and suppose that a base $\mathfrak{B}$ for $\mathfrak{U}$ is linearly ordered by inclusion. Then either $(X,\mathfrak{U})$ is metrizable or the intersection of every countable family of open subsets of X is open.

E EXAMPLE: A VERY INCOMPLETE SPACE (THE ORDINALS)

Let Ω_0 be the set of all ordinals less than the first uncountable ordinal Ω, and let $\mathfrak{J}$ be the order topology for Ω_0. Then there is a unique uniformity for Ω_0 whose topology is $\mathfrak{J}$ and Ω_0 is not complete relative to this uniformity. (Using the methods of problem **4.E** show that, if U is an open subset of $\Omega_0 \times \Omega_0$ which contains the diagonal, then for some x it is true that $(y,z) \in U$ whenever $y > x$ and $z > x$. Then show that a uniformity whose topology is $\mathfrak{J}$ must be identical with the relativized uniformity of the compact space $\Omega' = \{x: x \leq \Omega\}$.)

Note This property of Ω_0 was observed by Dieudonné [5]. Doss [1] has characterized topological spaces which, like Ω_0, have a unique uniformity.

F THE SUBBASE THEOREM FOR TOTAL BOUNDEDNESS

The uniform space analogue of Alexander's theorem 5.6 on compact subbases is: Let $(X,\mathfrak{U})$ be a uniform space such that for each member U of some subbase for $\mathfrak{U}$ there is a finite cover $A_1, \cdots, A_n$ of X such that $A_i \times A_i \subset U$ for each i. Then the space $(X,\mathfrak{U})$ is totally bounded.

Consequently the product of uniform spaces is totally bounded if and only if each coordinate space is totally bounded.

The Tychonoff product theorem 5.13 for completely regular spaces may be derived from the preceding proposition and 6.32.

G SOME EXTREMAL UNIFORMITIES

(a) If $(X,\mathfrak{J})$ is a Tychonoff space, then the uniformity of the Stone-Čech compactification of X, relativized to X, is the smallest uniformity such that each bounded real-valued continuous function is uniformly continuous.

(b) If $(X,\mathfrak{J})$ is a completely regular space, then there is a largest uniformity $\mathfrak{V}$ for X whose topology is $\mathfrak{J}$. This uniformity may be described alternately as the smallest which makes uniformly continuous each continuous map into a metric space, or each continuous map into a uniform space. Explicitly, V is a member of $\mathfrak{V}$ iff V is a neighborhood of the diagonal in $X \times X$ and there is a sequence $\{V_n, n \in \omega\}$ of symmetric neighborhoods of the diagonal such that $V_0 \subset V$ and $V_{n+1} \circ V_{n+1} \subset V_n$ for each n in ω.

Note These two constructions are examples of a method which has been used before. If F is an arbitrary family of functions on X, each member f mapping X into a uniform space Y_f, then there is a smallest uniformity which makes each f uniformly continuous (or equivalently, makes the natural map into $\times\{Y_f: f \in F\}$ uniformly continuous).

For further information on some extremal uniformities see Shirota [1].

H UNIFORM NEIGHBORHOOD SYSTEMS

A *uniform neighborhood system* for a set X is a correspondence V and an ordering $\geq$ such that the following conditions are satisfied:

(i) $V_a(x)$ is a subset of X to which x belongs, for each member a of an index set A and each point x of X;
(ii) the relation $\geq$ directs the index set A;
(iii) if $a \geq b$, then $V_a(x) \subset V_b(x)$ for all x;
(iv) for each member a of A there is b in A such that $y \in V_a(x)$ whenever $x \in V_b(y)$; and

(v) for each member a of A there is b in A such that $z \in V_a(x)$ whenever $y \in V_b(x)$ and $z \in V_b(y)$.

(a) If $(V, \geqq)$ is a uniform neighborhood system for X, then the family of all sets of the form $\{(x,y): y \in V_a(x)\}$, for a an arbitrary member of A is the base of a uniformity $\mathcal{U}$ for X. This uniformity is called the *uniformity of the system*. This uniformity has the property that: for each a in A, for some U in $\mathcal{U}$, $U[x] \subset V_a(x)$ for all x, and for each U in $\mathcal{U}$ for some a in A, $V_a(x) \subset U[x]$ for all x.

(b) Let $\mathcal{U}$ be a uniformity for X, and let $V_U(x) = U[x]$ for each member U of $\mathcal{U}$ and each member x of X. Then $\mathcal{U}$ is directed by $\subset$ and $(V, \subset)$ is a uniform neighborhood system for X whose uniformity is $\mathcal{U}$.

(c) Let P be the gage of a uniformity $\mathcal{U}$ for X, let A be the cartesian product of P and the set of positive real numbers, and direct A by agreeing that $(p,r) \geqq (q,s)$ iff $r \leqq s$ and $p(x,y) \geqq q(x,y)$ for all x and y in X. If $V_{p,r}(x) = \{y: p(x,y) < r\}$, then $(V, \geqq)$ is a uniform neighborhood system for X whose uniformity is $\mathcal{U}$.

Note It is evident from the foregoing that "indexed" neighborhoods may be used to discuss uniformity and that the theory so obtained is identical with that of uniform spaces. These facts are due to Weil [1].

I ÉCARTS AND METRICS

An *écart* for a set X is a non-negative real-valued function e on $X \times X$ such that

(i) $e(x,y) = 0$ iff $x = y$ and
(ii) for each positive number s there is a positive number r such that $e(x,z) < s$ whenever $e(x,y)$ and $e(y,z)$ are both less than r.

If e is an écart for X then there is a non-negative function p on $X \times X$ such that

(i) $p(x,y) = 0$ iff $x = y$;
(ii) $p(x,y) + p(y,z) \geqq p(x,z)$ for all x, y, and z in X; and
(iii) for each positive s there is a positive number r such that $p(x,y) < s$ whenever $e(x,y) < r$ and, similarly, $e(x,y) < s$ whenever $p(x,y) < r$.

If $e(x,y) = e(y,x)$ for all x and y then p may be taken to be a metric.

Note This is essentially Chittenden's metrization theorem (see 6.14). The "metrization" of a topological space by a function d satisfying all of the requirements for a metric except "$d(x,y) = d(y,x)$" has been investigated by Ribeiro [2] and by Balanzat [1].

The term "écart" has been used by some authors to mean a distance function taking values in a structure less restricted than that of the real numbers (for example, a partially ordered set). For treatments of uniformity based on ideas of this sort see Appert [1], Colmez [1], Cohen and Goffman [1], Gomes [1], Kalisch [1], and Lasalle [1].

J UNIFORM COVERING SYSTEMS

Let Φ be a collection of covers of a set X such that:

(i) if $\mathcal{A}$ and $\mathcal{B}$ are members of Φ, then there is a member of Φ which is a refinement of both $\mathcal{A}$ and $\mathcal{B}$;
(ii) if $\mathcal{A} \in \Phi$, then there is a member of Φ which is a star refinement of $\mathcal{A}$; and
(iii) if $\mathcal{A}$ is a cover of X and some refinement of $\mathcal{A}$ belongs to Φ, then $\mathcal{A}$ belongs to Φ.

Let $\mathcal{U}$ be the uniformity for X such that the family of all sets of the form $\bigcup \{A \times A : A \in \mathcal{A}\}$ for $\mathcal{A}$ in Φ is a base for $\mathcal{U}$. Then Φ is precisely the family of all covers of X which are uniform relative to $\mathcal{U}$.

Note Description of a uniformity by means of covers has been used very effectively by J. W. Tukey [1]; a very early use of this general sort was made by Alexandroff and Urysohn [2].

K TOPOLOGICALLY COMPLETE SPACES: METRIZABLE SPACES

A topological space $(X,\mathfrak{J})$ is called *metrically topologically complete* ifi there is a metric d for X such that (X,d) is complete and $\mathfrak{J}$ is the metric topology. A topological space $(X,\mathfrak{J})$ is an *absolute* G_δ iff it is metrizable and is a G_δ (a countable intersection of open sets) in every metric space in which it is topologically embedded. Then: A topological space is metrically topologically complete if and only if it is an absolute G_δ. The proof depends on a sequence of lemmas.

(a) Let (X,d) be a complete metric space, let U be an open subset of X, for x in U let $f(x) = 1/\text{dist}\,(x, X \sim U)$, and let $d^*(x,y) = d(x,y) + |f(x) - f(y)|$. Then d^* is a metric, U is a complete relative to d^*, and the d and d^* topologies for U are identical.

(b) A G_δ in a complete metric space is homeomorphic to a complete metric space. (If $U = \bigcap \{U_n : n \in \omega\}$ consider the map of U into the product of the complete metric spaces (U_n, d_n^*), where d_n^* is constructed from d and U_n as in (a).)

(c) If there is a homeomorphism of a dense subset Y of a Hausdorff space X onto a complete metric space Z, then Y is a G_δ in X. (For each integer n let U_n be the set of all points x of X such that the image of some neighborhood of x is of diameter less than $1/n$. Then the homeomorphism f can be extended continuously to a continuous map f^- of $\bigcap\{U_n: n \in \omega\}$ into Z and $f^{-1} \circ f^-$ must be the identity.)

Note These are classical results; (b) is due to Alexandroff [1] and to Hausdorff [2] and (c) is due to Sierpinski [2].

L TOPOLOGICALLY COMPLETE SPACES: UNIFORMIZABLE SPACES

A topological space $(X,\mathfrak{J})$ is said to be *topologically complete* iff there is a uniformity $\mathfrak{U}$ for X such that $(X,\mathfrak{U})$ is complete and $\mathfrak{J}$ is the uniform topology.

(a) If $\mathfrak{U}$ and $\mathfrak{V}$ are uniformities for X such that $\mathfrak{U} \subset \mathfrak{V}$, if $(X,\mathfrak{U})$ is complete, and if the topology of $\mathfrak{U}$ is identical with that of $\mathfrak{V}$, then $(X,\mathfrak{V})$ is complete. Hence a completely regular space is topologically complete iff it is complete relative to the largest uniformity whose topology is $\mathfrak{J}$.

(b) Let $(X,\mathfrak{U})$ be a complete uniform space, let F be an F_σ (a countable union of closed sets) and let $x \in X \sim F$. Then there is a continuous real-valued function on X which is positive on F and 0 at x. Consequently there is an open set V and a uniformity $\mathfrak{V}$ for V such that V contains F, $x \notin V$, $(V,\mathfrak{V})$ is complete, and the topology of $\mathfrak{V}$ is identical with the relativized topology of $\mathfrak{U}$. (Recall the device used in 6.**K**(a).)

(c) If $(X,\mathfrak{U})$ is a complete uniform space and Y is a subset of X which is the intersection of the members of a family of F_σ's, then Y, with the relativized uniform topology, is topologically complete. (See 6.**K**.)

(d) Each paracompact space X is topologically complete. (Consider the uniformity consisting of all neighborhoods of the diagonal. A Cauchy net which converges to no point of X must, for each point x, be eventually in the complement of some neighborhood of x, and the application of the even covering property of paracompact spaces leads to a contradiction.)

Note The problem of topological completeness has been studied by Dieudonné [6]; in particular he has shown that each metrizable space is topologically complete (this is a consequence of either (c) or (d) above). Shirota [2] has proved several interesting and profound theorems on topological completeness, in a direction connected with work of Hewitt [2]. See also Umegaki[1].

I conjecture that a completely regular space X is paracompact iff

(i) the family of all neighborhoods of the diagonal is a uniformity, and

(ii) X is topologically complete.

Neither (i) or (ii) is in itself sufficient to imply paracompactness. A non-paracompact space satisfying (i) is exhibited in 6.E. The condition (i) implies normality (if A and B are disjoint closed sets choose a symmetric U such that $U \circ U \subset (X \sim A) \times (X \sim A) \cup (X \sim B) \times (X \sim B)$ and consider $U[A]$ and $U[B]$; a stronger normality condition may be obtained by a similar argument, as shown by H. J. Cohen [1]). However, the product of uncountably many copies of the space of real numbers is complete and not normal (A. H. Stone [1]).

The F_σ condition encountered in (c) above is suggestive of the work of Smirnov [3] on normality.

M THE DISCRETE SUBSPACE ARGUMENT; COUNTABLE COMPACTNESS

(a) If a subset A of a uniform space $(X,\mathfrak{U})$ is not totally bounded, then there is a member U of $\mathfrak{U}$ and an infinite subset B of A such that $U[x]$ is disjoint from $U[y]$ for every pair of distinct points of B; equivalently, there is a pseudo-metric d in the gage of $\mathfrak{U}$ such that $d(x,y) \geq 1$ for distinct points x and y of B. (A set such as B might be called uniformly discrete.)

(b) A subset A of a topological space $(X,\mathfrak{J})$ is called *relatively countably compact* iff each sequence in A has a cluster point in X. Each relatively countably compact subset of a completely regular space $(X,\mathfrak{J})$ is totally bounded relative to the largest uniformity whose topology is $\mathfrak{J}$. If $(X,\mathfrak{J})$ is topologically complete a subset is relatively countably compact iff its closure is compact, and a closed subset is compact iff it is countably compact.

N INVARIANT METRICS

A pseudo-metric p for a set X is said to be *invariant* under the members of a family F of one-to-one maps of X onto itself, or simply *F-invariant*, iff $p(x,y) = p(f(x),f(y))$ for all x and y in X and all f in F.

A member U of a uniformity $\mathfrak{U}$ for X is called *F-invariant*, provided $(x,y) \in U$ iff $(f(x),f(y)) \in U$ for all f in F. Then: The family of F-invariant pseudo-metrics which are uniformly continuous on $X \times X$ generates the uniformity $\mathfrak{U}$ if and only if the family of F-invariant members of $\mathfrak{U}$ is a base. (See 6.12.)

Note This is a straightforward generalization of the metrization theorem for topological groups which is stated in the next problem.

O TOPOLOGICAL GROUPS: UNIFORMITIES AND METRIZATION

Let $(G,\mathfrak{J})$ be a topological group, and for each neighborhood U of the identity let $U_L = \{(x,y): x^{-1}y \in U\}$ and let $U_R = \{(x,y): xy^{-1} \in U\}$. Consider the following uniformities for G: the *left uniformity* $\mathcal{L}$ having as a base the family of all sets U_L with U a neighborhood of the identity, the *right uniformity* $\mathcal{R}$ with all U_R as a base, and the *two-sided uniformity* $\mathfrak{U}$ having $\mathcal{L} \cup \mathcal{R}$ as a subbase.

(a) The topology $\mathfrak{J}$ is the topology of each of $\mathcal{L}$, $\mathcal{R}$, and $\mathfrak{U}$.

(b) The uniformity $\mathcal{L}$ (respectively $\mathcal{R}$) is generated by the family of all left-invariant (right-invariant) pseudo-metrics which are continuous on $G \times G$. (See 6.**N**.)

(c) Let I be the family of all neighborhoods of the identity e which are invariant under inner automorphisms. Then I is a base for the neighborhood system of e iff the family of all pseudo-metrics which are both left and right invariant and are continuous on $G \times G$ generates a uniformity whose topology is $\mathfrak{J}$. (If U is an invariant neighborhood of e, then $U_L = U_R$, and this set is invariant under both left and right translation. If p is left and right invariant, then $p(e,y) = p(x^{-1}ex,x^{-1}yx)$.)

(d) Let G be the set of all real-valued functions of the form $g(x) = ax + b$ where $a \neq 0$. Then G is a group under composition and may be topologized by agreeing that g is near the identity iff a is near 1 and $|b|$ is near zero. For this group $\mathcal{L} \neq \mathcal{R}$ and there is no two-sided invariant metric. (The fact that $\mathcal{L} \neq \mathcal{R}$ follows directly from inspection of the defining bases. To see that no invariant metric exists show that, for each g, if $a \neq 1$, then there is f in G such that the constant coefficient of $f^{-1} \circ g \circ f$ is arbitrarily large.)

Note The existence of left-, right- or two-sided invariant metrics for G follows from the foregoing under the additional hypothesis that there is a countable base for the neighborhood system of e. The existence of left-invariant metrics is due to Birkhoff [2] and to Kakutani [1]. The two-sided invariant theorem is due to Klee [1].

It should be remarked that the requirement that a topological group be metrizable with a two-sided invariant metric is very stringent. In particular, a locally compact group of this sort has a Haar measure which is invariant under both right and left translation.

P ALMOST OPEN SUBSETS OF A TOPOLOGICAL GROUP

A subset A of a topological space X is *almost open* in X, or satisfies the *condition of Baire*, iff there is a meager set B such that the symmetric difference $(A \sim B) \cup (B \sim A)$ is open.

(a) A subset A is almost open in X iff there are meager sets B and C such that $(A \sim B) \cup C$ is open. Countable unions and complements of almost open sets are almost open. Every Borel set is almost open. (The family of *Borel* sets is the smallest family $\mathfrak{B}$ such that $\mathfrak{B}$ contains all open sets, and countable unions and complements of members of $\mathfrak{B}$ belong to $\mathfrak{B}$.)

(b) *Banach-Kuratowski-Pettis Theorem* If A contains a non-meager almost open subset of a topological group X, then AA^{-1} is a neighborhood of the identity element. (If A is non-meager so is X, and because X is a topological group each non-void open subset is also non-meager. For each almost open subset B of X let B^* be the union of all open sets U such that $U \cap (X \sim B)$ is meager. Then $(xB)^* = xB^*$ and $(B \cap C)^* = B^* \cap C^*$ if C is also almost open. Hence $xA^* \cap A^* = (xA \cap A)^*$ and if $xA^* \cap A^*$ is non-void, then $xA \cap A$ is non-void. Then $A^*(A^*)^{-1} = \{x\colon xA^* \cap A^* \textit{ is non-void}\} \subset \{x\colon xA \cap A \textit{ is non-void}\} = AA^{-1}$.)

(c) An almost open subgroup of a non-meager topological group X is either meager in X or open and closed in X.

(d) The requirement "almost open" cannot be omitted from theorem (c). There is a subgroup Y of the group X of real numbers such that the quotient X/Y is countably infinite, and since for each member Z of X/Y there is a homeomorphism of X onto itself carrying Y onto Z it follows that Y is not meager in X. (Let B be a Hamel base for X relative to the rational numbers, let C be a countably infinite subset of B, and let Y be the set of all finite linear rational combinations of members of $B \sim C$.)

Note For history and references on theorem (b) see Pettis [1]. The construction in (d) is not peculiar to the real numbers; a related phenomenon occurs in the much more general situation. The basic idea is due to Hausdorff; the sharpest known results in this direction are found in Pettis [2], where history and further references are also given.

Q COMPLETION OF TOPOLOGICAL GROUPS

Let $(G,\cdot,\mathfrak{J})$ be a topological group, let $\mathfrak{L}$ be its left uniformity, $\mathfrak{R}$ its right uniformity, and $\mathfrak{U}$ its two-sided uniformity ($\mathfrak{U}$ is the smallest uniformity which is larger than each of $\mathfrak{L}$ and $\mathfrak{R}$). It has been noted that $\mathfrak{J}$ is the topology of each of $\mathfrak{L}$, $\mathfrak{R}$, and $\mathfrak{U}$.

(a) $(G,\mathfrak{L})$ is complete iff $(G,\mathfrak{R})$ is complete. A net is Cauchy relative to $\mathfrak{U}$ iff it is Cauchy relative to each of $\mathfrak{L}$ and $\mathfrak{R}$. If $(G,\mathfrak{L})$ is complete so is $(G,\mathfrak{U})$. The uniform space $(G,\mathfrak{L})$ is complete, provided $(G,\mathfrak{U})$ is complete, and the group has the property: if $\{x_n, n \in D\}$ is a Cauchy net relative to $\mathcal{L}$, then $\{(x_n)^{-1}, n \in D\}$ is also a Cauchy net relative to $\mathcal{L}$. (Equivalently, $\mathcal{L}$ and $\mathcal{R}$ have the same Cauchy nets.) Left translation by a fixed member of the group is $\mathcal{L}$-uniformly continuous, right translation is $\mathcal{R}$-uniformly continuous, and inversion (x into x^{-1}) is $\mathfrak{U}$-uniformly continuous. Multiplication ((x,y) into xy) is usually not uniformly continuous.

(b) *Theorem* Let $(G,\cdot,\mathfrak{J})$ be a Hausdorff topological group, let $(H,\mathfrak{V})$ be a Hausdorff completion of the uniform space $(G,\mathfrak{U})$, and let $\mathcal{S}$ be the topology of $\mathfrak{V}$. Then the group operation $\cdot$ can be extended in a unique way such that $(H,\cdot,\mathcal{S})$ becomes a topological group and $\mathfrak{V}$ becomes its two-sided uniformity.

(c) The preceding theorem yields a topological group completion relative to the right uniformity, provided $\mathcal{L}$ and $\mathcal{R}$ have the same Cauchy nets. But in view of (a) this condition is necessary for the existence of "right completion." The condition is not always satisfied. For example, let G be the group of all homeomorphisms of the closed unit interval $[0,1]$ onto itself with composition for group operation and with the topology of the (right invariant) metric: $d(f,g) = \sup\{|f(x) - g(x)|: x \in [0,1]\}$. There is a sequence $\{f_n, n \in \omega\}$ in G which converges uniformly to a function which is not one to one, and the sequence $\{(f_n)^{-1}, n \in \omega\}$ is therefore not Cauchy relative to the left uniformity. The group G is already complete relative to the two-sided uniformity $\mathfrak{U}$, for $\mathfrak{U}$ is the uniformity of the metric: $d(x,y) + d(x^{-1},y^{-1})$.

(d) *Theorem* Let $(G,\cdot,\mathfrak{J})$ be a metrizable topological group, let d be a right invariant metric metrizing G, and let $d^*(x,y) = d(x,y) + d(x^{-1},y^{-1})$. Then the two-sided uniformity $\mathfrak{U}$ is the uniformity of the metric d^*. The uniform space $(G,\mathfrak{U})$ is complete iff G is complete relative to some metric whose topology is $\mathfrak{J}$. (Equivalently, iff G is a G_δ in each metrizable space in which it is topologically embedded.) If $\mathcal{L}$ and $\mathcal{R}$ have the same Cauchy sequences and G is complete relative to some metric whose topology is $\mathfrak{J}$, then G is complete relative to every right invariant metric whose topology is $\mathfrak{J}$. (See 6.**K** and 6.**P**.)

Note There are two important special cases in which "right-handed completion" may be accomplished. If there is a totally bounded neighborhood of the identity of the group, or if inversion (the map carrying x into x^{-1}) is uniformly continuous on some neighborhood of the identity, then each right Cauchy net is also a left Cauchy net and the two-sided completion yields also a right completion. These results may be proved directly without great difficulty; they are given in Bourbaki [1] and Weil [2]. The example of (c) is due to Dieudonné 3], and the result (d) is due to Klee [1].

The result of part (d)—the deduction of completeness from metric topological completeness—cannot be extended to non-metrizable groups. (See 7.M.)

R CONTINUITY AND OPENNESS OF HOMOMORPHISMS: THE CLOSED GRAPH THEOREM

Throughout this problem G and H will be Hausdorff topological groups, $\mathcal{U}$ will be the family of all neighborhoods of the identity in G, and $\mathcal{V}$ will be the corresponding family in H.

(a) *Closed graph theorem* Let G be a topological group, let H be a metrizable topological group which is complete relative to its right uniformity, and let f be a homomorphism of G into H such that

(i) the graph of f is a closed subset of $G \times H$, and
(ii) the closure of $f^{-1}[V]$ belongs to $\mathcal{U}$ whenever $V \in \mathcal{V}$.

Then f is continuous.

Dually, a homomorphism g of H into G is open if

(i)* the graph of g is a closed subset of $H \times G$, and
(ii)* the closure of $g[V]$ belongs to $\mathcal{U}$ whenever $V \in \mathcal{V}$.

(The proof is made by applying lemma 6.36 to the relations f^{-1} and g respectively. Use a right invariant metric for H. H is complete relative to each right invariant metric which metrizes H.)

(b) If in the preceding theorem it is assumed that H is a Lindelöf space (each open cover has a countable subcover) and G is non-meager, then condition (ii) is automatically satisfied; if further $g[H] = G$, then (ii)* is also automatically satisfied. If G and H are linear topological space, f and g are linear functions, $g[H] = G$, and G is non-meager, then (ii) and (ii)* are automatically satisfied. (If $V \in \mathcal{V}$, then $f[G]^- \subset Vf[G]$, and if H is Lindelöf, then $f[G]$ is covered by a countable number of translates of V by members of $f[G]$. The closures of inverses under f of these translates are mutually homeomorphic and must have non-void interiors if G is not meager. Hence $f^{-1}[V]^-$ contains an open set and $(f^{-1}[V^{-1}V])^- \supset (f^{-1}[V^{-1}]f^{-1}[V])^- \supset f^{-1}[V^{-1}]^-f^{-1}[V]^- = (f^{-1}[V]^-)^{-1}(f^{-1}[V]^-)$. It follows that $f^{-1}[V]^- \in \mathcal{U}$ for each V in $\mathcal{V}$ and a similar argument applies to g. In the linear topological space case it is possible to use scalar multiples instead of translates of members of $\mathcal{V}$.)

(c) If H is a locally compact topological group, then the closed graph theorem is valid; that is, (i) and (ii) of (a) imply continuity, and dually.

(This is a simpler result than that above. It depends on the lemma 6.A.)

Note The closed graph theorem for complete normed linear spaces is due to Banach [1;41]. Every known form of the theorem requires drastic countability or compactness assumptions on H. A counter example to a number of attractive conjectures may be constructed as follows. Let G be an arbitrary infinite dimensional complete normed linear space, and let H be G with the topology such that a base for the neighborhoods of 0 is the family of all convex sets which contain a line segment in every direction. The identity map g of H onto G is continuous and satisfies (i)* and (ii)* above (see 6.Ua). The space H has many pleasant properties: for example, it is complete, and the uniform boundedness theorem (6.Ub) holds for it. Nevertheless g is evidently not open.

S SUMMABILITY

Let f be a function whose domain includes a set A and whose values line in a complete abelian Hausdorff topological group G. Let $\mathcal{A}$ be the family of finite subsets of A, and for F in $\mathcal{A}$ let S_F be the sum of $f(a)$ for a in F. The family $\mathcal{A}$ is directed by $\supset$, and $\{S_F, F \in \mathcal{A}, \supset\}$ is a net in G. If this net converges to a member s of G, then f is said to be *summable over* A, s is defined to be the sum of f over A, and we write $s = \sum\{f(a): a \in A\} = \sum_A f$.

(a) *Cauchy criterion for summability* The function f is summable over A iff for each neighborhood U of 0 in G there is a finite subset B of A such that for every finite subset C of $A \sim B$ it is true that $\sum_C f \in U$. Hence a function summable over A is summable over each subset of A.

(b) If f and g are summable over A, then $f + g$ (where $(f + g)(x) = f(x) + g(x)$) is summable over A and $\sum_A(f + g) = \sum_A f + \sum_A g$.

(c) If f is defined and summable over A and $\mathcal{B}$ is a disjoint family of subsets of A which cover A, then $\sum_A f = \sum\{\sum\{f(b): b \in B\}: B \in \mathcal{B}\}$. However, from the existence of the iterated sum it is not possible to deduce summability over A. (See 2.G for a special case in which the existence of the iterated sum implies summability over A.)

T UNIFORMLY LOCALLY COMPACT SPACES

A uniform space $(X,\mathfrak{U})$ is *uniformly locally compact* iff there is a member U of $\mathfrak{U}$ such that $U[x]$ is compact for each x in X. In particular, each locally compact topological group is uniformly locally compact relative to its left and its right uniformity.

(a) Let $(X,\mathfrak{U})$ be a uniform space, let U be a member of $\mathfrak{U}$, let $U_0 = U$ and $U_n = U \circ U_{n-1}$ for each positive integer n. Then for each subset A of X the set $\bigcup\{U_n[A]: n \in \omega\}$ is both open and closed.

(b) If U is a closed neighborhood of the diagonal in $X \times X$, A is a compact subset of X, and $U \circ U[x]$ is compact for each x in A, then $U[A]$ is compact. ($U[A]$ is closed by 6.A.)

(c) A connected uniformly locally compact space $(X,\mathfrak{U})$ is σ-compact (that is, X is the union of a countable family of compact subsets).

(d) Each uniformly locally compact space is the union of a disjoint open family of σ-compact subspaces. Hence each such space is paracompact.

(e) Let $(X,\mathfrak{I})$ be a topological space. Then there is a uniformity $\mathfrak{U}$ whose topology is $\mathfrak{I}$ such that $(X,\mathfrak{U})$ is uniformly locally compact iff $(X,\mathfrak{I})$ is locally compact and paracompact. (See 5.28.)

Note Part (a) is essentially the chain argument of **5.T**. It may be noted that the propositions on components and connected sets of **5.T** cannot be extended to uniformly locally compact spaces.

U THE UNIFORM BOUNDEDNESS THEOREM

(a) Let X be a real linear topological space which is not meager in itself and let K be a closed convex subset of X such that $K = -K$ and K contains a line segment in each direction (that is, for each x in X there is a positive real number t such that $sx \in K$ if $0 \leq s \leq t$). Then K is a neighborhood of 0. (Show that K is not meager in X. Then by 6.**P**, $K - K$ is a neighborhood of 0 and convexity implies that $2K$ is a neighborhood of 0.)

(b) *Theorem* Let F be a family of continuous linear functions on a non-meager linear topological space X to a normed linear space Y and suppose that $\sup\{\|f(x)\|: f \in F\}$ is finite for each point x of X. Then for some neighborhood U of 0 in X it is true that $\sup\{\|f(x)\|: x \in U$ and $f \in F\}$ is finite. (Use the foregoing proposition to show that, if S is the unit sphere about 0 in Y, then $\bigcap\}f^{-1}[S]: f \in F\}$ is a neighborhood of 0 in X.)

Note Part (b) is the classic Banach-Steinhaus theorem. (Banach [1;80].) The formulation is clearly capable of some generalization; the basic idea of such generalization is that of proposition (a). In the terminology of the next chapter the conclusion of (b) can be stated: F is equicontinuous at 0.

V BOOLEAN σ-RINGS

A Boolean ring $(B,+,\cdot)$ is a *σ-ring* iff each countable subset has a least upper bound relative to the natural ordering of B (see 2.**K**). Natural examples of Boolean σ-rings are:

(i) The ring $(\mathcal{L},\Delta,\cap)$ where $\mathcal{L}$ is the family of all Lebesgue measurable subsets of $[0,1]$, or the ring $\mathcal{L}$ modulo the family of $\mathfrak{N}$ of all sets of measure zero is a σ-ring. (Here Δ is symmetric difference. The family $\mathfrak{N}$ is actually a σ-ideal, in the obvious sense.)

(ii) The ring $(\mathfrak{A}/\mathfrak{M},\Delta,\cap)$, where $\mathfrak{A}$ is the family of all Borel subsets of $[0,1]$ and $\mathfrak{M}$ is the subfamily consisting of meager Borel sets.

It is the purpose of this problem to exhibit a representation theorem of the type (ii) for an arbitrary Boolean σ-ring. Throughout $\mathfrak{B}$ will be the family of all compact open subsets of a locally compact Boolean space X. There is no loss in generality in restricting attention to rings of the type $(\mathfrak{B},\Delta,\cap)$. (See the Stone representation theorem, 5.S.)

(a) If $(\mathfrak{B},\Delta,\cap)$ is a Boolean σ-ring, then the closure of the union of a countable subfamily of $\mathfrak{B}$ is a member of $\mathfrak{B}$ (that is, the closure of the union of a countable family of compact open subsets of X is compact and open).

(b) Let $\mathfrak{A}$ be the smallest family of subsets of X such that $\mathfrak{B} \subset \mathfrak{A}$ and countable unions and symmetric differences of members of $\mathfrak{A}$ belong to $\mathfrak{A}$. Let $\mathfrak{M}$ be the family of all meager subsets of X. Then for each member A of $\mathfrak{A}$ there is a unique member B of $\mathfrak{B}$ such that $A \Delta B \in \mathfrak{M}$. (See 6.P(a).)

(c) *Theorem* The σ-ring $\mathfrak{A}$ is (additively) the direct sum of $\mathfrak{B}$ and the σ-ideal $\mathfrak{A} \cap \mathfrak{M}$. Hence $\mathfrak{B}$ is isomorphic to the Boolean σ-ring $\mathfrak{A}$ modulo the σ-ideal $\mathfrak{A} \cap \mathfrak{M}$.

Note The results of this problem are due to Loomis [1]. A space which has the property that the closure of an open set is open (such as the Stone space of a Boolean σ-ring which satisfies a countable chain condition) is sometimes called *extremally disconnected.* The space of real-valued bounded Borel functions on a compact space of this sort decomposes, in a way analogous to proposition (c), into continuous functions and functions vanishing outside a meager set. For this and other results see M. H. Stone [4] and also Dixmier [1].

Chapter 7

FUNCTION SPACES

This chapter is devoted to function spaces. That is, the elements of the spaces are functions on a fixed set X to a fixed topological or uniform space Y. Almost all of the development concerns spaces of functions which are continuous relative to a topology for X. Briefly, the purpose of the study is to define topologies and uniformities for sets of continuous functions, and to prove compactness, completeness, and continuity properties for the resulting spaces.

Most of the results of the chapter have their origins in the early theory of real variables. However, the theorems on joint continuity and the compact open topology are relatively recent; they are due primarily to Fox [1]. Further information on function spaces may be found in Arens [2], Bourbaki [1], Myers [2], and Tukey [1].

POINTWISE CONVERGENCE

One topology for a function space has already been investigated rather extensively. If F is a family of functions, each on a set X to a topological space Y, then F is contained in the product $Y^X = \times\{Y{:}x \in X\}$. The topology $\mathcal{P}$ of pointwise convergence (coordinatewise convergence, simple convergence) or simply the **pointwise topology** for F is the relativized product topology. A net $\{f_n, n \in D\}$ converges to g iff $\{f_n(x), n \in D\}$ converges to $g(x)$ for each x in X (see 3.4). A subbase for $\mathcal{P}$ is the family of all subsets of the form $\{f: f(x) \in U\}$, where x is a point of X and U is open in Y. For each point x of X there is a function e_x on F, which is called the evaluation at x (or the projection into the x-th coordinate space) which is defined by $e_x(f) = f(x)$ for all f in F. Evaluation at x is continuous and open relative to $\mathcal{P}$ (theorem 3.2), and $\mathcal{P}$ is the smallest topology for F such that each evaluation is continuous. A function g on a topological space to F is continuous relative to $\mathcal{P}$ iff $e_x \circ g$ is continuous for each point x of X (theorem 3.3). It is clear that the pointwise topology depends only on the family of functions and the topology of Y. A topology for X, if such is given, does not enter into the definitions or the theorems. If Y is Hausdorff or regular, then the space F inherits the same property (3.5 and 4.**A**), but in general Y may be locally compact or satisfy the first or second axiom of countability and F may fail to have these properties (3.6 and 5.19).

A characterization of those function spaces which are compact relative to the topology is an immediate consequence of the Tychonoff theorem, 5.13, on the product of compact spaces. Before stating the result let us agree, for convenience, that a family F of functions on a set X to a topological space Y is **pointwise closed** iff F is a closed subset of the product space Y^X. If A is a subset of X, then $F[A]$ is defined to be the set of all points $f(x)$ for x in A and f in F. If $x \in X$, then $F[\{x\}]$ is abbreviated to $F[x]$. If e_x is the evaluation at x, then clearly $e_x[F] = F[x]$.

1 THEOREM *In order that a family F of functions on a set X to a topological space Y be compact relative to the topology of pointwise convergence it is sufficient that*

(a) *F be pointwise closed in Y^X, and*
(b) *for each point x of X the set $F[x]$ has a compact closure.*

If Y is a Hausdorff space the conditions (a) *and* (b) *are also necessary.*

PROOF The family F is not only a subfamily of Y^X but is also contained in $\times\{F[x]^-: x \in X\}$. If condition (b) is satisfied, then this product is a compact subset of Y^X by the Tychonoff product theorem, and if F is pointwise closed, then F is compact. The sufficiency of (a) and (b) is then proved. If Y is a Hausdorff space and F is compact relative to the pointwise topology, then F is closed by 5.7. The set $F[x]$ is compact and closed because the evaluation at each point x is a continuous map of F into the Hausdorff space Y. ∎

The preceding theorem is more important than casual consideration of the topology of pointwise convergence might indicate. The pointwise topology is in many ways unnatural. For example, let X be a set and for each finite subset A of X let C_A be the characteristic function of A (that is, $C_A(x) = 1$ if $x \in A$ and $C_A(x) = 0$ if $x \notin A$). The family $\mathfrak{A}$ of all finite subsets of X is directed by $\supset$, and consequently $\{C_A, A \in \mathfrak{A}\}$ is a net of functions on X to the closed unit interval. This net converges to the function e which is identically one, because $\{x\} \in \mathfrak{A}$ for each point x, and if $A \supset \{x\}$, then $C_A(x) = 1$. Now a topology such that the characteristic function of a finite set is "near" the unit function is obviously unsuitable for many purposes. The more interesting topologies are those for which convergence is more restricted, that is, the larger topologies. But observe: if $(F,\mathfrak{J})$ is compact and $\mathfrak{J}$ is larger than the topology $\mathfrak{P}$ of pointwise convergence, then the identity map i of $(F,\mathfrak{J})$ onto $(F,\mathfrak{P})$ is continuous, and if $(F,\mathfrak{P})$ is a Hausdorff space, then i must be a homeomorphism. Consequently if $(F,\mathfrak{J})$ is compact, $(F,\mathfrak{P})$ is Hausdorff, and $\mathfrak{J}$ is larger than the pointwise topology, then $\mathfrak{J}$ is identical with the topology of pointwise convergence. This simple remark indicates the standard method of proving a function space F compact relative to a topology $\mathfrak{J}$. One first shows that F is compact relative to the topology of pointwise convergence and then proves that $\mathfrak{P}$-convergence of a net in F implies $\mathfrak{J}$-convergence. If Y is Hausdorff there can be no loss in restricting attention to these two propositions, for if either fails F is not compact relative to $\mathfrak{J}$.

It is sometimes convenient to consider pointwise convergence for points in a subset of the domain space. Suppose F is a family of functions, each on a set X to a topological space Y, and suppose that A is a subset of X. There is a natural map R of F into the product space Y^A, obtained by mapping each member f of F into its restriction to A: that is, $R(f) = f \mid A$ for each f in

F. The smallest topology $\mathcal{P}_A$ for F such that R is continuous evidently consists of the inverses under R of the open subsets Y^A. This topology is that of **pointwise convergence on A.** A subbase for $\mathcal{P}_A$ is the family of sets of the form $\{f: f(x) \in U\}$ for x in A and U open in Y, and a net $\{f_n, n \in D\}$ in F converges to g relative to $\mathcal{P}_A$ iff $\{f_n(x), n \in D\}$ converges to $g(x)$ for each x in A. The map R will be one to one iff, whenever f and g are distinct members of F, then for some point x of A it is true that $f(x) \neq g(x)$. A subset A of X for which this is the case is said **to distinguish members** of the family F.

2 THEOREM *Let F be a family of functions, each on a set X to a Hausdorff space Y, and let A be a subset of X. The family F with the topology $\mathcal{P}_A$ of pointwise convergence on A is a Hausdorff space if and only if A distinguishes members of F. If F is compact relative to the topology of pointwise convergence on X and if A distinguishes members of F, then $\mathcal{P}$ and $\mathcal{P}_A$ are identical.*

PROOF The product space Y^A is a Hausdorff space and, in view of the definition of $\mathcal{P}_A$, F with this topology will be Hausdorff iff the restriction map R is one to one. This is the case iff A distinguishes members of F. The identity map i of $(F,\mathcal{P})$ onto $(F,\mathcal{P}_A)$ is always continuous since $\mathcal{P}_A \subset \mathcal{P}$. If $(F,\mathcal{P})$ is compact and $(F,\mathcal{P}_A)$ is Hausdorff, then i is a homeomorphism and $\mathcal{P} = \mathcal{P}_A$. ∎

If the range space is a uniform space, then the topology of pointwise convergence is the topology of a uniformity.

If F is a family of functions on a set X to a uniform space $(Y,\mathcal{V})$, then F is a subset of the product $\times\{Y: x \in X\}$ and the relativized product uniformity is called the **uniformity of pointwise convergence** (or of simple convergence). This is sometimes abbreviated as the $\mathcal{P}$ uniformity. Its properties have already been studied (for example, 6.25).

If A is a subset of X, then the uniformity of pointwise convergence on A, or simply the $\mathcal{P}_A$ uniformity, is defined to be the smallest uniformity which makes the restriction map R of F into the family of all functions on A to Y uniformly continuous. The following simple facts about this uniformity are listed without proof.

3 THEOREM *Let F be a family of functions on a set X to a uniform space $(Y,\mathcal{V})$ and let A be a subset of X. Then the uniformity of pointwise convergence on A has the properties:*

(a) *The family of all sets of the form $\{(f,g): (f(x),g(x)) \in V\}$ for V in $\mathcal{V}$ and x in A is a subbase for the $\mathcal{P}_A$ uniformity.*
(b) *The topology of the $\mathcal{P}_A$ uniformity is the topology of pointwise convergence on A.*
(c) *A net $\{f_n, n \in D\}$ is a Cauchy net if and only if $\{f_n(x), n \in D\}$ is a Cauchy net for each x in A.*
(d) *If $(Y,\mathcal{V})$ is complete and $R[F]$ is closed in Y^A relative to pointwise convergence on A, then F is complete relative to the $\mathcal{P}_A$ uniformity.*

COMPACT OPEN TOPOLOGY AND JOINT CONTINUITY

Given a topology for a family F of functions on a topological space X to a topological space Y one might reasonably ask whether $f(x)$ is continuous simultaneously in f and in x. Stated somewhat more formally, the question is: for which topologies for F is the map $F \times X$ which carries (f,x) onto $f(x)$ continuous, if $F \times X$ is given the product topology? This section is devoted to a brief examination of this question. It turns out that there is a particular function space topology which is related to this problem, and we begin by defining this topology and establishing some elementary properties. The section is devoted entirely to topological questions; connections with a uniformity for function spaces will be established later. Throughout the section F will be a family of functions, each on a topological space X to a topological space Y.

For convenience, for each subset K of X and each subset U of Y, define $W(K,U)$ to be the set of all members of F which carry K into U; that is, $W(K,U) = \{f: f[K] \subset U\}$. The family of all sets of the form $W(K,U)$, for K a compact subset of X and U open in Y, is a subbase for the **compact open** topology for F. The family of finite intersections of sets of the form $W(K,U)$ is then a base for the compact open topology; each member of this base is the form $\bigcap \{W(K_i,U_i): i = 0, 1, \cdots, n\}$, where each K_i

is a compact subset of X and each U_i is an open subset of Y. The fact that each set consisting of a single point is compact makes comparison with the pointwise topology simple.

4 THEOREM *The compact open topology* $\mathcal{C}$ *contains the topology* $\mathcal{P}$ *of pointwise convergence. The space* $(F,\mathcal{C})$ *is a Hausdorff space if the range space* Y *is Hausdorff, and is regular if* Y *is regular and the members of* F *are continuous.*

PROOF For each x in X and each open subset U of Y the set $W(\{x\},U) = \{f: f(x) \in U\}$ belongs to $\mathcal{C}$ because $\{x\}$ is compact. Hence $\mathcal{P} \subset \mathcal{C}$, for the family of all sets of this form is a subbase for the pointwise topology $\mathcal{P}$. If Y is a Hausdorff space, then $(F,\mathcal{P})$ is also a Hausdorff space, by 3.5, and if U and V are disjoint $\mathcal{P}$-neighborhoods of members of F they are also $\mathcal{C}$-neighborhoods. Therefore $(F,\mathcal{C})$ is Hausdorff.

Finally, assume that Y is regular; it must be shown that each neighborhood of each member f of F contains a closed neighborhood. It is sufficient to prove that each neighborhood of f which belongs to a subbase for $\mathcal{C}$ contains a closed neighborhood, for each neighborhood of f contains a finite intersection of neighborhoods belonging to the subbase. Suppose that $f \in W(K,U)$ where K is compact and U is an open subset of Y. Then $f[K]$ is compact, and since Y is regular there is by 5.10 a closed neighborhood V of $f[K]$ such that $V \subset U$. Surely $f \in W(K,V) \subset W(K,U)$ and evidently $W(K,V)$ is a neighborhood of f. It remains to show that $W(K,V)$ is closed. But $W(K,V)$ is the intersection of the sets $W(\{x\},V)$ for x in K, and each of the sets $W(\{x\},V)$ is $\mathcal{P}$-closed and hence $\mathcal{C}$-closed. ∎

There is no hope of showing that, if Y is normal or satisfies the first or second axiom of countability, then $(F,\mathcal{C})$ has these properties, for if X is discrete the only compact sets are finite and hence $\mathcal{C}$ is identical with the topology of pointwise convergence. The product of normal spaces or spaces satisfying one of the countability axioms may fail to have the corresponding property and hence F with the topology $\mathcal{C}$ also may fail to have the property.

Let P be the map of $F \times X$ into Y which carries (f,x) into $f(x)$. Each topology for F gives rise to a product topology for

$F \times X$, and one may ask whether P is continuous relative to this product topology. A topology for F is said to be **jointly continuous** iff the map P of $F \times X$ into Y is continuous. It is very easy to see that the topology of pointwise convergence is usually not jointly continuous. The discrete topology is jointly continuous, for if U is an open subset of Y, then $P^{-1}[U] = \{(f,x): f(x) \in U\} = \bigcup\{\{f\} \times f^{-1}[U]: f \in F\}$, which is the union of open sets (assuming that F is a family of continuous functions). If a topology for F is jointly continuous, then each larger topology is also jointly continuous. Consequently the natural problem is to find the smallest jointly continuous topology, if such exists. It turns out that there is generally no such smallest topology; however, a slight relaxation of the conditions for joint continuity yields a precise description of the compact open topology. A topology for a family F of functions is **jointly continuous on a set** A iff the map P is continuous on $F \times A$, where $P(f,x) = f(x)$. (Caution: This does not mean that P is continuous at the points of $F \times A$; the condition is that the restriction $P \mid (F \times A)$ be continuous.) A topology for F is **jointly continuous on compacta** iff it is jointly continuous on each compact subset of the domain space. Each member f of such a family is necessarily continuous on each compact set K (that is, $f \mid K$ is continuous).

5 Theorem *Each topology which is jointly continuous on compacta is larger than the compact open topology* $\mathfrak{C}$. *If* X *is regular or Hausdorff and each member of* F *is continuous on every compact subset of* X, *then* $\mathfrak{C}$ *is jointly continuous on compacta.*

Proof Suppose a topology $\mathfrak{J}$ for F is jointly continuous on compacta, U is an open subset of Y, K is a compact subset of X, and P is the map such that $P(f,x) = f(x)$. It must be shown that $W(K,U)$ is $\mathfrak{J}$-open, where $W(K,U) = \{f: f[K] \subset U\}$. The set $V = (F \times K) \cap P^{-1}[U]$ is open in $F \times K$ because $\mathfrak{J}$ is jointly continuous on compacta. If $f \in W(K,U)$, then $\{f\} \times K \subset V$, and since $\{f\} \times K$ is compact there is a $\mathfrak{J}$-neighborhood N of f such that $N \times K \subset P^{-1}[U]$ by theorem 5.12. In other words, each member of the $\mathfrak{J}$-neighborhood N of f is a member of the compact open neighborhood $W(K,U)$. It follows that $W(K,U)$ is $\mathfrak{J}$-open and the first statement of the theorem is proved. To prove the second assertion, suppose K is a compact subset of X, $x \in K$, U is open in Y, and $(f,x) \in P^{-1}[U]$. Then, since f is continuous on K, there is a compact set M which is a neighborhood of x in K such that $f[M] \subset U$ (recall that X is either Hausdorff or regular). Then $W(M,U) \times M$ is a neighborhood of (f,x) in $F \times K$ and is contained in $P^{-1}[U]$. Joint continuity on K follows. ▮

It may be noticed that, if X is locally compact, then a topology is jointly continuous on compacta iff it is jointly continuous. Hence, if X is a locally compact regular space, then the compact open topology for a family of continuous functions is the smallest jointly continuous topology.

If a topology $\mathfrak{I}$ for a family F is jointly continuous on compacta, then $\mathfrak{I} \supset \mathcal{C} \supset \mathcal{P}$, where $\mathcal{C}$ is the compact open topology and $\mathcal{P}$ is the pointwise. If $(F,\mathfrak{I})$ is compact and the range space is Hausdorff, then $(F,\mathcal{P})$ is Hausdorff and consequently $\mathfrak{I} = \mathcal{C} = \mathcal{P}$. This fact shows the necessity of one of the conditions given for $\mathcal{C}$-compactness in the next theorem. The result is given in a rather curious form in order to be directly applicable to the later problem.

6 THEOREM *Let X be a topological space which is either regular or Hausdorff, let Y be a Hausdorff space, let C be the family of all unctions on X to Y which are continuous on each compact subset of X, and let $\mathcal{C}$ and $\mathcal{P}$ be respectively the compact open and the pointwise topologies. Then a subfamily F of C is $\mathcal{C}$-compact if and only if*

(a) *F is $\mathcal{C}$-closed in C,*
(b) *$F[x]$ has a compact closure for each member x of X, and*
(c) *the topology $\mathcal{P}$ for the $\mathcal{P}$-closure of F in Y^X is jointly continuous on compacta.*

PROOF Suppose F is $\mathcal{C}$-compact. The space $(C,\mathcal{C})$ is Hausdorff because Y is Hausdorff and hence F is $\mathcal{C}$-closed in C. Evaluation at a point x is $\mathcal{P}$-continuous, hence $\mathcal{C}$-continuous, and the image $F[x]$ of F is therefore compact. The topologies $\mathcal{C}$ and $\mathcal{P}$ for F coincide because F is $\mathcal{C}$-compact and $\mathcal{P}$-Hausdorff, hence F is

$\mathcal{P}$-closed in Y^X, and by 7.5 the topology $\mathcal{C}$ (and hence $\mathcal{P}$) for F is jointly continuous on compacta. This completes the proof that conditions (a), (b), and (c) are necessary.

Assuming conditions (a), (b), and (c), let F^- be the $\mathcal{P}$-closure of F in Y^X. Condition (b) states that $F[x]^-$ is compact for each x, and since F^- is a closed subset of the $\mathcal{P}$-compact set $\times\{F[x]^-: x \in X\}$ it follows that F^- is $\mathcal{P}$-compact. By (c) the topology $\mathcal{P}$ for F^- is jointly continuous on compacta. Consequently each member of F^- is continuous on each compactum and $F^- \subset C$. Theorem 5.5 implies that the topology $\mathcal{P}$ for F^- is larger than $\mathcal{C}$, and hence these two topologies for F^- coincide. By (a) the family F is $\mathcal{C}$-closed in C and is hence $\mathcal{C}$ (and $\mathcal{P}$) closed in the subset F^- of C; in fact, $F^- = F$, and F is $\mathcal{C}$-compact. ∎

7 *Notes* The family C of all functions which are continuous on every compact subset coincides with the family of all continuous functions if the space is either locally compact or satisfies the first axiom of countability (see theorem 7.13 and the discussion preceding it). It is usually the family of all continuous functions which is of interest; however, the mathematical structure (and not my whim) is responsible for the appearance of the class C. The class also shows up a little later in a discussion of completeness.

The relation between the compact open topology and joint continuity was first studied by Fox [1], who showed that the compact open topology for a family of continuous functions is smaller than each jointly continuous topology and is itself jointly continuous if the domain space is locally compact. For proof of the fact that there is generally no smallest jointly continuous topology see Arens [2].

UNIFORM CONVERGENCE

This section is devoted to the study of a uniformity for a family F of functions on a set X to a uniform space $(Y,\mathcal{V})$. The uniformity is independent of any topology which may be assigned to the set X, but one of the principal results is that the family of functions continuous relative to a topology for X is closed in the space of all functions on X to Y. That is, the uniform limit of continuous functions is continuous.

The uniformity of uniform convergence is the largest which will be considered and the uniformity of pointwise convergence is the smallest. Both of these may be considered as special instances of uniform convergence on the members of a family $\mathcal{A}$ of sets. This concept is investigated briefly; a uniformity is constructed for each family $\mathcal{A}$ of subsets of X, and the elementary properties are derived.

Let F be a family of functions on a set X to a uniform space $(Y,\mathcal{V})$. For each member V of $\mathcal{V}$ let $W(V)$ be the set * of all pairs (f,g) such that $(f(x),g(x)) \in V$ for each x in X. Then $W(V)[f]$ is the set of all g such that $g(x) \in V[f(x)]$ for every x in X. It is easy to see that $W(V^{-1}) = (W(V))^{-1}$, $W(U \cap V) = W(U) \cap W(V)$, and $W(U \circ V) \supset W(U) \circ W(V)$ for all members U and V of $\mathcal{V}$. Consequently the family of all sets $W(V)$ for V in $\mathcal{V}$ is a base for a uniformity $\mathcal{U}$ for F by theorem 6.2. The family $\mathcal{U}$ is the **uniformity of uniform convergence,** or simply the **u.c. uniformity.** The topology of $\mathcal{U}$ is the **topology of uniform convergence,** or the **u.c. topology.**

It is clear that $\mathcal{U}$ is larger than the uniformity of pointwise convergence, for if y is an arbitrary member of X and $V \in \mathcal{V}$, then $\{(f,g): (f(x),g(x)) \in V \textit{ for all } x \textit{ in } X\} \subset \{(f,g): (f(y),g(y)) \in V\}$, and hence each member of the defining base for $\mathcal{U}$ is a subset of a member of the defining subbase for the pointwise uniformity. It follows that the u.c. topology is larger than the pointwise. It is also easy to see directly that uniform convergence implies pointwise convergence, for a net $\{f_n, n \in D\}$ in F converges to g relative to the u.c. topology iff the net is eventually in $W(V)[g]$ for each V in $\mathcal{V}$, and this is true iff there is some m in D such that, when $n \geq m$, then $f_n(x) \in V[g(x)]$ for all x in X. The following theorem lists other elementary properties of the uniformity $\mathcal{U}$.

8 THEOREM *Let F be the family of all functions on a set X to a uniform space* $(Y,\mathcal{V})$ *and let $\mathcal{U}$ be the uniformity of uniform con-*

* The set $W(V)$ may be described very simply in terms of the usual notation for relations: $W(V) = \{(f,g): g \circ f^{-1} \subset V\}$. This statement is clear since $g \circ f^{-1}$ is precisely the set of all pairs $(f(x),g(x))$ with x in X. It is also clear that $W(V) = \{(f,g): g \subset V \circ f\}$ and $W(V)[f] = \{g: g \subset V \circ f\} = \{g: g(x) \in V[f(x)] \textit{ for each } x \textit{ in } X\}$.

vergence. Then:

(a) *The uniformity* $\mathfrak{U}$ *is generated by the family of all pseudo-metrics of the form* $d^*(f,g) = \sup \{d(f(x),g(x)) : x \in X\}$, *where d is a bounded member of the gage of* $(Y,\mathcal{V})$.

(b) *A net* $\{f_n, n \in D\}$ *in F converges uniformly to g if and only if it is a Cauchy net relative to* $\mathfrak{U}$ *and* $\{f_n(x), n \in D\}$ *converges to* $g(x)$ *for each x in X.*

(c) *If* $(Y,\mathcal{V})$ *is complete so is the uniform space* $(F,\mathfrak{U})$.

PROOF To prove part (a) observe that the family of all sets of the form $\{(y,z): d(y,z) \leqq r\}$, for r positive and for d a bounded member of the gage of $\mathcal{V}$, is a base for $\mathcal{V}$. This is true because for each pseudo-metric e the pseudo-metric $d = \min [1,e]$ is bounded and has the same uniformity. But $\{(f,g): d^*(f,g) \leqq r\} = \{(f,g): d(f(x),g(x)) \leqq r \textit{ for each } x \textit{ in } X\} = W(\{(y,z): d(y,z) \leqq r\})$, where W is the correspondence used above in defining the u.c. uniformity. It follows that d^* belongs to the gage of $\mathfrak{U}$ and that pseudo-metrics of this form generate the gage.

One half of the proposition (b) is obvious, and it is only necessary to show that, if a Cauchy net $\{f_n, n \in D\}$ converges pointwise to g, then it converges uniformly to g. Let V be an arbitrary closed symmetric member of $\mathcal{V}$, and choose m in D such that, if $n \geqq m$ and $p \geqq m$, then $f_p(x) \in V[f_n(x)]$ for each x in X. Such a choice is possible because the net is assumed to be Cauchy relative to $\mathfrak{U}$. Since $V[f_n(x)]$ is closed and $f_p(x)$ converges to $g(x)$ it follows that $g(x) \in V[f_n(x)]$ and hence $f_n(x) \in V[g(x)]$ for each $n \geqq m$ and every x in X, and (b) is established. Proposition (c) is an immediate consequence of (b) and of the fact that the product of complete spaces is complete. ∎

The following theorem states the principal properties of $\mathfrak{U}$ for a family of continuous functions.

9 THEOREM *Let F be the family of all continuous functions on a topological space X to a uniform space* $(Y,\mathcal{V})$, *and let* $\mathfrak{U}$ *be the uniformity of uniform convergence. Then:*

(a) *The family F is closed in the space of all functions on X to Y, and consequently* $(F,\mathfrak{U})$ *is complete if* $(Y,\mathcal{V})$ *is complete.*

(b) *The topology of uniform convergence is jointly continuous.*

PROOF Proposition (a) is proved by showing that the set of non-continuous functions is an open subset of the space G of all functions on X to Y. If f is not continuous at a point x of X there is a member V of $\mathcal{V}$ such that $f^{-1}[V[f(x)]]$ is not a neighborhood of x. Choose a symmetric member W of $\mathcal{V}$ such that $W \circ W \circ W \subset V$. It will be proved that if g is a function such that $(g(y),f(y)) \in W$ for each y, then $g^{-1}[W[g(x)]]$ is not a neighborhood of x and hence g is not continuous. It will follow that $G \sim F$ is open relative to the topology of uniform convergence. If $(g(y),f(y)) \in W$ for each y, then $g \subset W \circ f$ and $g^{-1} \subset f^{-1} \circ W^{-1} = f^{-1} \circ W$ and hence $g^{-1} \circ W \circ g \subset f^{-1} \circ W \circ W \circ W \circ f \subset f^{-1} \circ V \circ f$. Therefore $g^{-1}[W[g(x)]]$ is a subset of $f^{-1}[V[f(x)]]$ and is not a neighborhood of x.

The proof of (b) remains. To show continuity of the map of $F \times X$ into Y at a point (f,x) it is only necessary to verify that for V in $\mathcal{V}$, if $y \in f^{-1}[V[f(x)]]$ and $g(z) \in V[f(z)]$ for all z, then $g(y) \in V[f(y)] \subset V \circ V[f(x)]$. ∎

A number of useful uniformities are constructed by considering uniform convergence on each of a family $\mathcal{A}$ of subsets of the domain space. Explicitly, if F is a family of functions on a set X to a uniform space $(Y,\mathcal{V})$ and $\mathcal{A}$ is a family of subsets of X, then the uniformity of **uniform convergence on members** of $\mathcal{A}$, abbreviated $\mathcal{U} \mid \mathcal{A}$, has for a subbase the family of all sets of the form $\{(f,g): (f(x),g(x)) \in V \textit{ for all } x \textit{ in } A\}$, for V in $\mathcal{V}$ and A in $\mathcal{A}$. This uniformity may be described in another way. For each A in $\mathcal{A}$ let R_A be the map which carries f into the restriction of f to A; that is, $R_A(f) = f|A$. Then R_A carries F into a family of functions on A to Y, this family may be assigned the uniformity of uniform convergence, and the uniformity $\mathcal{U} \mid \mathcal{A}$ may be described as the smallest which makes each R_A uniformly continuous.

The preceding propositions on uniform convergence imply corresponding results about the $\mathcal{U} \mid \mathcal{A}$ uniformity. The simple proofs are omitted.

10 Theorem *Let X be a topological space, let $(Y,\mathcal{V})$ be a uniform space, let $\mathcal{A}$ be a family of subsets of X which covers X, let G be the family of all functions on X to Y, and let F be the family of all* *functions which are continuous on each member of* $\mathcal{A}$. *Then:*

(a) *The uniformity* $\mathcal{U} \mid \mathcal{A}$ *of uniform convergence on members of* $\mathcal{A}$ *is larger than the uniformity of pointwise convergence and smaller than that of uniform convergence on* X.

(b) *A net* $\{f_n, n \in D\}$ *converges to* g *relative to the topology of* $\mathcal{U} \mid \mathcal{A}$ *if and only if it is a Cauchy net (relative to* $\mathcal{U} \mid \mathcal{A}$*) and converges to* g *pointwise.*

(c) *If* $(Y,\mathcal{V})$ *is complete, then* G *is complete relative to* $\mathcal{U} \mid \mathcal{A}$.

(d) *The family* F *is closed in* G *relative to the topology of* $\mathcal{U} \mid \mathcal{A}$, *and hence if* $(Y,\mathcal{V})$ *is complete so is* $(F,\mathcal{U} \mid \mathcal{A})$.

(e) *The topology of* $\mathcal{U} \mid \mathcal{A}$ *for* F *is jointly continuous on each member of* $\mathcal{A}$.

It should be emphasized that the family of continuous functions may fail to be complete relative to $\mathcal{U} \mid \mathcal{A}$. If $\mathcal{A}$ is the family of all sets $\{x\}$ for x in X, then $\mathcal{U} \mid \mathcal{A}$ is simply the uniformity of pointwise convergence, and the family of continuous functions is generally not complete relative to this uniformity. If $\mathcal{A}$ is such that continuity on each member of $\mathcal{A}$ implies continuity on X, then proposition (d) above shows $\mathcal{U} \mid \mathcal{A}$ completeness of the family of continuous functions on X to a complete space. In particular, this is the case if there is a neighborhood of each point of X which belongs to $\mathcal{A}$.

UNIFORM CONVERGENCE ON COMPACTA

In this section two distinct lines of investigation will be combined. Suppose that F is a family of continuous functions on a topological space X to a uniform space $(Y,\mathcal{V})$. The uniformity of **uniform convergence on compacta** is the uniformity $\mathcal{U} \mid \mathcal{C}$, where $\mathcal{C}$ is the family of all compact subsets of X. The topology of $\mathcal{U} \mid \mathcal{C}$ is sometimes called the **topology of compact convergence.** It will be proved that this topology is identical with the compact open topology which is constructed from the topology of X and the topology of the uniformity $\mathcal{V}$. Thus the uniformity $\mathcal{U} \mid \mathcal{C}$ depends on the uniformity $\mathcal{V}$ for Y, but the topology of $\mathcal{U} \mid \mathcal{C}$ depends only on the topology of $\mathcal{V}$. The uniformity $\mathcal{U} \mid \mathcal{C}$ is particularly useful in case the space X has a "rich" supply of compact sets, and the section concludes with a brief examination of spaces satisfying a "richness" condition.

11 THEOREM *Let F be a family of continuous functions on a topological space X to a uniform space $(Y,\mathcal{V})$. Then the topology of uniform convergence on compacta is the compact open topology.*

PROOF Let K be a compact subset of X, U an open subset of Y, let $f \in F$, and suppose that $f[K] \subset U$. Then $f[K]$ is compact and by 6.33 there is V in $\mathcal{V}$ such that $V[f[K]] \subset U$. It is then clear that, if g is a function such that $g(x) \in V[f(x)]$ for each x in K, then $g[K] \subset U$ also. Consequently each set of the form $\{f: f[K] \subset U\}$ is open relative to the topology of $\mathcal{U} \mid \mathcal{C}$, and the compact open topology is therefore smaller than that of $\mathcal{U} \mid \mathcal{C}$.

To prove the converse it must be shown that for each compact subset K of X, each V in $\mathcal{V}$, and each continuous f there are compact subsets $K_1, \cdots K_n$ of X and open subsets $U_1, \cdots U_n$ of Y such that $f[K_i] \subset U_i$, and if $g[K_i] \subset U_i$ for each i then $g(x) \in V[f(x)]$ for each x in K. Choose a closed symmetric member W of $\mathcal{V}$ such that $W \circ W \circ W \subset V$, choose $x_1, \cdots x_n$ in K such that the sets $W[f(x_i)]$ cover $f[K]$, let $K_i = K \cap f^{-1}[W[f(x_i)]]$, and let U_i be the interior of $W \circ W[f(x_i)]$. If $g[K_i] \subset U_i$ for each i, then: for each x in K there is i such that $x \in K_i$, hence $g(x) \in W \circ W[f(x_i)]$, and since $f(x) \in W[f(x_i)]$ it follows that $(g(x),f(x)) \in W \circ W \circ W \subset V$. ∎

If the uniform space $(Y,\mathcal{V})$ is complete and $\mathcal{A}$ is a family of subsets of the topological space X then the family of all functions on X to Y which are continuous on each member of $\mathcal{A}$ is $\mathcal{U} \mid \mathcal{A}$-complete, according to 7.10. In order that the family of all continuous functions be complete it is then sufficient that $\mathcal{A}$ satisfy the condition: a function is continuous whenever it is continuous on each member of $\mathcal{A}$. If f is a function on X to Y and B is a subset of Y, then this condition would be implied by: if $A \cap f^{-1}[B]$ is closed for each member A of $\mathcal{A}$, then $f^{-1}[B]$ is closed. In particular, the space of all continuous functions on X to Y is complete relative to uniform convergence on compacta if X satisfies the condition: if a subset A of X intersects each closed compact set in a closed set, then A is closed. Such a topological space is called a ***k*-space.** It is clear that the family $\mathcal{C}$ of closed compact sets determines the topology of a k-space entirely, for A is closed iff $A \cap C \in \mathfrak{C}$ for each C in $\mathfrak{C}$. By complementation it follows that a subset U of a k-space is open iff $U \cap C$ is open in C for each closed compact set C.

The following is evident in view of the definition of k-space and the remarks preceding.

12 THEOREM *The family of all continuous functions on a k-space to a complete uniform space is complete relative to uniform convergence on compacta.*

The two most important examples of k-spaces are given in the following.

13 THEOREM *If X is a Hausdorff space which is either locally compact or satisfies the first axiom of countability, then X is a k-space.*

PROOF In each case the proof proceeds by assuming that B is a non-closed subset of X and showing that for some closed compact set C the intersection $B \cap C$ is not closed. Suppose x is an accumulation point of B which does not belong to B. If X is locally compact there is a compact neighborhood U of x and the intersection $B \cap U$ is not closed because x is an accumulation point but not a member. If X satisfies the first axiom of countability, then there is a sequence $\{y_n, n \in \omega\}$ in $B \sim \{x\}$ which converges to x, and the set which is the union of $\{x\}$ and the set of all points y_n is clearly compact, but its intersection with B is not closed. ∎

COMPACTNESS AND EQUICONTINUITY

This is the first of two sections devoted to the problem of finding conditions for compactness of a family of functions relative to the compact open topology. The conclusion desired is topological, and the sharpest results are obtained from purely topological premises. However, the arguments are simpler for uniformities and the discussion of this section concerns maps into a uniform space. The last section of the chapter treats the purely topological problem.

Let F be a family of maps of a topological space X into a uniform space $(Y,\mathcal{V})$. The family F is **equicontinuous at a point** x if and only if for each member V of $\mathcal{V}$ there is a neighborhood U of x such that $f[U] \subset V[f(x)]$ for every member f of F. Equivalently, F is equicontinuous at x iff $\bigcap \{f^{-1}[V[f(x)]]: f \in F\}$ is a neighborhood of x for each V in $\mathcal{V}$. Roughly speaking, F is equicontinuous at x iff there is a neighborhood of x whose image under every member of F is small.

14 THEOREM *If F is equicontinuous at x, then the closure of F relative to the topology $\mathcal{P}$ of pointwise convergence is also equicontinuous at x.*

PROOF If V is a closed member of the uniformity of Y, then the class of all functions f which satisfy the condition $f[U] \subset V[f(x)]$ is evidently closed relative to the topology $\mathcal{P}$ of pointwise convergence because it is identical with $\bigcap \{\{f: (f(y),f(x)) \in V\}: y \in U\}$. It follows that the pointwise closure of F is equicontinuous. ▮

A family F of functions is **equicontinuous** iff it is equicontinuous at every point. In view of the preceding theorem the closure of an equicontinuous family relative to the topology of pointwise convergence is also equicontinuous; in particular the members of the closure are continuous functions. The topology of pointwise convergence has other noteworthy properties for equicontinuous families.

15 THEOREM *If F is an equicontinuous family, then the topology of pointwise convergence is jointly continuous and hence coincides with the topology of uniform convergence on compacta.*

PROOF To prove that the map of $F \times X$ into Y is continuous at (f,x) let V be a member of the uniformity of Y and let U be a neighborhood of x such that $g[U] \subset V[g(x)]$ for all g in F. If g is a member of the $\mathcal{P}$-neighborhood $\{h: h(x) \in V[f(x)]\}$ of f and $y \in U$, then $g(y) \in V[g(x)]$ and $g(x) \in V[f(x)]$. Consequently $g(y) \in V \circ V[f(x)]$, and joint continuity follows. Each jointly continuous topology is larger than the compact open by 7.5, and the compact open topology coincides with that of uniform convergence on compacta by 7.11. ▮

The preceding theorem implies that an equicontinuous family of functions is compact relative to the topology of uniform convergence on compacta if it is compact relative to the pointwise topology $\mathcal{P}$, and the Tychonoff product theorem gives sufficient conditions for $\mathcal{P}$-compactness. In this way equicontinuity together with certain other conditions implies compactness of a family of functions. An implication in the reverse direction, from compactness to equicontinuity, is shown in the following theorem.

16 Theorem *If a family F of functions on a topological space X to a uniform space $(Y,\mathcal{V})$ is compact relative to a jointly continuous topology, then F is equicontinuous.*

PROOF Suppose that x is a fixed point of X and V is a symmetric member of $\mathcal{V}$. The theorem will follow if it is shown that there is a neighborhood U of x such that $g[U] \subset V \circ V[g(x)]$ for each g in F. Because the topology for F is jointly continuous there is for each member f of F a neighborhood G of f and a neighborhood W of x such that $G \times W$ maps into $V[f(x)]$. If $g \in G$ and $w \in W$, then $g(x)$ and $g(w)$ both belong to $V[f(x)]$ and hence $g(w) \in V \circ V[g(x)]$. That is, $g[W] \subset V \circ V[g(x)]$ for each g in G. Because F is compact there is a finite family $G_1, \cdots, G_n$ covering F and corresponding neighborhoods $W_1, \cdots, W_n$ of x such that $g[W_i] \subset V \circ V[g(x)]$ for each g in G_i. If we let U be the intersection of the neighborhoods W_i of x, it is clear that $g[U] \subset V \circ V[g(x)]$ for every g in F. ▮

The Ascoli theorem for locally compact spaces is an immediate consequence of the preceding results. It is obtained from 7.6 by replacing the condition "the pointwise topology $\mathcal{P}$ for the $\mathcal{P}$-closure of F is jointly continuous on compacta" by "the family F is equicontinuous." The latter condition implies the former (7.14 and 7.15) and compactness implies equicontinuity by 7.16. (A proof which does not depend on 7.6 is also simple to construct.)

17 Ascoli theorem *Let C be the family of all continuous functions on a regular locally compact topological space to a Hausdorff uniform space, and let C have the topology of uniform convergence* *on compacta. Then a subfamily F of C is compact if and only if*

(a) *F is closed in C,*
(b) *F[x] has a compact closure for each member x of X, and*
(c) *the family F is equicontinuous.*

A form of the Ascoli theorem is true for families of functions on a k-space (a space such that a set is closed whenever its intersection with every closed compact set is closed). A variant of the notion of equicontinuity is required. A family F of functions is **equicontinuous on a set** A iff the family of all restrictions of members of F to A is an equicontinuous family. A family which is equicontinuous at every point of A is equicontinuous on A, but the converse proposition is false. However, a family which is equicontinuous on A is equicontinuous at each point of the interior of A.

The proof of the following theorem is omitted. It is a straightforward application of 7.6, the results of this section and the fact that a function on a k-space is continuous if it is continuous on each compact set.*

18 ASCOLI THEOREM *Let C be the family of all continuous functions on a k-space X which is either Hausdorff or regular to a Hausdorff uniform space Y, and let C have the topology of uniform convergence on compacta. Then a subfamily F of C is compact if and only if*

(a) *F is closed in C,*
(b) *the closure of F[x] is compact for each x in X, and*
(c) *F is equicontinuous on every compact subset of X.*

* *EVEN CONTINUITY*

This section is devoted to the proof of an Ascoli theorem for topological spaces. The pattern of attack is much the same as the foregoing except that a topological concept replaces the (uni-

* It is evident that the condition "X is a k-space" may be omitted from the hypothesis of the theorem if the family C of continuous functions is replaced by the family of all functions which are continuous on each compact set. However, the same result may be obtained by applying the given theorem to X with the topology $\mathfrak{J}$ such that a set A is $\mathfrak{J}$-closed iff $A \cap B$ is closed for every closed compact set B.

form) concept of equicontinuity. The connections between the two concepts are discussed briefly at the end of the section.

Let F be a family of functions, each on a topological space X to a topological space Y. The concept of even continuity can be described intuitively by the statement: for each x in X, y in Y, and f in F, if $f(x)$ is near y, then f maps points near x into points near y. Explicitly, the family F is **evenly continuous** iff for each x in X, each y in Y, and each neighborhood U of y there is a neighborhood V of x and a neighborhood W of y such that $f[V] \subset U$ whenever $f(x) \in W$. The close connection between this definition and joint continuity may be emphasized by the restatement: F is evenly continuous iff for each x in X and y in Y and for each neighborhood U of y there are neighborhoods V of x and W of y such that $\{f: f \in F \textit{ and } f(x) \in W\} \times V$ is carried into U by the natural map. The crucial property of evenly continuous families is easily demonstrated.

19 THEOREM *Let F be an evenly continuous family of functions on a topological space X to a regular space Y and let $\mathcal{P}$ be the topology of pointwise convergence. Then the $\mathcal{P}$-closure F^- of F is evenly continuous and $\mathcal{P}$ is jointly continuous on F^-.*

PROOF The latter statement of the theorem is evident from the second formulation of the definition of even continuity, since $\{f: f \in F \textit{ and } f(x) \in W\}$ is $\mathcal{P}$-open whenever W is open in Y. To show that the $\mathcal{P}$-closure of F is evenly continuous suppose $x \in X$, $y \in Y$ and U is a neighborhood of y. Because Y is regular it may be supposed that U is closed. Let V be a neighborhood of x and W an open neighborhood of y such that, if $f \in F$ and $f(x) \in W$, then $f[V] \subset U$, and suppose that $\{g_n, n \in D\}$ is a net in F which converges pointwise to g and $g(x) \in W$. Then $\{g_n(x), n \in D\}$ is eventually in W; hence for each z in V it is true that $\{g_n(z), n \in D\}$ is eventually in U and therefore $g(z) \in U$. This shows that $g[V] \subset U$. ∎

Sufficient conditions for compactness of an evenly continuous family of functions are more or less self-evident in view of the preceding result and 7.6. The following proposition shows the necessity of the conditions given in the Ascoli theorem.

20 THEOREM *If a family F of continuous functions on a topological space X to a regular Hausdorff space Y is compact relative to a jointly continuous topology, then F is evenly continuous.*

PROOF The identity map of the compact space F into F with the topology of pointwise convergence is continuous, and since the latter topology is Hausdorff, the two topologies coincide. The pointwise topology for F is therefore jointly continuous. Suppose that $x \in X$, $y \in Y$, and U is an open neighborhood of y. Let W be a closed neighborhood of y such that $W \subset U$, and observe that the set K of all members f of F such that $f(x) \in W$ is pointwise closed and hence compact. If P is the function such that $P(f,x) = f(x)$, then the compact set $K \times \{x\}$ is contained in $P^{-1}[U]$, and since P is continuous there is a neighborhood V of x such that $K \times V \subset P^{-1}[U]$ by 5.12. That is, if $v \in V$ and $f(x) \in W$, then $f(v) \in U$. ∎

21 ASCOLI THEOREM *Let C be the family of all continuous functions on a regular locally compact space X to a regular Hausdorff space Y, and let C have the compact open topology. Then a subset F of C is compact if and only if*

(a) *F is closed in C,*
(b) *the closure of $F[x]$ is compact for each x in X, and*
(c) *F is evenly continuous.*

PROOF If F is compact relative to the compact open topology conditions (a), (b), and (c) follow from 7.6 and 7.20. If F satisfies (a), (b), and (c), then the pointwise closure of F is an evenly continuous family on which the pointwise topology is jointly continuous, by 7.19. Compactness follows from 7.6. ∎

The foregoing theorem can be extended to k-spaces in the same fashion that 7.17 was extended. A family F of functions is **evenly continuous on a set** A iff the family of all restrictions of members of F to A is evenly continuous. With this definition the Ascoli theorem (21) can be proved for k-spaces X if condition (c) is replaced by "F is evenly continuous on each compact subset of X." The straightforward proof of this fact is omitted.

The section is concluded with two propositions which clarify the relation between even continuity and equicontinuity.

22 THEOREM *An equicontinuous family of functions on a topological space to a uniform space is evenly continuous.*

PROOF Suppose that F is an equicontinuous family of functions on X to Y, that $x \in X$ and $y \in Y$, and that U is a neighborhood of y. Then one may assume that U is the sphere of d-radius r about y, where d is a pseudo-metric belonging to the gage of Y and $r > 0$. Since F is equicontinuous at x there is a neighborhood V of x such that, if $z \in V$, then $d(f(x),f(z)) < r/2$ for all f in F. Consequently, if $z \in V$ and $f(x)$ belongs to the sphere of d-radius $r/2$ about y, then $f(z) \in U$. ∎

In a certain sense equicontinuity is the result of "uniformizing" even continuity with respect to the range space, and, as might be expected, equicontinuity may be deduced from even continuity in the presence of a suitable compactness condition.

23 THEOREM * *If F is an evenly continuous family of functions on a topological space X to a uniform space Y, and x is a point of X such that $F[x]$ has a compact closure, then F is equicontinuous at x.*

PROOF Suppose d is a member of the gage of Y and $r > 0$. For each y in $F[x]^-$ there are neighborhoods W of y and V of x such that, if $f(x) \in W$, then $f[V]$ is contained in the sphere of d-radius $r/2$ about y. Because $F[x]^-$ is compact, there is a finite number of neighborhoods W_i of points y_i of $F[x]^-$ and corresponding neighborhoods V_i of x, for $i = 1, \cdots, n$, such that the family of all W_i covers $F[x]^-$, and such that, if $f(x) \in W_i$, then $f[V_i]$ is a subset of the sphere of d-radius $r/2$ about y_i. Consequently, if $T = \bigcap \{V_i: i = 0, 1, \cdots, n\}$ and $f \in F$, then $f(x)$ belongs to W_i for some i, and since $f[T]$ is a subset of some sphere of d-radius $r/2$, $d(f(x),f(y)) < r$ for each y in T. Hence F is equicontinuous. ∎

Notes The results of this section are due to A. P. Morse and myself. Another form of the Ascoli theorem for topological spaces has been obtained by Gale [1].

* This theorem is false if the condition "$F[x]$ has a compact closure" is replaced by "$F[x]$ is totally bounded".

PROBLEMS

A EXERCISE ON THE TOPOLOGY OF POINTWISE CONVERGENCE

The set of all continuous real-valued functions on a Tychonoff space X is dense, relative to the topology of pointwise convergence, in the set of all real-valued functions on X.

B EXERCISE ON CONVERGENCE OF FUNCTIONS

Let f be a continuous real-valued function on the closed unit interval $[0,1]$ such that $f(0) = f(1) = 0$ and f is not identically zero. Let $g_n(x) = f(x^n)$ for each non-negative integer n. Then $\{g_n, n \in \omega\}$ converges pointwise (but not uniformly) to the function h which is identically zero. The union of $\{h\}$ and the set of all g_n is compact relative to the pointwise topology but is not compact relative to the topology of uniform convergence.

C POINTWISE CONVERGENCE ON A DENSE SUBSET

Let F be an equicontinuous family of functions on a topological space X to a uniform space and let A be a dense subset of X. Then the uniformity of pointwise convergence on X is identical with the uniformity of pointwise convergence on A.

D THE DIAGONAL PROCESS AND SEQUENTIAL COMPACTNESS

Prior to the proof of the Tychonoff product theorem the diagonal process, as outlined below, was the standard method of proving compactness of a family of functions. Recall that a topological space is called sequentially compact if each sequence in the space has a subsequence which converges to a point of the space.

(a) The product of a countable number of sequentially compact topological spaces is sequentially compact. (Suppose $\{Y_m, m \in \omega\}$ is a sequence of sequentially compact spaces and $\{f_n, n \in \omega\}$ is a sequence in the product $\times\{Y_m: m \in \omega\}$. Choose an infinite subset A_0 of ω such that $\{f_n(0), n \in A_0\}$ converges to a point of Y_0, and continue inductively, choosing an infinite subset A_{k+1} of A_k such that $\{f_n(k + 1), n \in A_{k+1}\}$ converges to a point of Y_{k+1}. If N_k is the k-th member of A_k, then $\{f_{N_k}, k \in \omega\}$ is the required subsequence.)

(b) Let Y be a sequentially compact uniform space, let X be a separable topological space, and let F be an equicontinuous family of functions on X to Y which is closed in Y^X relative to the topology of pointwise convergence. Then F is sequentially compact relative to the pointwise topology (or the compact open topology). (Use 7.C and observe that each Cauchy sequence in Y has a limit point.)

Note Some very beautiful results on countable compactness of function spaces have been obtained recently by Grothendieck [1]. His results apply directly to some interesting linear topological space problems.

E DINI'S THEOREM

If a monotonically increasing net $\{f_n, n \in D\}$ of continuous real-valued functions on a topological space X converges pointwise to a continuous function f, then the net converges to f uniformly on compacta. (This is a straightforward compactness argument. If C is a compact subset of X let $A_n = \{(x,y): x \in C \text{ and } f_n(x) \leq y \leq f(x)\}$ and observe that the intersection of the sets A_n for n in D is simply the graph of $f \mid C$.)

F CONTINUITY OF AN INDUCED MAP

Let X and Y be sets, let $\mathfrak{A}$ and $\mathfrak{B}$ be families of subsets of X and of Y respectively, let F be the family of all functions on X to a uniform space $(Z,\mathfrak{U})$, and let G be the family of all functions on Y to $(Z,\mathfrak{U})$. If T is a map of X into Y the *induced map* T^* of G into F is defined by $T^*(g) = g \circ T$ for g in G. If for each member A of $\mathfrak{A}$ the set $T[A]$ is contained in some member of $\mathfrak{B}$, then T^* is uniformly continuous relative to the uniformities $\mathfrak{U} \mid \mathfrak{A}$ for F and $\mathfrak{U} \mid \mathfrak{B}$ for G (uniform convergence on members of $\mathfrak{A}$ and of $\mathfrak{B}$ respectively). In particular T^* is always uniformly continuous relative to the uniformity of uniform convergence and is continuous relative to that of pointwise convergence if $\mathfrak{B}$ covers Y. If X and Y are topological spaces and T is continuous, then T^* is uniformly continuous relative to uniform convergence on compacta.

Note The continuity of certain other naturally induced maps has been studied by Arens and Dugundji [2].

G UNIFORM EQUICONTINUITY

A family F of functions on a uniform space $(X,\mathfrak{U})$ to a uniform space $(Y,\mathfrak{V})$ is *uniformly equicontinuous* iff for each member V of $\mathfrak{V}$ there is U in $\mathfrak{U}$ such that $(f(x),f(y)) \in V$ whenever $f \in F$ and $(x,y) \in U$.

(a) A family F is uniformly equicontinuous iff it is uniformly jointly continuous, in the sense that the natural map of $F \times X$ into Y is uniformly continuous when the uniformity of F is that of uniform convergence and $F \times X$ has the product uniformity.

(b) The pointwise closure of a uniformly equicontinuous family is uniformly equicontinuous.

(c) If X is compact and F is equicontinuous, then F is uniformly equicontinuous.

Note The proofs of the foregoing propositions require no new methods. A more detailed treatment of the subject is given in Arens [2] and in Bourbaki [1].

H EXERCISE ON THE UNIFORMITY $\mathfrak{U} | \mathfrak{A}$

Let X be a set, let $\mathfrak{A}$ be a cover of X which is directed by $\supset$ (that is, for A and B in $\mathfrak{A}$ there is C in $\mathfrak{A}$ such that $C \supset A \cup B$), let $(Y,\mathfrak{V})$ be a uniform space, and let F be the family of functions on X to Y with the uniformity $\mathfrak{U} | \mathfrak{A}$ of uniform convergence on members of $\mathfrak{A}$. Finally, suppose that S is a net in F and that for each member A of $\mathfrak{A}$ there is given a subnet $\{S \circ T_A(m), m \in E_A\}$ of S which converges to a member s of F uniformly on A. Give an explicit formula for a subnet of S which converges to s relative to the topology of $\mathfrak{U} | \mathfrak{A}$.

I CONTINUITY OF EVALUATION

If F is a family of functions on a set X to a set Y, then X is mapped by evaluation into a family G of functions on F to Y; explicitly, the evaluation $E(x)$ at a point x of X is defined by $E(x)(f) = f(x)$ for all f in F. Let $(X,\mathfrak{U})$ and $(Y,\mathfrak{V})$ be uniform spaces and let G have the uniformity of uniform convergence on members of a family $\mathfrak{A}$ of subsets of F. Then the evaluation map E of X into G is continuous if each member of $\mathfrak{A}$ is equicontinuous, and evaluation is uniformly continuous if each member of $\mathfrak{A}$ is uniformly equicontinuous.

J SUBSPACES, PRODUCTS, AND QUOTIENTS OF k-SPACES

(a) There are Tychonoff spaces which are not k-spaces, and since every Tychonoff space can be embedded in a compact Hausdorff space it follows that not every subspace of a k-space is a k-space. (See the example 2.E.)

(b) The product of uncountably many copies of the real line is not a k-space. (Let A be the subset of the product consisting of all members x such that for some non-negative integer n each coordinate of x is equal to n except for a set of at most n indices, and on this set x is zero. Then A is not closed, but $A \cap C$ is compact for each compact set C.)

(c) Let X be a k-space, let R be an equivalence relation on X, and let X/R have the quotient topology. If X/R is a Hausdorff space, then it is a k-space.

K THE k-EXTENSION OF A TOPOLOGY

Let $(X,\mathfrak{J})$ be a Hausdorff space. The *k-extension of* $\mathfrak{J}$ is defined to be the family $\mathfrak{J}_k$ of all subsets U of X such that $U \cap C$ is open in C for every compact set C (equivalently, A is $\mathfrak{J}_k$-closed iff $A \cap C$ is $\mathfrak{J}$-compact for every $\mathfrak{J}$-compact set C).

(a) If C is a $\mathfrak{J}$-compact subset of X, then the relativization of $\mathfrak{J}$ to C is identical with that of $\mathfrak{J}_k$. Consequently a set is $\mathfrak{J}$-compact iff it is $\mathfrak{J}_k$-compact.

(b) The space $(X,\mathfrak{J}_k)$ is a k-space.

(c) A function on X is $\mathfrak{J}_k$-continuous iff it is $\mathfrak{J}$-continuous on every compact subset of X.

(d) The topology $\mathfrak{J}_k$ is the largest topology which agrees with $\mathfrak{J}$ on compact sets (in the sense that the relativization to a compact set is identical with the relativization of $\mathfrak{J}$).

L CHARACTERIZATION OF EVEN CONTINUITY

A family F of functions on a topological space X to a topological space Y is evenly continuous if and only if for each net $\{(f_n,x_n), n \in D\}$ in $F \times X$ such that $\{x_n, n \in D\}$ converges to x and $\{f_n(x), n \in D\}$ converges to y it is true that $\{f_n(x_n), n \in D\}$ converges to y.

M CONTINUOUS CONVERGENCE

Let F be a family of continuous functions, each on a space X to a space Y. A net $\{f_n, n \in D\}$ *converges continuously* to a member f of F iff it is true that $\{f_n(x_n), n \in D\}$ converges to $f(x)$ whenever $\{x_n, n \in D\}$ is a net in X converging to a point x.

(a) A topology $\mathfrak{J}$ for F is jointly continuous iff a net in F converges continuously to a member f whenever it $\mathfrak{J}$-converges to f.

(b) If a sequence in F converges to f relative to the compact open topology, then it converges to f continuously.

(c) Suppose that X satisfies the first axiom of countability and that F, with the compact open topology $\mathfrak{C}$, also satisfies this axiom. Then $\mathfrak{C}$ is jointly continuous and a sequence in F $\mathfrak{C}$-converges to a member f iff it converges continuously to f.

N THE ADJOINT OF A NORMED LINEAR SPACE

Let X be a real normed linear space and let X^*, its adjoint, be the space of all continuous real-valued linear functions on X. The *norm topology* for X^* is defined by: $\|f\| = \sup \{|f(x)|: \|x\| \leqq 1\}$. The topology of pointwise convergence for X^* is called the w^*-topology.

A subset F of X^* is called *w^*-bounded* iff for each member x of X the set of all $f(x)$ with f in F is bounded.

(a) The space X^* is not complete relative to the w^*-topology unless every linear function on X is continuous. (See 3.**W**. Assume that there are enough continuous linear functionals on X to distinguish points—this fact is a consequence of the Hahn-Banach theorem, Banach [1;27].)

(b) *Theorem* (*Alaoglu*) The unit sphere in X^* is compact relative to the w^*-topology. Hence each norm bounded w^*-closed subset of X^* is w^*-compact. (The unit sphere is a closed subset of the product $\times\{[-\|x\|, \|x\|]: x \in X\}$.)

(c) The space X^* with the w^*-topology is paracompact and hence topologically complete. (See 5.**Y** and 6.**L**.)

(d) If a subset F of X^* is equicontinuous, then its w^*-closure is equicontinuous. If F is equicontinuous, then the w^*-closure of F is w^*-compact. If the w^*-closure of F is w^*-compact, then F is w^*-bounded. (Observe that F is equicontinuous iff it is norm bounded.)

(e) If X is non-meager, and in particular if X is complete, then each w^*-bounded subset F of X^* is equicontinuous. (Apply 6.**U**(**b**), or apply 6.**U**(**a**) to the set $\{x: |f(x)| \leq 1$ *for each* f *in* $F\}$.)

(f) The hypothesis "X is non-meager" cannot be omitted from (e). (Consider the space X of all real sequences which are zero save on a finite set, with the norm $\|x\| = \sum\{|x_n|: n \in \omega\}$. If $f_n(x) = nx_n$, then the sequence $\{f_n, n \in \omega\}$ converges to zero relative to the w^*-topology.)

Note The principal results of this problem are more or less classical and certain of them may clearly be extended to less restricted situations. However, the equivalences resulting from (d) and (e) do not hold for an arbitrary complete linear topological space. In connection with (f) it is interesting to note that a w^*-compact *convex* subset of the adjoint of a normed linear space X is always equicontinuous; the proof of this fact is not entirely trivial.

O TIETZE EXTENSION THEOREM *

(a) Let X be a normal topological space, let A be a closed subset, and let f be a continuous function on A to the closed interval $[-1,1]$. Then f has a continuous extension g which carries X into $[-1,1]$. (Let $C = \{x: f(x) \leq -1/3\}$ and let $D = \{x: f(x) \geq 1/3\}$. By Urysohn's lemma there is f_1 on X to $[-1/3, 1/3]$ such that f_1 is $-1/3$ on C and $1/3$ on

* This theorem occurs here because the proof requires the fact that the uniform limit of continuous functions is continuous. In all honesty I should admit that there are three problems in earlier chapters where the same fact is used.

D. Evidently $|f(x) - f_1(x)| \leqq 2/3$ for all x in A. The same sort of argument may be applied to the function $f - f_1$.)

Note Dugundji [1], Dowker [3], and Hanner [1] have proved interesting extensions of Tietze's theorem.

P DENSITY LEMMA FOR LINEAR SUBSPACES OF $C(X)$

Let X be a topological space, let $C(X)$ be the space of all bounded continuous real-valued functions on X, and let $C(X)$ have the topology of uniform convergence (equivalently, norm $C(X)$ by $\|f\| = \sup \{|f(x)|: x \in X\}$). A subset L of $C(X)$ is said to have the *two-set property* iff for closed disjoint subsets A and B of X and for each closed interval $[a,b]$ there is a member f of L such that f maps X into $[a,b]$, f is a on A, and f is b on B. Each linear subspace of $C(X)$ which has the two-set property is dense in $C(X)$. (If g is an arbitrary member of $C(X)$ and dist $(g,L) > 0$ choose h in L such that dist (g,L) is approximately $\|g - h\|$. If $k = g - h$, then dist $(k,L) =$ dist (g,L) which is approximately $\|k\|$. Show that there is a member f of L such that $\|k - f\| \leqq 2\|k\|/3$.)

Q THE SQUARE ROOT LEMMA FOR BANACH ALGEBRAS *

A real (or complex) *Banach algebra* is an algebra A over the real (complex) numbers together with a norm such that A is a complete normed linear space and multiplication satisfies the condition: $\|xy\| \leqq \|x\| \|y\|$. (In terms of the usual operator norm the algebra A can be described as a Banach space with an associative multiplication such that multiplication on the left by a fixed element x is a linear operator of norm at most $\|x\|$.) Throughout the following, A is a fixed (real or complex) Banach algebra.

A function f on D to a normed linear space is *absolutely summable* iff $\sum\{\|f(n)\|: n \in D\}$ exists.

(a) Each function in A which is absolutely summable is summable. If

$\{x_n, n \in \omega\}$ and $\{y_m, m \in \omega\}$ are absolutely summable, then $\{x_n y_m, (m,n) \in \omega \times \omega\}$ is absolutely summable, and

$$\sum\{x_n: n \in \omega\}\sum\{y_m: m \in \omega\} = \sum\{x_n y_m: (m,n) \in \omega \times \omega\}.$$

(The usefulness of this result lies in the fact that the last sum may be computed by grouping the summands in a more or less arbitrary fashion. See 6.S.)

* This proposition is given here essentially as a preliminary to the Stone-Weierstrass theorem. However, the lemma is of some importance in a more general situation and is consequently stated for an arbitrary Banach algebra.

(b) Let a_n be the n-th binomial coefficient in the expansion of $(1 - t)^{1/2}$ about $t = 0$. Then $a_0 = 1$, a_n is negative for n positive, $\sum\{a_n: n \in \omega\} = 0$, and $\sum\{a_n a_{p-n}: n \in \omega \text{ and } n \leq p\}$ is 1, -1 and 0 for $p = 0$, $p = 1$, and $p > 1$, respectively. (Alternatively, one may define the coefficients a_n recursively so that the last stated relation is satisfied. After verifying that $a_n < 0$ for n positive observe that the partial sums $\sum\{a_n t^n: n < p\}$ are monotonically decreasing in n and bounded below by $(1 - t)^{1/2}$ for $0 \leq t < 1$—hence also for $t = 1$.)

(c) If the algebra has a unit u and if $\| x - u \| \leq 1$, then there is an element y in the algebra such that $x = y^2$. Explicitly, y may be taken to be $\sum\{a_n(u - x)^n: n \in \omega\}$, where a_n is defined as in (b). (Here it is assumed that $x^0 = u$. The element y may also be written in the form: $y = \sum\{a_n[(u - x)^n - u]: n \geq 1\}$. In this form it is clear that y is the limit of polynomials in x and that these polynomials may be taken to be without constant coefficients.)

Note It is evident that a great deal more information can be obtained by means of the methods sketched above (for example, if $\| x \| < 1$, then $\sum\{x^n: n \in \omega\}$ is the multiplicative inverse of $u - x$). For a systematic treatment of Banach algebras see Loomis [2] and Hille [1].

R THE STONE-WEIERSTRASS THEOREM

(a) Let X be a compact topological space, let $C(X)$ be the algebra of all continuous real-valued functions on X, and let $C(X)$ have the norm: $\|f\| = \sup\{|f(x)|: x \in X\}$. Then a subalgebra R of $C(X)$ is dense in $C(X)$ if it has the *two-point property:* for distinct points x and y of X and for each pair a and b of real numbers there is f in R such that $f(x) = a$ and $f(y) = b$.

In particular R is dense if the constant functions belong to R and R distinguishes points (in the sense that, if $x \neq y$, then $f(x) \neq f(y)$ for some f in R).

The proof is accomplished by a sequence of lemmas.

(i) If $f \in R$, then $|f|$ belongs to the closure R^- of R, where $|f|(x) = |f(x)|$. (Take the square root of f^2 using 7.P.)

(ii) If f and g belong to a subalgebra, then $\max[f,g]$ and $\min[f,g]$ belong to its closure. (Here $\max[f,g](x) = \max[f(x),g(x)]$. Observe that $\max[a,b] = [(a + b) + |a - b|]/2$ and $\min[a,b] = [(a + b) - |a - b|]/2$.)

(iii) If the subalgebra R has the two-point property, $f \in C(X)$, $x \in X$, and $e > 0$, then there is g in R^- such that $g(x) = f(x)$ and $g(y) < f(y) + e$ for all y in X. (Using compactness of X, take the minimum of a suitably chosen finite family of functions.)

The theorem now follows from (iii) by taking the maximum of a properly chosen finite family of functions.

(b) If X is a topological space and the family $C(X)$ of all continuous real-valued functions on X is given the topology of uniform convergence on compacta, then each subalgebra of $C(X)$ which has the two-point property is dense in $C(X)$.

Note This is unquestionably the most useful known result on $C(X)$. The corresponding theorem for complex-valued functions is false (consider, for example, the functions which are continuous on the unit disk in the plane and are analytic in its interior). See M. H. Stone [5] for a more detailed discussion.

S STRUCTURE OF $C(X)$

Throughout this problem X, Y, and Z will be compact Hausdorff spaces and $C(X)$, $C(Y)$, and $C(Z)$ will be the algebras of all continuous real-valued functions on X, Y, and Z, respectively. A *real homomorphism* of an elgebra is a homomorphism into the real numbers.

(a) For each continuous function F on X to Y let F^* be the induced map of $C(Y)$ into $C(X)$ defined by $F^*(h) = h \circ F$ for all h in $C(Y)$. Then

(i) F^* is a homomorphism of $C(Y)$ into $C(X)$;
(ii) F maps X onto Y iff F^* is an isomorphism of $C(Y)$ onto a subalgebra of $C(X)$ which contains the unit;
(iii) F is one to one iff F^* maps $C(Y)$ onto $C(X)$;
(iv) if G is a continuous map of Y into Z, then $(G \circ F)^* = F^* \circ G^*$; and
(v) if F is a topological map of X onto Y, then $(F^{-1})^* = (F^*)^{-1}$.

(b) The topology of $C(X)$ is entirely determined by the algebraic operations. In detail: $f \geqq g$ iff $f - g$ is the square of an element of $C(X)$, and $\|f\| = \inf \{k: -ku \leqq f \leqq ku\}$ where u is the function which is identically one. If ϕ is a real homomorphism of $C(X)$, then $|\phi(f)| \leqq \|f\|$ and, unless ϕ is identically zero, $\phi(u) = 1$.

(c) Let S be the set of all real homomorphisms ϕ of $C(X)$ such that $\phi(u) = 1$, let S have the topology of pointwise convergence, and let E be the evaluation map of X into S (that is, $E(x)(f) = f(x)$). Then E is a topological map of X onto S. (Show that S is compact, use the Stone-Weierstrass theorem to show that the evaluation map D of $C(X)$ into $C(S)$ is an isomorphism of $C(X)$ onto $C(S)$, verify that $E^* = D^{-1}$, and use (a).)

(d) The space X is metrizable if and only if $C(X)$ is separable. (This result is not needed for the rest of the problem; it is given simply as an exercise in the use of (c).)

(e) If H is a homomorphism of $C(Y)$ into $C(X)$ which carries the unit of $C(Y)$ into the unit of $C(X)$, then there is a unique continuous map F of X into Y such that $H = F^*$. (The homomorphism H induces a map of the real homomorphisms on $C(X)$ into real homomorphisms on $C(Y)$.)

(f) Let R be a closed subalgebra of $C(X)$ such that $u \in R$, let F be the map of X into $\times\{f[X]: f \in R\}$ which is defined by $F(x)_f = f(x)$, and let Y be the range of F. Then R is the range of the induced isomorphism F^* of $C(Y)$ into $C(X)$.

(g) Let I be a closed ideal in $C(X)$ and let $Z = \{x: f(x) = 0$ *for all* f *in* $I\}$. Then I is the set of all members of $C(X)$ which vanish identically on Z. (If Z is empty, then there is a member of I which vanishes at no point and therefore has an inverse. Consider the subalgebra $C + I$, where C is the set of constant functions. Because Z is nonvoid the set $C + I$ is closed, and (f) may be applied.)

Notes Quite a bit is known about the structure of $C(X)$. Further information and references are given in a review of the subject by S. B. Myers [1]. See also Hewitt [2].

The line of attack outlined in the preceding problem is not the only one possible—the fundamental facts (the Stone-Weierstrass theorem, the Tychonoff product theorem, and the Tietze theorem) may be used in various ways to yield the desired results. However, the pattern used above is, in part, an example of a general method. To each member of a certain collection of objects (in this case compact Hausdorff spaces X) there is associated another object (in this case the Banach algebras $C(X)$). Moreover, to each of a specified class of maps of the original objects (continuous maps in the case at hand) there is assigned an induced map satisfying certain conditions (for example (iv) and (v) of (a)). In this case the induced maps "go in the direction opposite" that of the inducing maps—such a correspondence is called *contravariant.* The assignment of the Stone-Čech compactification of a Tychonoff space, together with the obvious induced maps, furnishes an example where the induced map is in the same direction as the original—a *covariant* correspondence.

This general method of investigation has been used most successfully by Eilenberg and Steenrod in their axiomatic treatment of homology theory [1]. The method itself was first studied by Eilenberg and MacLane. The study of objects and maps might be called the galactic theory, continuing the analogy whereby the study of a topological space is called global.

T COMPACTIFICATION OF GROUPS; ALMOST PERIODIC FUNCTIONS

It is natural to attempt to map a topological group into a dense subgroup of a compact topological group in somewhat the same way that a Tychonoff space is embedded in its Stone-Čech compactification. A topological embedding is usually impossible—a complete group is closed in each Hausdorff group in which it is topologically and isomorphically embedded. However, a number of interesting results can be obtained; the propositions that follow are intended to be an introduction to these. The development is motivated by the observation: If ϕ is a continuous homomorphism of a topological group G into a compact group H and if g is a continuous real-valued function on H, then $g \circ \phi$ has the property that the set of all left translates is totally bounded (relative to the uniformity of uniform convergence).

Throughout it is assumed that G is a fixed topological group. For each bounded real-valued function f on the group G and each x in G the *left translate* of f by x, $L_x(f)$, is defined by: $L_x(f)(y) = f(x^{-1}y)$. The space of bounded real functions is metrized by $d(f,g) = \sup\{|f(x) - g(x)|: x \in X\}$ and the *left orbit* X_f of a function f is defined to be the closure, relative to the metric topology, of the set of all left translates of f. A function f is called *left almost periodic* iff X_f is compact.

Let A be the set of all continuous left almost periodic functions on G. Then for each x in G the left translation L_x maps A into A. Topologize the space of all maps of A into A by pointwise convergence, and let $\alpha[G]$ be the closure relative to this topology of the set of left translations.

(a) *Lemma* Let (X,d) be a compact metric space and let K be the group (under composition) of all isometries of K into itself. Then the topology (for K) of uniform convergence on X is the topology of the metric: $d^*(R,S) = \sup\{d(R(x),S(x)): x \in X\}$ and this is identical with the topology of pointwise convergence on X. The group K with this topology is a compact topological group.

(b) $\alpha[G]$ is compact. (Observe that $\alpha[G] \subset \times\{X_f: f \in A\}$.)

(c) Each member of $\alpha[G]$ is an isometry which carries each left orbit X_f onto itself. The natural map of $\alpha[G]$ into the product space $\times\{K_f: f \in A\}$, where K_f is the group of isometries of X_f, is a topological isomorphism. Hence $\alpha[G]$ is a topological group.

(d) If A is given the topology of pointwise convergence on G and $\alpha[G]$ (a subset of A^A) has the resulting product topology, then the two topologies for $\alpha[G]$ coincide. Hence $R_n \to R$ in $\alpha[G]$ iff $R_n(f)(x) \to R(f)(x)$ for all f in A and all x in G.

(e) The map L of G into $\alpha[G]$ which carries a member x of G into L_x is a continuous homomorphism. The smallest topology for G which makes L continuous is identical with the smallest topology which makes each member f of A continuous. ($\alpha[G]$ may also be described as the completion, relative to the smallest uniformity which makes each f in A uniformly continuous, of G modulo the subgroup of members of G which are not distinguished from the identity by members of A.)

(f) If g is a continuous real function on $\alpha[G]$, then $g \circ L \in A$. If $f \in A$ and g is the function on $\alpha[G]$ defined by $g(R) = R^{-1}\ (f)(e)$, then $f = g \circ L$ and g is continuous. The family of continuous real functions on $\alpha[G]$ is isometric (and isomorphic) to A.

(g) If ϕ is a continuous homomorphism of G into a compact topological group H, then there is a continuous homomorphism θ of $\alpha[G]$ into H such that $\phi = \theta \circ L$. (More generally, for H arbitrary ϕ induces a natural homomorphism θ of $\alpha[G]$ into $\alpha[H]$ such that $\theta \circ L = L \circ \phi$. See the definition of α.)

There are several obvious corollaries to the preceding development; for example, a function is left periodic iff it is right periodic, and the class A is a Banach algebra which is isomorphic to the algebra of all continuous functions on the compact group $\alpha[G]$.

(h) The term "almost periodic" is derived from an alternate description of the class A. A member x of G is called a *left e-period* of a real function f iff $|f(x^{-1}y) - f(y)| < e$ for all y in G. Let A_e be the set of all left e-periods of a continuous function f. Then the following are equivalent:

(i) There is a homomorphism ϕ of G into a compact group H and a continuous real-valued function h on H such that $g = h \circ \phi$.
(ii) The set of left translates of f is totally bounded relative to the uniformity of uniform convergence.
(iii) For each positive number e there is a finite subset B of G such that $G = BA_e$.

(The connection between (ii) and (iii) is clarified by observing that $|L_x(f)(z) - L_y(f)(z)| < e$ for all z iff $y^{-1}x$ is a left e-period.)

Notes The results above are due primarily to Weil [2]. The equivalence of parts (ii) and (iii) of (h) is a classical theorem of Bochner. Loomis [2] investigates almost periodic functions by showing first that the set of all left almost periodic functions on a group satisfies the conditions which characterize a Banach algebra of functions, and then defining $\alpha[G]$ to be the set of real homomorphisms of this Banach algebra.

Proposition (a) suggests the general problem of topologizing a homeomorphism group in such a fashion as to obtain a topological group. For results in this direction and for references see Arens [3] and Dieudonné [4].

Appendix

ELEMENTARY SET THEORY

This appendix is devoted to elementary set theory. The ordinal and cardinal numbers are constructed and the most commonly used theorems are proved. The non-negative integers are defined and Peano's postulates are proved as theorems.

A working knowledge of elementary logic is assumed, but acquaintance with formal logic is not essential. However, an understanding of the nature of a mathematical system (in the technical sense) helps to clarify and motivate some of the discussion. Tarski's excellent exposition [1] describes such systems very lucidly and is particularly recommended for general background.

This presentation of set theory is arranged so that it may be translated without difficulty into a completely formal language.* In order to facilitate either formal or informal treatment the introductory material is split into two sections, the second of which is essentially a precise restatement of part of the first. It may be omitted without loss of continuity.

The system of axioms adopted is a variant of systems of Skolem and of A. P. Morse and owes much to the Hilbert-Bernays-von Neumann system as formulated by Gödel. The formulation used here is designed to give quickly and naturally a foundation for mathematics which is free from the more obvious paradoxes.

* That is, it is possible to write the theorems in terms of logical constants, logical variables, and the constants of the system, and the proofs may be derived from the axioms by means of rules of inference. Of course, a foundation in formal logic is necessary for this sort of development. I have used (essentially) Quine's meta-axioms for logic [1] in making this kind of presentation for a course.

For this reason a finite axiom system is abandoned and the development is based on eight axioms and one axiom scheme * (that is, all statements of a certain prescribed form are accepted as axioms).

It has been convenient to state as theorems many propositions which are essentially preliminary to the desired results. This clutters up the list of theorems, but it permits omission of many proofs and abbreviation of others. Most of the devices used are more or less evident from the statements of the definitions and theorems.

THE CLASSIFICATION AXIOM SCHEME

Equality is always used in the sense of logical identity; '1 + 1 = 2' is to mean that '1 + 1' and '2' are names of the same object. Besides the usual axioms for equality an unrestricted substitution rule is assumed: in particular the result of changing a theorem by replacing an object by its equal is again a theorem.

There are two primitive (undefined) constants besides '=' and the other logical constants. The first of those is 'ε,' which is read 'is a member of' or 'belongs to.' The second constant is denoted, rather strangely, '{··: ···}' and is read 'the class of all ·· such that ···.' It is the *classifier*. A remark on the use of the term 'class' may clarify matters. The term does not appear in any axiom, definition, or theorem, but the primary interpretation † of these statements is as assertions about classes (aggregates, collections). Consequently the term 'class' is used in the discussion to suggest this interpretation.

Lower case Latin letters are (logical) variables. The difference between a constant and variable lies entirely in the substitution rules. For example, the result of replacing a variable in a theorem by another variable which does not occur in the theorem is

* Actually, an axiom scheme for definition is also assumed without explicit statement. That is, statements of a certain form, which in particular involve one new constant and are either an equivalence or an identity, are accepted as definitions and are treated in precisely the same fashion as theorems. The axiom scheme of definition is in the fortunate position of being justifiable in the sense that, if the definitions conform with the prescribed rules, then no new contradictions and no real enrichment of the theory results. These results are due to S. Lésniewski.

† Presumably other interpretations are also possible.

again a theorem, but there is no such substitution rule for constants.

I Axiom of extent * *For each* x *and each* y *it is true that* $x = y$ *if and only if for each* z, $z \varepsilon x$ *when and only when* $z \varepsilon y$.

Thus two classes are identical iff every member of each is a member of the other. We shall frequently omit 'for each x' or 'for each y' in the statement of a theorem or definition. If a variable, for example 'x,' occurs and is not preceded by 'for each x' or 'for some x' it is understood that 'for each x' is to be prefixed to the theorem or definition in question.

The first definition assigns a special name to those classes which are themselves members of classes. The reason for this dichotomy among classes is discussed a little later.

1 Definition x *is a set iff for some* y, $x \varepsilon y$.

The next task is to describe the use of the classifier. The first blank in the classifier constant is to be occupied by a variable and the second by a formula, for example $\{x: x \varepsilon y\}$. We accept as an axiom the statement: $u \varepsilon \{x: x \varepsilon y\}$ iff u is a set and $u \varepsilon y$. More generally, each statement of the following form is supposed to be an axiom: $u \varepsilon \{x: \cdots x \cdots\}$ iff u is a set and $\cdots u \cdots$. Here '$\cdots x \cdots$' is supposed to be a formula and '$\cdots u \cdots$' is supposed to be the formula which is obtained from it by replacing every occurrence of 'x' by 'u.' Thus $u \varepsilon \{x: x \varepsilon y \text{ and } z \varepsilon x\}$ iff u is a set and $u \varepsilon y$ and $z \varepsilon u$.

This axiom scheme is precisely the usual intuitive construction of classes except for the requirement 'u is a set.' This requirement is very evidently unnatural and is intuitively quite undesirable. However, without it a contradiction may be constructed simply on the basis of the axiom of extent. (See theorem 39 and the discussion preceding it.) This complication, which necessitates a good bit of technical work on the existence of sets, is simply the price paid to avoid obvious inconsistencies. Less obvious inconsistencies may very possibly remain.

* One is tempted to make this the definition of equality, thus dispensing with one axiom and with all logical presuppositions about equality. This is perfectly feasible. However, there would be no unlimited substitution rule for equality and one would have to assume as an axiom: If $x \varepsilon z$ and $y = x$, then $y \varepsilon z$

THE CLASSIFICATION AXIOM SCHEME (*Continued*)

A precise statement of the classification axiom scheme requires a description of formulae. It is agreed that: *

(a) The result of replacing 'α' and 'β' by variables is, for each of the following, a formula.

$$\alpha = \beta \qquad \alpha \varepsilon \beta$$

(b) The result of replacing 'α' and 'β' by variables and 'A' and 'B' by formulae is, for each of the following, a formula

if A, then B A iff B it is false that A
A and B A or B
for every α, A for some α, A
$\beta \varepsilon \{\alpha: A\}$ $\{\alpha: A\} \varepsilon \beta$ $\{\alpha: A\} \varepsilon \{\beta: B\}$

Formulae are constructed recursively, beginning with the primitive formulae of (a) and proceeding via the constructions permitted by (b).

II Classification axiom-scheme *An axiom results if in the following 'α' and 'β' are replaced by variables, 'A' by a formula $\mathfrak{A}$ and 'B' by the formula obtained from $\mathfrak{A}$ by replacing each occurrence of the variable which replaced α by the variable which replaced β:*
For each β, $\beta \varepsilon \{\alpha: A\}$ if and only if β is a set and B.

ELEMENTARY ALGEBRA OF CLASSES

The axioms already stated permit the deduction of a number of theorems directly from logical results. The deduction is straightforward and only an occasional proof is given.

2 DEFINITION $x \cup y = \{z: z \varepsilon x$ or $z \varepsilon y\}$.

3 DEFINITION $x \cap y = \{z: z \varepsilon x$ and $z \varepsilon y\}$.

* This circuitous sort of language is unfortunately necessary. Using the convention o quotation marks for names, for example 'Boston' is the name of Boston, if $\mathfrak{A}$ is a formula and $\mathfrak{B}$ is a formula, then '$\mathfrak{A} \to \mathfrak{B}$' is not a formula. For example, if $\mathfrak{A}$ is '$x = y$' and $\mathfrak{B}$ is '$y = z$', then ''$x = y$' $\to$ '$y = z$'' is not a formula. Formulae (for example '$x = y$') contain no quotation marks. Instead of '$\mathfrak{A} \to \mathfrak{B}$' we want to discuss the result of replacing 'α' by $\mathfrak{A}$ and 'β' by $\mathfrak{B}$ in '$\alpha \to \beta$.' This sort of circumlocution can be avoided by using Quine's corner convention

The class $x \cup y$ is the *union* of x and y, and $x \cap y$ is the *intersection* of x and y.

4 THEOREM *$z \in x \cup y$ if and only if $z \in x$ or $z \in y$, and $z \in x \cap y$ if and only if $z \in x$ and $z \in y$.*

PROOF From the classification axiom $z \in x \cup y$ iff $z \in x$ or $z \in y$ and z is a set. But in view of the definition 1 of set, $z \in x$ or $z \in y$ and z is a set iff $z \in x$ or $z \in y$. A similar argument proves the corresponding result for intersection. ▮

5 THEOREM *$x \cup x = x$ and $x \cap x = x$.*

6 THEOREM *$x \cup y = y \cup x$ and $x \cap y = y \cap x$.*

7 THEOREM* *$(x \cup y) \cup z = x \cup (y \cup z)$ and $(x \cap y) \cap z = x \cap (y \cap z)$.*

These theorems state that union and intersection are, in the usual sense, commutative and associative operations. The distributive laws follow.

8 THEOREM *$x \cap (y \cup z) = (x \cap y) \cup (x \cap z)$ and $x \cup (y \cap z) = (x \cup y) \cap (x \cup z)$.*

9 DEFINITION *$x \notin y$ if and only if it is false that $x \in y$.*

10 DEFINITION $\sim x = \{y : y \notin x\}$.

The class $\sim x$ is the *complement* of x.

11 THEOREM $\sim(\sim x) = x$.

12 THEOREM (DE MORGAN) *$\sim(x \cup y) = (\sim x) \cap (\sim y)$ and $\sim(x \cap y) = (\sim x) \cup (\sim y)$.*

PROOF Only the first of the two statements will be proved. For each z, $z \in \sim(x \cup y)$ iff z is a set and it is false that $z \in x \cup y$, because of the classification axiom and the definition 10 of complement. Using theorem 4, $z \in x \cup y$ iff $z \in x$ or $z \in y$. Consequently, $z \in \sim(x \cup y)$ iff z is a set and $z \notin x$ and $z \notin y$; that is, iff $z \in \sim x$ and $z \in \sim y$. Using 4 again, $z \in \sim(x \cup y)$ iff $z \in (\sim x)$

* There would be no necessity for parentheses if the constant '$\cup$' occurred first in the definition; that is, '$\cup xy$' instead of '$x \cup y$.' In this case the first part of the theorem would read: $\cup \cup xyz = \cup x \cup yz$.

$\cap (\sim y)$. Hence $\sim(x \cup y) = (\sim x) \cap (\sim y)$ because of the axiom of extent. ▮

13 Definition $x \sim y = x \cap (\sim y)$.

The class $x \sim y$ is the *difference* of x and y, or the *complement* of y relative to x.

14 Theorem $x \cap (y \sim z) = (x \cap y) \sim z$.

The proposition '$x \cup (y \sim z) = (x \cup y) \sim z$' is unlikely, although at this stage it is impossible to exhibit a counter example. To be a little more precise, the negation of the proposition cannot be proved on the basis of the axioms so far assumed; it is possible to make a model for this initial part of the system such that $x \notin y$ for each x and each y (there are no sets). The negation of the proposition can be proved on the basis of axioms which will presently be assumed.

15 Definition $0 = \{x: x \neq x\}$.

The class 0 is the *void class*, or *zero*.

16 Theorem $x \notin 0$.

17 Theorem $0 \cup x = x$ *and* $0 \cap x = 0$.

18 Definition $\mathcal{U} = \{x: x = x\}$.

The class $\mathcal{U}$ is the *universe*.

19 Theorem $x \in \mathcal{U}$ *if and only if* x *is a set.*

20 Theorem $x \cup \mathcal{U} = \mathcal{U}$ *and* $x \cap \mathcal{U} = x$.

21 Theorem $\sim 0 = \mathcal{U}$ *and* $\sim \mathcal{U} = 0$.

22 Definition* $\bigcap x = \{z:$ for each y, if $y \in x$, then $z \in y\}$.

23 Definition $\bigcup x = \{z:$ for some y, $z \in y$ and $y \in x\}$.

The class $\bigcap x$ is the *intersection* of the members of x. Note that the members of $\bigcap x$ are members of members of x and may or may not belong to x. The class $\bigcup x$ is the *union* of the mem-

* A bound variable notation for the intersection of the members of a family is not needed in this appendix, and consequently a notation is adopted which is simpler than that used in the rest of the book.

bers of x. Observe that a set z belongs to $\bigcap x$ (or to $\bigcup x$) iff z belongs to every (respectively, to some) member of x.

24 THEOREM $\bigcap 0 = \mathcal{U}$ *and* $\bigcup 0 = 0$.

PROOF $z \varepsilon \bigcap 0$ iff z is a set and z belongs to each member of 0. Since (theorem 16) there is no member of 0, $z \varepsilon \bigcap 0$ iff z is a set, and by 19 and the axiom of extent $\bigcap 0 = \mathcal{U}$. The second assertion is also easy to prove. ∎

25 DEFINITION $x \subset y$ *iff for each* z, *if* $z \varepsilon x$, *then* $z \varepsilon y$.

A class x is a *subclass* of y, or is *contained in* y, or y *contains* x iff $x \subset y$. It is absolutely essential that '$\subset$' not be confused with 'ε.' For example, $0 \subset 0$ but it is false that $0 \varepsilon 0$.

26 THEOREM $0 \subset x$ *and* $x \subset \mathcal{U}$.

27 THEOREM $x = y$ *iff* $x \subset y$ *and* $y \subset x$.

28 THEOREM *If* $x \subset y$ *and* $y \subset z$, *then* $x \subset z$.

29 THEOREM $x \subset y$ *iff* $x \cup y = y$.

30 THEOREM $x \subset y$ *iff* $x \cap y = x$.

31 THEOREM *If* $x \subset y$, *then* $\bigcup x \subset \bigcup y$ *and* $\bigcap y \subset \bigcap x$.

32 THEOREM *If* $x \varepsilon y$, *then* $x \subset \bigcup y$ *and* $\bigcap y \subset x$.

The preceding definitions and theorems are used very frequently—often without explicit reference.

EXISTENCE OF SETS

This section is concerned with the existence of sets and with the initial steps in the construction of functions and other relations from the primitives of set theory.

III Axiom of subsets *If* x *is a set there is a set* y *such that for each* z, *if* $z \subset x$, *then* $z \varepsilon y$.

33 THEOREM *If* x *is a set and* $z \subset x$, *then* z *is a set.*

PROOF According to the axiom of subsets, if x is a set there is y such that, if $z \subset x$, then $z \varepsilon y$, and hence by the definition 1, z is a set. (Observe that this proof does not use the full strength of the axiom of subsets since the argument does not require that y be a set.) ∎

34 THEOREM $0 = \bigcap \mathcal{U}$ *and* $\mathcal{U} = \bigcup \mathcal{U}$.

PROOF If $x \in \bigcap \mathcal{U}$, then x is a set and since $0 \subset x$ it follows from 33 that 0 is a set. Then $0 \in \mathcal{U}$ and each member of $\bigcap \mathcal{U}$ belongs to 0. It follows that $\bigcap \mathcal{U}$ has no members. Clearly (that is, theorem 26) $\bigcup \mathcal{U} \subset \mathcal{U}$. If $x \in \mathcal{U}$, then x is a set and by the axiom of subsets there is a set y such that, if $z \subset x$, then $z \in y$. In particular $x \in y$, and since $y \in \mathcal{U}$ it follows that $x \in \bigcup \mathcal{U}$. Consequently $\mathcal{U} \subset \bigcup \mathcal{U}$ and equality follows. ∎

35 THEOREM *If* $x \neq 0$, *then* $\bigcap x$ *is a set.*

PROOF If $x \neq 0$, then for some y, $y \in x$. But y is a set and since $\bigcap x \subset y$ by 32 it follows from 33 that $\bigcap x$ is a set. ∎

36 DEFINITION $2^x = \{y : y \subset x\}$.

37 THEOREM $\mathcal{U} = 2^{\mathcal{U}}$.

PROOF Every member of $2^{\mathcal{U}}$ is a set and consequently belongs to $\mathcal{U}$. Each member of $\mathcal{U}$ is a set and is contained in $\mathcal{U}$ (theorem 26) and hence belongs to $2^{\mathcal{U}}$. ∎

38 THEOREM *If x is a set, then 2^x is a set, and for each y, $y \subset x$ iff $y \in 2^x$.*

It is interesting to notice that the existence of sets is not provable on the basis of the axioms so far enunciated, but it is possible to prove that there is a class which is not a set. Letting R be $\{x : x \notin x\}$, by the classifier axiom $R \in R$ iff $R \notin R$ and R is a set. It follows that R is not a set. Observe that, if the classifier axiom did not contain the "is a set" qualification, then an outright contradiction, $R \in R$ iff $R \notin R$, would result. This is the Russell paradox. A consequence of this argument is that $\mathcal{U}$ is not a set, because $R \subset \mathcal{U}$ and 33 applies. (The regularity axiom will imply that $R = \mathcal{U}$; this axiom also yields a different proof that $\mathcal{U}$ is not a set.)

39 THEOREM $\mathcal{U}$ *is not a set.*

40 DEFINITION $\{x\} = \{z\colon \text{if } x \in \mathcal{U}, \text{ then } z = x\}$.

Singleton x is $\{x\}$.

This definition is an example of a technical device which is very convenient. If x is a set, then $\{x\}$ is a class whose only member is x. However, if x is not a set, then $\{x\} = \mathcal{U}$ (these statements are theorems 41 and 43). Actually, the primary interest is in the case where x is a set, and for this case the same result is given by the more natural definition $\{z\colon z = x\}$. However, it simplifies statements of results greatly if computations are arranged so that $\mathcal{U}$ is the result of applying the computation outside its pertinent domain.

41 THEOREM *If x is a set, then, for each y, $y \in \{x\}$ iff $y = x$.*

42 THEOREM *If x is a set, then $\{x\}$ is a set.*

PROOF If x is a set $\{x\} \subset 2^x$ and 2^x is a set. ▌

43 THEOREM $\{x\} = \mathcal{U}$ *if and only if x is not a set.*

PROOF If x is a set, then $\{x\}$ is a set and consequently is not equal to $\mathcal{U}$. If x is not a set, then $x \notin \mathcal{U}$ and $\{x\} = \mathcal{U}$ by the definition. ▌

44 THEOREM *If x is a set, then $\bigcap\{x\} = x$ and $\bigcup\{x\} = x$; if x is not a set, then $\bigcap\{x\} = 0$ and $\bigcup\{x\} = \mathcal{U}$.*

PROOF Use 34 and 41. ▌

IV Axiom of union *If x is a set and y is a set so is $x \cup y$.*

45 DEFINITION $\{xy\} = \{x\} \cup \{y\}$.

The class $\{xy\}$ is an *unordered pair.*

46 THEOREM *If x is a set and y is a set, then $\{xy\}$ is a set and $z \in \{xy\}$ iff $z = x$ or $z = y$; $\{xy\} = \mathcal{U}$ if and only if x is not a set or y is not a set.*

47 THEOREM *If x and y are sets, then $\bigcap\{xy\} = x \cap y$ and $\bigcup\{xy\} = x \cup y$; if either x or y is not a set, then $\bigcap\{xy\} = 0$ and $\bigcup\{xy\} = \mathcal{U}$.*

ORDERED PAIRS: RELATIONS

This section is devoted to the properties of ordered pairs and relations. The crucial property for ordered pairs is theorem 55: if x and y are sets, then $(x,y) = (u,v)$ iff $x = u$ and $y = v$.

48 DEFINITION $(x,y) = \{\{x\}\{xy\}\}$.

The class (x,y) is an *ordered pair.*

49 THEOREM *(x,y) is a set if and only if x is a set and y is a set; if (x,y) is not a set, then $(x,y) = \mathcal{U}$.*

50 THEOREM *If x and y are sets, then $\bigcup(x,y) = \{xy\}$, $\bigcap(x,y) = \{x\}$, $\bigcup\bigcap(x,y) = x$, $\bigcap\bigcap(x,y) = x$, $\bigcup\bigcup(x,y) = x \cup y$ and $\bigcap\bigcup(x,y) = x \cap y$.*

If either x or y is not a set, then $\bigcup\bigcap(x,y) = 0$, $\bigcap\bigcap(x,y) = \mathcal{U}$, $\bigcup\bigcup(x,y) = \mathcal{U}$, and $\bigcap\bigcup(x,y) = 0$.

51 DEFINITION 1st coord $z = \bigcap\bigcap z$.

52 DEFINITION 2nd coord $z = (\bigcap\bigcup z) \cup ((\bigcup\bigcup z) \sim \bigcup\bigcap z)$.

These definitions will be used, with one exception, only in the case where z is an ordered pair. The *first coordinate* of z is 1st coord z and the *second coordinate* of z is 2nd coord z.

53 THEOREM 2nd coord $\mathcal{U} = \mathcal{U}$.

54 THEOREM *If x and y are sets* 1st coord $(x,y) = x$ *and* 2nd coord $(x,y) = y$. *If either of x and y is not a set, then* 1st coord $(x,y) = \mathcal{U}$ *and* 2nd coord $(x,y) = \mathcal{U}$.

PROOF If x and y are sets, then the equality for 1st *coord* is immediate from 50 and 51. The equality for 2nd *coord* reduces to showing that $y = (x \cap y) \cup ((x \cup y) \sim x)$, by 50 and 52. It is straightforward to see that $(x \cup y) \sim x = y \sim x$ and by the distributive law $(y \cap x) \cup (y \cap \sim x)$ is $y \cap (x \cup \sim x) = y \cap \mathcal{U} = y$. If either x or y is not a set, then, using 50 it is easy to compute 1st *coord* (x,y) and 2nd *coord* (x,y). ∎

55 THEOREM *If x and y are sets and $(x,y) = (u,v)$, then $x = u$ and $y = v$.*

56 DEFINITION *r is a relation if and only if for each member z of r there is x and y such that $z = (x,y)$.*

A *relation* is a class whose members are ordered pairs.

57 DEFINITION $r \circ s = \{u$: for some x, some y and some z, $u = (x,z)$, $(x,y) \in s$ and $(y,z) \in r\}$.

The class $r \circ s$ is the *composition* of r and s.

To avoid excessive notation we agree that $\{(x,z): \cdots\}$ is to be identical with $\{u$: *for some x, some z, $u = (x,z)$ and* $\cdots\}$. Thus $r \circ s = \{(x,z)$: *for some y, $(x,y) \in s$ and $(y,z) \in r$*$\}$.

58 THEOREM $(r \circ s) \circ t = r \circ (s \circ t)$.

59 THEOREM $r \circ (s \cup t) = (r \circ s) \cup (r \circ t)$ *and* $r \circ (s \cap t) \subset (r \circ s) \cap (r \circ t)$.

60 DEFINITION $r^{-1} = \{(x,y): (y,x) \in r\}$.

If r is a relation r^{-1} is the *relation inverse to r*.

61 THEOREM $(r^{-1})^{-1} = r$.

62 THEOREM $(r \circ s)^{-1} = s^{-1} \circ r^{-1}$.

FUNCTIONS

Intuitively, a function is to be identical with the class of ordered pairs which is its graph. All functions are single-valued, and consequently two distinct ordered pairs belonging to a function must have different first coordinates.

63 DEFINITION *f is a function if and only if f is a relation and for each x, each y, each z, if $(x,y) \in f$ and $(x,z) \in f$, then $y = z$.*

64 THEOREM *If f is a function and g is a function so is $f \circ g$.*

65 DEFINITION domain $f = \{x$: for some y, $(x,y) \in f\}$.

66 DEFINITION range $f = \{y$: for some x, $(x,y) \in f\}$.

67 THEOREM domain $\mathcal{U} = \mathcal{U}$ *and* range $\mathcal{U} = \mathcal{U}$.

PROOF If $x \in \mathcal{U}$, then $(x,0)$ and $(0,x)$ belong to $\mathcal{U}$ and hence x belongs to *domain* $\mathcal{U}$ and *range* $\mathcal{U}$. ∎

68 DEFINITION $f(x) = \bigcap\{y\colon (x,y) \in f\}$.

Hence $z \in f(x)$ if z belongs to the second coordinate of each member of f whose first coordinate is x.

The class $f(x)$ is the *value* of f at x or the *image* of x under f. It is to be noticed that if x is a subset of *domain* f, then $f(x)$ is not $\{y\colon$ *for some* z, $z \in x$ *and* $y = f(z)\}$.

69 THEOREM *If* $x \notin$ domain f, *then* $f(x) = \mathcal{U}$; *if* $x \in$ domain f, *then* $f(x) \in \mathcal{U}$.

PROOF If $x \notin$ *domain* f, then $\{y\colon (x,y) \in f\} = 0$, and $f(x) = \mathcal{U}$ (theorem 24). If $x \in$ *domain* f, then $\{y\colon (x,y) \in f\} \neq 0$ and (theorem 35) $f(x)$ is a set. ▮

The foregoing theorem does not require that f be a function.

70 THEOREM *If* f *is a function, then* $f = \{(x,y)\colon y = f(x)\}$.

71 THEOREM * *If* f *and* g *are functions, then* $f = g$ *if and only if* $f(x) = g(x)$ *for each* x.

The two following axioms † further delineate the class of all sets.

V Axiom of substitution *If* f *is a function and* domain f *is a set, then* range f *is a set.*

VI Axiom of amalgamation *If* x *is a set so is* $\bigcup x$.

72 DEFINITION $x \times y = \{(u,v)\colon u \in x$ *and* $v \in y\}$.

The class $x \times y$ is the *cartesian product* of x and y.

73 THEOREM *If* u *and* y *are sets so is* $\{u\} \times y$.

PROOF Clearly one can construct a function (namely, $\{(w,z)\colon$ $w \in y$ *and* $z = (u,w)\}$) whose domain is y and whose range is $\{u\} \times y$. Then apply the axiom of substitution. ▮

* This theorem would not be true if $f(x)$ had been defined to be the union of the second coordinates of the members of f with first coordinate x. For then, if $y \in \mathcal{U}$ and $y \notin$ *domain* f, then $f(y) = 0$, and, if $g = f \cup \{(y,0)\}$, then $g(x) = f(x)$ for each x and f is not equal to g.

† These two axioms may be replaced by the single axiom: if f is a function and domain f is a set, then $\bigcup$ range f is a set. (In the bound variable notation used earlier in the book this can be stated very naturally: if d is a set and $x(a)$ is a set for each a in d, then $\bigcup\{x(a)\colon a \in d\}$ is a set.) To obtain V and VI from the above one may proceed roughly as follows: For V, given f make a new function whose members are of the form $(x,\{f(x)\})$. For VI, given x consider the function whose members are of the form (u,u) with u in x.

74 THEOREM *If x and y are sets so is $x \times y$.*

PROOF Let f be the function such that *domain* $f = x$ and $f(u) = \{u\} \times y$ for u in x. (There is a unique function of this sort; namely, $f = \{(u,z)\colon u \in x \text{ and } z = \{u\} \times y\}$.) Because of the axiom of substitution, *range* f is a set. By a direct computation range $f = \{z\colon \text{for some } u,\ u \in x \text{ and } z = \{u\} \times y\}$. Consequently $\bigcup$ *range* f, which by the axiom of amalgamation is a set, is $x \times y$. ▮

75 THEOREM *If f is a function and* domain *f is a set, then f is a set.*

PROOF For $f \subset (\text{domain } f) \times (\text{range } f)$. ▮

76 DEFINITION $y^x = \{f\colon f \text{ is a function, domain } f = x \text{ and range } f \subset y\}$.

77 THEOREM *If x and y are sets so is y^x.*

PROOF If $f \in y^x$, then $f \subset x \times y$, which is a set, and hence $f \in 2^{x \times y}$ (theorem 38) and $2^{x \times y}$ is a set. Since $y^x \subset 2^{x \times y}$ it follows from the axiom of subsets that y^x is a set. ▮

For convenience, three more definitions are made.

78 DEFINITION *f is on x if and only if f is a function and $x =$* domain *f.*

79 DEFINITION *f is to y if and only if f is a function and* range *$f \subset y$.*

80 DEFINITION *f is onto y if and only if f is a function and* range *$f = y$.*

WELL ORDERING

Many of the results of this section are not needed in the development of the integers, ordinals, and cardinals which follows. They are included here because they are interesting in themselves and because the methods are simplified forms of the constructions used later.

Since the basic constructive results have now been proved we are able to assume a somewhat less pedestrian pace.

81 Definition *$x\,r\,y$ if and only if $(x,y) \in r$.*

If $x\,r\,y$, then x is *r-related* to y or x *r-precedes* y.

82 Definition *r connects x if and only if when u and v belong to x either $u\,r\,v$ or $v\,r\,u$ or $v = u$.*

83 Definition *r is transitive in x if and only if, when u, v, and w are members of x and $u\,r\,v$ and $v\,r\,w$, then $u\,r\,w$.*

If r is transitive in x, then r *orders* x. The terminology 'u r-precedes v' is especially suggestive if u and v belong to x and r orders x.

84 Definition *r is asymmetric in x if and only if, when u and v are members of x and $u\,r\,v$, then it is not true that $v\,r\,u$.*

Restated, if $u \in x$ and $v \in x$ and u r-precedes v, then v does not r-precede u.

85 Definition *$x \neq y$ if and only if it is false that $x = y$.*

86 Definition *z is an r-first member of x if and only if $z \in x$ and if $y \in x$, then it is false that $y\,r\,z$.*

87 Definition *r well-orders x if and only if r connects x and if $y \subset x$ and $y \neq 0$, then there is an r-first member of y.*

88 Theorem *If r well-orders x, then r is transitive in x and r is asymmetric in x.*

PROOF If $u \in x$, $v \in x$, $u\,r\,v$, and $v\,r\,u$, then $\{uv\} \subset x$ and consequently there is an r-first member z of $\{uv\}$. Either $z = u$ or $z = v$, and hence it is either false that $v\,r\,u$ or that $u\,r\,v$. This contradiction shows that r is asymmetric in x. If r fails to be transitive in x, then for some members u, v, and w of x it is true that $u\,r\,v$, $v\,r\,w$, and $w\,r\,u$, since r connects x. But then $\{u\} \cup \{v\} \cup \{w\}$ fails to have an r-first member. ▮

89 Definition *y is an r-section of x if and only if $y \subset x$, r well-orders x, and for each u and v such that $u \in x$, $v \in y$, and $u\,r\,v$ it is true that $u \in y$.*

That is, a subset y of x is an r-section of x iff r well-orders x and no member of $x \sim y$ r-precedes a member of y.

90 THEOREM *If $n \neq 0$ and each member of n is an r-section of x, then $\bigcup n$ and $\bigcap n$ are r-sections of x.*

91 THEOREM *If y is an r-section of x and $y \neq x$, then $y =$ $\{u\colon u \in x$ and $u\, r\, v\}$ for some v in x.*

PROOF If y is an r-section of x and $y \neq x$, then there is an r-first member v of $x \sim y$. If $u \in x$ and $u\, r\, v$, then, since v is the r-first member of $x \sim y$, $u \notin x \sim y$ and hence $u \in y$. Therefore $\{u\colon u \in x$ *and* $u\, r\, v\} \subset y$. On the other hand, if $u \in y$, then since $v \notin y$ and y is an r-section, it is false that $v\, r\, u$ and hence it is true that $u\, r\, v$. The required equality follows. ∎

92 THEOREM *If x and y are r-sections of z, then $x \subset y$ or $y \subset x$.*

93 DEFINITION * *f is r-s order preserving if and only if f is a function, r well-orders* domain f, *s well-orders* range f, *and $f(u)\, s\, f(v)$ whenever u and v are members of* domain f *such that $u\, r\, v$.*

94 THEOREM *If x is an r-section of y and f is an r-r order-preserving function on x to y, then for each u in x it is false that $f(u)\, r\, u$.*

PROOF It must be shown that $\{u\colon u \in x$ *and* $f(u)\, r\, u\}$ is void. If not there is an r-first member v of this class. Then $f(v)\, r\, v$, and if $u\, r\, v$, then $u\, r\, f(u)$ or $u = f(u)$. Since $f(v)\, r\, v$, then $f(v)\, r\, f(f(v))$ or $f(v) = f(f(v))$, but since f is r-r order preserving $f(f(v))\, r\, f(v)$ and this is a contradiction. ∎

Thus an r-r order-preserving function on an r-section cannot map a member of its domain into an r-predecessor.

A proof such as that of theorem 94 which depends on considering the r-first element for which the theorem fails is a *proof by induction.*

95 DEFINITION *f is a* 1-1 *function if and only if both f and f^{-1} are functions.*

This is the equivalent to the statement that f is a function and if x and y are distinct members of *domain* f, then $f(x) \neq f(y)$.

96 THEOREM *If f is r-s order preserving, then f is a* 1-1 *function and f^{-1} is s-r order preserving.*

* In this appendix there is no need to consider order-preserving functions (as in chapter 0) whose domain and range are not well-ordered. For the sake of simplicity the earlier terminology is modified.

PROOF If $f(u) = f(v)$, then it is impossible that $u\,r\,v$ or $v\,r\,u$, for in this case $f(u)\,s\,f(v)$ or $f(v)\,s\,f(u)$. Hence $u = v$ and f is 1-1. If $f(u)\,s\,f(v)$, then $u \neq v$, and if $v\,r\,u$, then $f(v)\,s\,f(u)$, which is a contradiction. Therefore f^{-1} is s-r order preserving. ▌

97 THEOREM *If f and g are r-s order preserving,* domain f *and* domain g *are r-sections of x and* range f *and* range g *are s-sections of y, then $f \subset g$ or $g \subset f$.*

PROOF By theorem 92 either *domain* $f \subset$ *domain* g or *domain* $g \subset$ *domain* f, and the theorem will follow if it is proved that $f(u) = g(u)$ for all u belonging to the domain of both f and g. If the class $\{z: z \in (\text{domain } f) \cap (\text{domain } g)$ *and* $g(z) \neq f(z)\}$ is not empty there is an r-first member u. Then $f(u) \neq g(u)$ and it may be supposed that $f(u)\,s\,g(u)$. Since *range* g is an s-section, $g(v) = f(u)$ for some v in x and $v\,r\,u$ because g^{-1} is order preserving. But u is the r-first point at which the functions differ, and therefore $f(v) = g(v) = f(u)$ which is a contradiction. ▌

98 DEFINITION *f is r-s order preserving in x and y if and only if r well-orders x, s well-orders y, f is r-s order preserving,* domain f *is an r-section of x, and* range f *is an s-section of y.*

According to theorem 97, if f and g are both r-s order preserving in x and y, then $f \subset g$ or $g \subset f$.

99 THEOREM *If r well-orders x and s well-orders y, then there is a function f which is r-s order preserving in x and y such that either* domain $f = x$ *or* range $f = y$.

PROOF Let $f = \{(u,v)$: *$u \in x$, and for some function g which is r-s order preserving in x and y, $u \in$* domain g *and* $(u,v) \in g\}$. Because of the preceding theorem, f is a function, and it is easy to see that its domain is an r-section of x and its range is an s-section of y. Hence f is r-s order preserving in x and y and it remains to show that either *domain* $f = x$ or *range* $f = y$. If not, there is an r-first member u of $x \sim$ *(domain f)* and an s-first member v of $y \sim$ *(range f)*, and the function $f \cup \{(u,v)\}$ is easily seen to be r-s order preserving in x and y. Then $(u,v) \in f$ by definition of f and hence $u \in$ *domain* f. This is a contradiction. ▌

In one case it is possible to state which of the alternatives in the conclusion of the preceding theorem occurs, for if x is a set and y is not, then it is impossible that range $f = y$ because of the axiom of substitution.

100 THEOREM *If r well-orders x, s well-orders y, x is a set, and y is not a set, then there is a unique r-s order-preserving function in x and y whose domain is x.*

ORDINALS

In this section the ordinal numbers are defined and the fundamental properties established. Another axiom is assumed before beginning the discussion of ordinals.

It is a priori possible that there are classes x and y such that x is the only member of y and y is the only member of x. More generally, it is possible that there is a class z whose members exist by taking in each other's laundry, in the sense that every member of z consists of members of z. The following axiom explicitly denies this possibility by requiring that each non-void class z have at least one member whose elements do not belong to z.

VII Axiom of regularity *If $x \neq 0$ there is a member y of x such that $x \cap y = 0$.*

101 THEOREM $x \notin x$.

PROOF If $x \in x$, then x is a non-void set and is the only member of $\{x\}$. By the axiom of regularity there is y in $\{x\}$ such that $y \cap \{x\} = 0$, and necessarily $y = x$. But then $y \in y \cap \{x\}$, which is a contradiction. ▌

102 THEOREM *It is false that $x \in y$ and $y \in x$.*

PROOF If $x \in y$ and $y \in x$, then both x and y are sets and are the only members of $\{z: z = x \text{ or } z = y\}$. Applying the axiom of regularity to the latter class leads to a contradiction, just as in the proof of the preceding theorem. ▌

Of course, this theorem may be generalized to more than two sets. The axiom of regularity actually implies another strong result, intuitively stated as follows: it is impossible that there be a sequence such that $x_{n+1} \varepsilon x_n$ for each n. A precise statement of the result must be deferred.

103 DEFINITION $E = \{(x,y): x \varepsilon y\}$.

The class E is the ε-*relation*. Notice that if $x \varepsilon y$ and y is not a set, then $(x,y) = \mathfrak{U}$, by theorem 49, and $(x,y) \notin E$.

104 THEOREM *E is not a set.*

PROOF If $E \varepsilon \mathfrak{U}$, then $\{E\} \varepsilon \mathfrak{U}$ and $(E, \{E\}) \varepsilon E$. Recall that $(x,y) = \{\{x\}\{xy\}\}$ and, if (x,y) is a set, $z \varepsilon (x,y)$ iff $z = \{x\}$ or $z = \{xy\}$. Consequently $E \varepsilon \{E\} \varepsilon \{\{E\}\{E\{E\}\}\} \varepsilon E$. But if $a \varepsilon b \varepsilon c \varepsilon a$, then, upon application of the axiom of regularity to $\{x: x = a \text{ or } x = b \text{ or } x = c\}$, a contradiction results. ▮

An informal discussion of the structure of the first few ordinals may be conceptually enlightening.* The first ordinal will be 0, the next $1 = 0 \cup \{0\}$, the next $2 = 1 \cup \{1\}$, and the next $3 = 2 \cup \{2\}$. Observe 0 is the only member of 1, that 0 and 1 are the only members of 2, and 0, 1, and 2 are the only members of 3. Each ordinal preceding 3 is not only a member but also a subset of 3. Ordinals are defined so that this very special sort of structure results.

105 DEFINITION † *x is full iff each member of x is a subset of x.*

In other words, x is full iff each member of a member of x is a member of x.

The following definition is due to R. M. Robinson.

106 DEFINITION *x is an ordinal if and only if E connects x and x is full.*

That is, given two members of x, one is a member of the other, and each member of a member of x belongs to x.

107 THEOREM *If x is an ordinal E well-orders x.*

* The discussion is not precisely accurate, in that it has not been proved that 0 is a set; in fact, with the axioms at our disposal this is not provable. The existence of sets (and hence the fact that 0 is a set) results from the axiom of infinity, which is stated at the beginning of the next section.

† The term 'complete' is usually used instead of 'full,' but 'complete' has been used earlier in a different sense.

PROOF If u and v are members of x and $u\ E\ v$, then (theorem 102) it is false that $v\ E\ u$ and hence E is asymmetric in x. If y is a non-void subset of x, then by the axiom of regularity there is u in y such that $u \cap y = 0$. Then no member of y belongs to u and u is the E-first member of y. ▮

108 THEOREM *If x is an ordinal, $y \subset x$, $y \neq x$, and y is full, then $y \varepsilon x$.*

PROOF If $u\ E\ v$ and $v\ E\ y$, then $u\ E\ y$ because y is full. Hence y is an E-section of x. Consequently there is a member v of x such that $y = \{u: u \varepsilon x \text{ and } u\ E\ v\}$ by theorem 91. Since every member of v is a member of x, $y = \{u: u \varepsilon v\}$ and $y = v$. ▮

109 THEOREM *If x is an ordinal and y is an ordinal, then $x \subset y$ or $y \subset x$.*

PROOF The class $x \cap y$ is full and by the preceding theorem either $x \cap y = x$ or $x \cap y \varepsilon x$. In the first case $x \subset y$. If $x \cap y \varepsilon x$, then $x \cap y \not\varepsilon y$ since in this case $x \cap y \ \varepsilon\ x \cap y$. Since $x \cap y \not\varepsilon y$ the preceding theorem implies that $x \cap y = y$ and therefore $y \subset x$. ▮

110 THEOREM *If x is an ordinal and y is an ordinal, then $x \varepsilon y$ or $y \varepsilon x$ or $x = y$.*

111 THEOREM *If x is an ordinal and $y \varepsilon x$, then y is an ordinal.*

PROOF It is clear that E connects y because x is full and E connects x. The relation E is transitive on y because E well-orders x and $y \subset x$. Consequently if $u\ E\ v$ and $v\ E\ y$, then $u\ E\ y$ and hence y is full. ▮

112 DEFINITION $R = \{x: x \text{ is an ordinal}\}$.

113 THEOREM* *R is an ordinal and R is not a set.*

PROOF The last two theorems show that E connects R and that R is full; hence R is an ordinal. If R were a set, then $R \varepsilon R$ and this is impossible. ▮

In view of theorem 110, R is the only ordinal which is not a set.

* This theorem is essentially the statement of the Burali-Forti paradox—historically the first of the paradoxes of intuitive set theory.

114 THEOREM *Each E-section of R is an ordinal.*

PROOF If an E-section x of R is not equal to R, then by 91 there is a member v of R such that $x = \{u: u \varepsilon R \text{ and } u \varepsilon v\}$. Since each member of v is an ordinal, $x = \{u: u \varepsilon v\} = v$. ▮

115 DEFINITION *x is an ordinal number if and only if $x \varepsilon R$.*

116 DEFINITION *$x < y$ if and only if $x \varepsilon y$.*

117 DEFINITION *$x \leqq y$ if and only if $x \varepsilon y$ or $x = y$.*

118 THEOREM *If x and y are ordinals, then $x \leqq y$ if and only if $x \subset y$.*

119 THEOREM *If x is an ordinal, then $x = \{y: y \varepsilon R$ and $y < x\}$.*

120 THEOREM *If $x \subset R$, then $\bigcup x$ is an ordinal.*

PROOF That E connects $\bigcup x$ follows from theorems 110 and 111, and that $\bigcup x$ is full follows from the fact that members of x are full. ▮

It is not hard to see that if x is a subset of R, then the ordinal $\bigcup x$ is the first ordinal which is greater than or equal to each member of x, and that $\bigcup x$ is a set iff x is a set. These results will not be needed, however.

121 THEOREM *If $x \subset R$ and $x \neq 0$, then $\bigcap x \varepsilon x$.*

Indeed, in this case $\bigcap x$ is the E-first member of x.

122 DEFINITION $x + 1 = x \cup \{x\}$.

123 THEOREM *If $x \varepsilon R$, then $x + 1$ is the E-first member of $\{y: y \varepsilon R$ and $x < y\}$.*

PROOF It is easy to verify that E connects $x + 1$ and that $x + 1$ is full and is hence an ordinal. If there is u such that $x < u$ and $u < x + 1$, then since x is a set and $u \varepsilon x \cup \{x\}$ either $u \varepsilon x$ and $x \varepsilon u$ or $u = x$ and $x \varepsilon u$. Both of these conclusions are impossible (theorems 101 and 102) and the desired conclusion is established. ▮

124 THEOREM *If $x \varepsilon R$, then $\bigcup(x + 1) = x$.*

125 DEFINITION $f | x = f \cap (x \times \mathcal{U})$.

This definition will be used only in case f is a relation. In this case $f \mid x$ is a relation and is called the *restriction* of f to x.

126 THEOREM *If f is a function, $f \mid x$ is a function whose domain is $x \cap$ (domain f) and $(f \mid x)(y) = f(y)$ for each y in domain $f \mid x$.*

The final theorem on ordinals asserts that (intuitively) it is possible to define a function on an ordinal so that its value at any member of its domain is given by applying a predetermined rule to the earlier values of the function. A little more precisely, given g it is possible to find a unique function f on an ordinal such that $f(x) = g(f \mid x)$ for each ordinal number x. The value $f(x)$ is then completely determined by g and the values of f at ordinal numbers preceding x. Application of this theorem is *defining a function by transfinite induction.* The proof is similar to that of theorem 99 and the same sort of preliminary lemma is needed.

127 THEOREM *Let f be a function such that* domain f *is an ordinal and $f(u) = g(f \mid u)$ for u in* domain f. *If h is also a function such that* domain h *is an ordinal and $h(u) = g(h \mid u)$ for u in* domain h, *then $h \subset f$ or $f \subset h$.*

PROOF Since both *domain f* and *domain h* are ordinals it may be assumed that *domain $f \subset$ domain h* (either this or the reverse inclusion follows from 109) and it remains to be proved that $f(u) = h(u)$ for u in *domain f*. Assuming the contrary, let u be the E-first member of *domain f* such that $f(u) \neq h(u)$. Then $f(v) = h(v)$ for each ordinal v preceding u and hence $f \mid u = h \mid u$. Then $f(u) = g(f \mid u) = h(u)$ and this is a contradiction. ▌

128 THEOREM *For each g there is a unique function f such that* domain f *is an ordinal and $f(x) = g(f \mid x)$ for each ordinal number x.*

PROOF Let $f = \{(u,v)$: *$u \in R$ and there is a function h such that* domain h *is an ordinal, $h(z) = g(h \mid z)$ for z in* domain h *and* $(u,v) \in h\}$. From the preceding theorem it follows that f is a function, and it is evident that the domain of f is an E-section of R and is hence an ordinal. Moreover, if h is a function on an ordinal such that $h(z) = g(h|z)$ for z in *domain* h, then $h \subset f$, and if $z \varepsilon$ *domain* h, then $f(z) = g(f|z)$.

Finally, suppose $x \varepsilon R \sim$ (*domain* f). Then $f(x) = \mathcal{U}$ by theorem 69 and since *domain* f is a set f is a set (theorem 75). If $g(f \mid x) = g(f) = \mathcal{U}$, then the equality $f(x) = g(f \mid x)$ follows. Otherwise $g(f)$ is a set (theorem 69 again). In this case if y is the E-first member of $R \sim$ (*domain* f) and $h = f \cup \{(y,g(f))\}$, then the domain of h is an ordinal and $h(z) = g(h \mid z)$ for z in *domain* h. Hence $h \subset f$ and $y \varepsilon$ *domain* f which is a contradiction. Consequently, $g(f) = \mathcal{U}$ and the theorem is proved. ▌

The mechanics of this theorem deserves a little comment. If *domain* f is not R, then $g(f) = \mathcal{U}$ and $f(x) = \mathcal{U}$ for each ordinal number x such that *domain* $f \leqq x$. If $g(0) = \mathcal{U}$, then $f = 0$.

INTEGERS *

In this section the integers are defined and Peano's postulates are derived as theorems. The real numbers may be constructed from the integers (see Landau [1]) by use of these postulates and the two facts: the class of integers is a set (theorem 138), and it is possible to define a function on the integers by induction (theorem 0.13; this fact may also be derived as a corollary to 128). Another axiom is needed.

VIII Axiom of infinity *For some* y, y *is a set*, $0 \varepsilon y$ *and* $x \cup \{x\} \varepsilon y$ *whenever* $x \varepsilon y$.

In particular 0 is a set because 0 is contained in a set.

129 DEFINITION *x is an integer if and only if x is an ordinal and E^{-1} well-orders x.*

130 DEFINITION *x is an E-last member of y if and only if x is an E^{-1}-first member of y.*

131 DEFINITION $\omega = \{x: x \text{ is an integer}\}$.

132 THEOREM *A member of an integer is an integer.*

PROOF A member of an integer x is an ordinal and is a subset of x and x is well-ordered by E^{-1}. ▌

* **Non-negative integers.**

133 THEOREM *If $y \in R$ and x is an E-last member of y, then $y = x + 1$.*

PROOF By theorem 123, $x + 1$ is the E-first member of $\{z: z \in R \text{ and } x < z\}$. Then $x + 1 \leqq y$ because $y \in R$ and $x < y$. Since x is the E-last member of y and $x < x + 1$, it is false that $x + 1 < y$. ∎

134 THEOREM *If $x \in \omega$, then $x + 1 \in \omega$.*

135 THEOREM $0 \in \omega$ *and if $x \in \omega$, then* $0 \neq x + 1$.

That is, 0 is the successor of no integer.

136 THEOREM *If x and y are members of ω and $x + 1 = y + 1$, then $x = y$.*

PROOF By theorem 124, if $x \in R$, then $\bigcup(x + 1) = x$. ∎

The following theorem is the *principle of mathematical induction.*

137 THEOREM *If $x \subset \omega$, $0 \in x$ and $u + 1 \in x$ whenever $u \in x$, then $x = \omega$.*

PROOF If $x \neq \omega$ let y be the E-first member of $\omega \sim x$, and notice that $y \neq 0$. Since $y \subset y + 1$ and $y + 1$ is an integer there is an E-last member u of y, and clearly $u \in x$. Then $y = u + 1$ by theorem 133 and hence $y \in x$. This is a contradiction. ∎

Theorems 134, 135, 136, and 137 are Peano's axioms for integers. The next theorem implies that ω is a set.

138 THEOREM $\omega \in R$.

PROOF By the axiom of infinity there is a set y such that $0 \in y$ and, if $x \in y$, then $x + 1 \in y$. By mathematical induction (that is, the previous theorem) $\omega \cap y = \omega$, and hence ω is a set because $\omega \subset y$. Since ω consists of ordinal numbers, E connects ω and ω is full because each member of an integer is an integer. ∎

THE CHOICE AXIOM

We now state the last axiom and derive two powerful consequences.

139 Definition *c is a choice function if and only if c is a function and $c(x) \varepsilon x$ for each member x of* domain c.

Intuitively, a choice function is a simultaneous selection of a member from each set belonging to domain c.

The following is a strong form of Zermelo's postulate or the axiom of choice.

IX Axiom of choice *There is a choice function c whose domain is $\mathcal{U} \sim \{0\}$.*

The function c selects a member from every non-void set.

140 Theorem *If x is a set there is a* 1-1 *function whose range is x and whose domain is an ordinal number.*

PROOF The plan of proof is to construct, by transfinite induction, a function satisfying the requirements of the theorem. Let g be the function such that $g(h) = c(x \sim \textit{range } h)$ for each set h, where c is a choice function satisfying the axiom of choice. Applying theorem 128 there is a function f such that *domain* f is an ordinal and $f(u) = g(f \mid u)$ for each ordinal number u. Then $f(u) = c(x \sim \textit{range } (f \mid u))$, and if $u \;\varepsilon$ *domain* f, then $f(u) \;\varepsilon\; x \sim$ *range* $(f \mid u)$. Now f is 1-1, for $f(v) = f(u)$ and $u < v$, then $f(v) \;\varepsilon$ *range* $(f \mid v)$, which contradicts the fact that $f(v) \;\varepsilon\; x \sim$ *range* $(f \mid v)$. Since f is 1-1 it is impossible that *domain* $f = R$, for in this case f^{-1} is a function whose domain is a subclass of x and is hence a set, then *range* f^{-1} is a set because of the axiom of substitution and R is not a set. Consequently *domain* $f \;\varepsilon\; R$. Because *domain* $f \notin$ *domain* f, f (*domain* f) $= \mathcal{U}$ and therefore $c\ (x \sim \textit{range } f) = \mathcal{U}$. Since the domain of c is $\mathcal{U} \sim \{0\}$, $x \sim$ *range* $f = 0$. It follows immediately that f is a function satisfying the requirements of the theorem. ∎

141 Definition *n is a nest if and only if, whenever x and y are members of n, then $x \subset y$ or $y \subset x$.*

The next result is a lemma which is needed for the proof of theorem 143.

142 Theorem *If n is a nest and each member of n is a nest, then $\bigcup n$ is a nest.*

PROOF If $x \varepsilon m$, $m \varepsilon n$, $y \varepsilon p$, and $p \varepsilon n$, then either $m \subset p$ or $p \subset m$ because n is a nest. Suppose $m \subset p$. Then $x \varepsilon p$ and $y \varepsilon p$ and since p is a nest, $x \subset y$ or $y \subset x$. ▮

The following theorem is the *Hausdorff maximal principle.* It asserts the existence of a maximal nest in any set. The proof is closely related to that of 140.

143 THEOREM *If x is a set there is a nest n such that $n \subset x$ and if m is a nest, $m \subset x$, and $n \subset m$, then $m = n$.*

PROOF The proof is by transfinite induction; intuitively we select a nest and then a larger nest, and "keep going," knowing that, because R is not a set, the set of all nests which are contained in x will be exhausted before the class R of ordinals. For each h let $g(h) = c(\{m\colon m \textit{ is a nest}, m \subset x \textit{ and for } p \textit{ in } \text{range } h, p \subset m \textit{ and } p \neq m\})$, where c is a choice function satisfying the axiom of choice. (Intuitively select $g(h)$ to be a nest in x containing properly each previously selected nest.) By theorem 128 there is a function f such that *domain* f is an ordinal and $f(u) = g(f \mid u)$ for each ordinal number u. From the definition of g it follows that, if $u \varepsilon$ *domain* f, then $f(u) \subset x$ and $f(u)$ is a nest, and if u and v are members of *domain* f and $u < v$, then $f(u) \subset f(v)$ and $f(u) \neq f(v)$. Consequently f is 1-1, f^{-1} is a function and, since x is a set, *domain* $f \varepsilon R$. Since $f(\textit{domain } f) = \mathcal{U}$, $g(f) = \mathcal{U}$; consequently there is no nest m which is contained in x and properly contains each member of *range* f. Finally, $\bigcup(\textit{range } f)$ is a nest which contains every member of *range* f, and consequently there is no nest m which is contained in x and properly contains $\bigcup(\textit{range } f)$. ▮

CARDINAL NUMBERS

In this section cardinal numbers are defined and the most commonly used properties are proved. The proofs lean heavily on the earlier results.

144 DEFINITION *$x \approx y$ if and only if there is a* 1-1 *function f with* domain $f = x$ *and* range $f = y$.

If $x \approx y$, then x is *equivalent* to y, or x and y are *equipollent.*

145 Theorem $x \approx x$.

146 Theorem *If* $x \approx y$, *then* $y \approx x$.

147 Theorem *If* $x \approx y$ *and* $y \approx z$, *then* $x \approx z$.

148 Definition *x is a cardinal number if and only if x is an ordinal number and, if $y \varepsilon R$ and $y < x$, then it is false that $x \approx y$.*

That is, a cardinal number is an ordinal number which is not equivalent to any smaller ordinal.

149 Definition $C = \{x\colon x \text{ is a cardinal number}\}$.

150 Theorem *E well-orders C.*

151 Definition $P = \{(x,y)\colon x \approx y \text{ and } y \varepsilon C\}$.

The class P consists of all pairs (x,y) such that x is a set and y is a cardinal number equivalent to x. For each set x the cardinal number $P(x)$ is the *power* of x or the *cardinal* of x.

The basic facts needed for the following sequence of results have already been demonstrated.

152 Theorem *P is a function,* domain $P = \mathfrak{U}$ *and* range $P = C$.

PROOF Theorem 140 is the essential step. ▮

153 Theorem *If x is a set, then* $P(x) \approx x$.

154 Theorem *If x and y are sets, then $x \approx y$ if and only if* $P(x) = P(y)$.

155 Theorem $P(P(x)) = P(x)$.

PROOF If x is not a set, then $P(x) = \mathfrak{U}$ by theorem 69 and $P(\mathfrak{U}) = \mathfrak{U}$. ▮

156 Theorem *$x \varepsilon C$ if and only if x is a set and* $P(x) = x$.

157 Theorem *If $y \varepsilon R$ and $x \subset y$, then* $P(x) \leqq y$.

PROOF By theorem 99 there is a 1-1 function f which is E-E order preserving in x and R, such that *domain* $f = x$ or *range* $f = R$. Since x is a set and R is not, *domain* $f = x$. By theorem 94, $f(u) \leqq u$ for u in x and consequently x is equivalent to an ordinal less than or equal to y. ▮

158 THEOREM *If y is a set and $x \subset y$, then $P(x) \leqq P(y)$.*

The following is the Schroeder-Bernstein theorem. It can be proved directly without the axiom of choice (theorem 0.20).

159 THEOREM *If x and y are sets, $u \subset x$, $v \subset y$, $x \approx v$, and $y \approx u$, then $x \approx y$.*

PROOF Using 158, $P(x) = P(v) \leqq P(y) = P(u) \leqq P(x)$. ▮

160 THEOREM *If f is a function and f is a set, then $P(\text{range } f) \leqq P(\text{domain } f)$.*

PROOF If f is on x onto y and c is a choice function satisfying the choice axiom there is a function g such that *domain* $g = y$ and $g(v) = c(\{u : v = f(u)\})$ for v in y. Consequently y is equivalent to a subset of x. ▮

The following classic theorem is due to Cantor.

161 THEOREM *If x is a set, then $P(x) < P(2^x)$.*

PROOF The function, whose domain is x and whose value at a member u of x is $\{u\}$, is 1-1 and hence x is equivalent to a subset of 2^x and $P(x) \leqq P(2^x)$. If $P(x) = P(2^x)$ there is a 1-1 function f whose domain is x and range is 2^x. Then there is a member u of x such that $f(u) = \{v : v \varepsilon x \text{ and } v \not\varepsilon f(v)\}$. But then $u \varepsilon f(u)$ iff $u \not\varepsilon f(u)$, which is a contradiction. ▮

The foregoing is structurally similar to that of the Russell paradox.

162 THEOREM *C is not a set.*

PROOF If C is a set, then $\bigcup C$ is a set, $P(2^{\bigcup C}) \varepsilon C$ and hence $P(2^{\bigcup C}) \subset \bigcup C$. Therefore $P(2^{\bigcup C}) \leqq P(\bigcup C)$, which is a contradiction. ▮

After some preliminaries we divide the cardinals into two classes, the finite cardinals and the infinite cardinals, and prove a few special properties for each class.

163 THEOREM *If $x \varepsilon \omega$, $y \varepsilon \omega$ and $x + 1 \approx y + 1$, then $x \approx y$.*

PROOF If f is a 1-1 function on $x + 1$ onto $y + 1$ there is a 1-1 function g on $x + 1$ onto $y + 1$ such that $g(x) = y$; for example, let g be $(f \sim \{(x,f(x))\} \cup \{(f^{-1}(y),y)\}) \cup \{(f^{-1}(y),f(x))\} \cup \{(x,y)\}$. Then $g \mid x$ is a 1-1 function on x onto y. ∎

164 THEOREM $\omega \subset C$.

PROOF The proof is by induction. Apply the preceding theorem to the first integer which is equivalent to a smaller integer to obtain a contradiction, thus proving that each integer is a cardinal number. ∎

165 THEOREM $\omega \,\varepsilon\, C$.

PROOF If $\omega \approx x$ and $x \,\varepsilon\, \omega$, then $x \subset x + 1 \subset \omega$, and hence $P(x + 1) = P(x)$. This contradicts the preceding theorem, which states that each integer is a cardinal number. ∎

166 DEFINITION *x is finite if and only if $P(x) \,\varepsilon\, \omega$.*

167 THEOREM *x is finite if and only if there is r such that r well-orders x and r^{-1} well-orders x.*

PROOF If $P(x) \,\varepsilon\, \omega$, then E and E^{-1} well-order $P(x)$, and since $x \approx P(x)$ there is no difficulty finding r such that both r and r^{-1} well-order x. Conversely, if r and r^{-1} well-order x, then by 99 there is a 1-1 function f which is r-E order preserving in x and R such that either $domain\ f = x$ or $range\ f = R$. If $\omega \subset range\ f$, then r^{-1} does not well-order x because ω has no E last element. Consequently $range\ f \,\varepsilon\, \omega$, $domain\ f = x$, and the theorem follows. ∎

Each of the following sequence of theorems on finite sets can be proved by induction on the power of a set or by constructing a well ordering and applying 167. Examples of both sorts of proof will be given.

168 THEOREM *If x and y are finite so is $x \cup y$.*

PROOF If both r and r^{-1} well-order x and both s and s^{-1} well-order y, then, using r for points of x, s for points of $y \sim x$, and letting each member of $y \sim x$ follow every point of x, one can construct an ordering of the required type for $x \cup y$. ∎

169 THEOREM *If x is finite and each member of x is finite, then $\bigcup x$ is finite.*

PROOF One may proceed by induction on $P(x)$. Explicitly, consider the set s of all integers u such that, if $P(x) = u$ and each member of x is finite, then $\bigcup x$ is finite. Clearly 0 belongs to the set s. If $u \in s$, $P(x) = u + 1$, and each member of x is finite, then one may split x into two subsets, one of which has power u and one of which is a singleton. The induction hypothesis and the preceding theorem then show that $\bigcup x$ is finite. Hence $s = \omega$. ▮

170 THEOREM *If x and y are finite so is $x \times y$.*

PROOF The class $x \times y$ is the union of the members of a finite class, the members being of the form $\{v\} \times y$ for v in x. ▮

171 THEOREM *If x is finite so is 2^x.*

PROOF If y is an integer, then the subsets of $y + 1$ can be divided into two classes: those which are subsets of y, and those which are the union of a subset of y and $\{y\}$. This gives the necessary basis for an inductive proof of the theorem. ▮

172 THEOREM *If x is finite, $y \subset x$ and $P(y) = P(x)$, then $x = y$.*

PROOF It is sufficient to consider the case where x is an integer. Suppose $y \subset x$, $y \neq x$, $P(y) = x$, and $x \in \omega$. Then $x \neq 0$ and hence $x = u + 1$ for some integer u. Because $y \neq x$ there is a subset of u which is equivalent to y and hence $P(y) \leqq u$. But $P(y) = x = u + 1$, and this contradicts the fact that each integer is a cardinal number. ▮

The property of theorem 172, that a finite set is equivalent to no proper subset, actually characterizes finite sets.

173 THEOREM *If x is a set and x is not finite, then there is a subset y of x such that $y \neq x$ and $x \approx y$.*

PROOF Since x is a set and is not finite, $\omega \subset P(x)$. There is a function f on $P(x)$ such that $f(u) = u + 1$ for u in ω, and for $f(u) = u$ for u in $P(x) \sim \omega$. The function f is 1-1 and *range* $f = P(x) \sim \{0\}$. Since $P(x) \approx x$ the theorem follows. ▮

174 THEOREM *If $x \in R \sim \omega$, then $P(x + 1) = P(x)$.*

PROOF Clearly $P(x) \leqq P(x + 1)$. Since x is not finite there is a subset u of x such that $u \neq x$ and $u \approx x$. Consequently there is a 1-1 function f on $x + 1$ such that $f(y) \varepsilon u$ for y in x and $f(x) \varepsilon x \sim u$. Hence $P(x + 1) \leqq P(x)$. ▮

The principal remaining theorem depends on an order which will be assigned to the cartesian product $R \times R$. An intuitive description of this order may be instructive. It is to be a well ordering, and on $\omega \times \omega$ it is to have the property that the class of all predecessors of a member (x,y) of $\omega \times \omega$ is finite (a generalization of this fact is the key to the usefulness of the order). Picture $\omega \times \omega$ as a subset of the Euclidean plane and divide $\omega \times \omega$ into classes, putting in the same class pairs (x,y) and (u,v) such that the maximum of x and y is identical with the maximum of u and v. Each class then consists of two sides of a square, and the ordering is arranged so that points on smaller squares precede points on large squares. For points on the sides of the same square the ordering proceeds along the upper edge and to the right, up to but not including the corner point, and then along the right-hand edge upward, ending with the corner point.

If x and y are ordinals, the larger of them is $x \cup y$. This motivates the following definition.

175 DEFINITION $\max[x,y] = x \cup y$.

176 DEFINITION $\ll = \{z$: *for some* (u,v) *in* $R \times R$ *and some* (x,y) *in* $R \times R$, $z = ((u,v),(x,y))$, *and* $\max[u,v] < \max[x,y]$, *or* $\max[u,v] = \max[x,y]$ *and* $u < x$, *or* $\max[u,v] = \max[x,y]$ *and* $u = x$ *and* $v < y\}$.

177 THEOREM $\ll$ *well-orders* $R \times R$.

The proof is a straightforward but tedious application of the definition and the fact that $<$ well-orders R.

178 THEOREM *If* $(u,v) \ll (x,y)$, *then* $(u,v) \varepsilon (\max[x,y] + 1) \times (\max[x,y] + 1)$.

PROOF Surely $\max[u,v] \leqq \max[x,y]$, and hence $\max[u,v] \subset \max[x,y]$. Since the ordinals u and v are subsets of $\max[x,y]$ they are members of $\max[x,y] + 1$. ▮

179 THEOREM *If* $x \varepsilon C \sim \omega$, *then* $P(x \times x) = x$.

PROOF We proceed by induction, supposing x to be the first member of $C \sim \omega$ for which the theorem fails. There is by 99 a function f which is $\ll$-E order preserving in $x \times x$ and R, such that either *domain* $f = x \times x$ or *range* $f = R$. Since $x \times x$ is a set and R is not, *domain* $f = x \times x$. We show that, if $(u,v) \in x \times x$, then $f((u,v)) < x$, and the theorem follows. By the preceding theorem the class of all predecessors of (u,v) is a subset of $(\max [u,v] + 1) \times (\max [u,v] + 1)$. If $x = \omega$, then both u and v are finite because $\max [u,v] < x$; by 170, $(\max [u,v] + 1) \times (\max [u,v] + 1)$ is finite, hence $f((u,v))$ has only a finite number of predecessors and $f((u,v)) < x$. If $x \neq \omega$, and $\max [u,v]$ is not finite, then by 174, $P(\max [u,v] + 1) = P(\max [u,v]) < x$ and hence $P(f((u,v))) < x$ and $f((u,v)) < x$. ∎

180 Theorem *If x and y are members of C, one of which fails to belong to ω, then* $P(x \times y) = \max [P(x),P(y)]$.

The members of $C \sim \omega$ are called *infinite*, or *transfinite*, cardinal numbers.

There are many important and useful theorems on cardinal numbers which have not been given in the preceding list; see, for example, Fraenkel [1] for further information and references. This discussion will be concluded with a brief statement on one of the classic unsolved problems of set theory.

181 Theorem *There is a unique $<$-$<$ order-preserving function with domain R and range $C \sim \omega$.*

PROOF There is, by 99, a unique $<$-$<$ order-preserving function f in R and $C \sim \omega$ such that either *domain* $f = R$ or *range* $f = C \sim \omega$. Since every E-section of R and of $C \sim \omega$ is a set and neither R nor $C \sim \omega$ is a set, it is impossible that *domain* $f \neq R$ or *range* $f \neq C \sim \omega$. ∎

The unique $<$-$<$ order-preserving function whose existence is guaranteed by the previous theorem is usually denoted by $\aleph$. Thus $\aleph(0)$ (or $\aleph_0$) is ω. The next cardinal $\aleph_1$ is also denoted by Ω; it is the first uncountable ordinal. Since $P(2^{\aleph_0}) > \aleph_0$ it follows that $P(2^{\aleph_0}) \geq \aleph_1$. The equality of these two cardinals is an extremely attractive conjecture. It is called the *hypothesis of the continuum*. The *generalized hypothesis of the continuum is* the statement: if x is an ordinal number, then $P(2^{\aleph_x}) = \aleph_{x+1}$. Neither hypothesis has been proved or disproved. However,

Gödel [1] has proved the beautiful metamathematical theorem: If, on the basis of the hypothesis of the continuum, a contradiction is constructed, then a contradiction may be found without assuming the hypothesis of the continuum. The same situation prevails with respect to the generalized hypothesis of the continuum and the axiom of choice.

Bibliography

A. D. Alexandroff

[1] *On the extension of a Hausdorff space to an H-closed space*, C. R. (Doklady) Acad. Sci. U.R.S.S., N.S. **37** (1942) 118–121

P. Alexandroff

[1] *Sur les ensembles de la première classe et les ensembles abstraits*, C. R. Acad. Sci. Paris **178** (1924) 185–187

P. Alexandroff and H. Hopf

[1] *Topologie I*, Berlin (1935)

P. Alexandroff and P. Urysohn

[1] *Mémoire sur les espaces topologiques compactes*, Verh. Akad. Wetensch. Amsterdam **14** (1929) 1–96

[2] *Une condition necessaire et suffisante pour qu'une class* ($\mathfrak{L}$) *soit une class* ($\mathfrak{D}$), C. R. Acad. Sci. Paris **177** (1923) 1274–1277

A. Appert

[1] *Écart partielement ordonné et uniformité*, C. R. Acad. Sci. Paris **224** (1947) 442–444

A. Appert and Ky Fan

[1] *Espaces topologiques intermédiares. Problème de la distanciation* Actualités Sci. Ind. **1121** Paris (1951)

R. Arens

[1] *Note on convergence in topology*, Math. Mag. **23** (1950) 229–234

[2] *A topology for spaces of transformations*, Ann. of Math. (2) **47** (1946) 480–495

[3] *Topologies for homeomorphism groups*, Amer. J. Math. **68** (1946) 593–610

R. Arens and J. Dugundji

[1] *Remark on the concept of compactness*, Portugaliae Math. **9** (1950) 141–143

[2] *Topologies for function spaces*, Pacific J. Math. **1** (1951) 5–31

N. Aronszajn

[1] *Quelques remarques sur les relations entre les notions d'écart régulier et de distance*, Bull. Amer. Math. Soc. **44** (1938) 653–657

[2] *Über ein Urbildproblem*, Fund. Math. **17** (1931) 92–121

M. Balanzat

[1] *On the metrization of quasi-metric spaces*, Gaz. Mat., Lisboa 12, no. 50 (1951) 91–94

S. Banach

[1] *Théorie des opérations linéaires*, Warsaw (1932)

E. G. Begle

[1] *A note on S-spaces*, Bull. Amer. Math. Soc. **55** (1949) 577–579

R. H. Bing

[1] *Metrization of topological spaces*, Canadian J. Math. **3** (1951) 175–186

G. Birkhoff

[1] *Lattice Theory* (Revised Ed.), A.M.S. Colloquium Publ. XXV, New York (1948)

[2] *A note on topological groups*, Compositio Math. **3** (1936) 427–430

[3] *Moore-Smith convergence in general topology*, Ann. of Math. (2) **38** (1937) 39–56

N. Bourbaki

[1] *Topologie générale*, Actualités Sci. Ind. **858** (1940), **916** (1942), **1029** (1947), **1045** (1948), **1084** (1949), Paris

[2] *Intégration*, Actualités Sci. Ind. **1175**, Paris (1952)

[3] *Espace vectoriels topologiqùe*, Actualités Sci. Ind. **1189** Paris (1953)

N. Bourbaki and J. Dieudonné

[1] *Note de tératopologie II*, Revue Scientifique **77** (1939) 180–181

E. Čech

[1] *On bicompact spaces*, Ann. of Math. (2) **38** (1937) 823–844

C. Chevalley

[1] *Theory of Lie Groups I*, Princeton (1946)

E. W. Chittenden

[1] *On the metrization problem and related problems in the theory of abstract sets*, Bull. Amer. Math. Soc. **33** (1927) 13–34

H. J. Cohen

[1] *Sur un problème de M. Dieudonné*, C. R. Acad. Sci. Paris **234** (1952) 290–292

L. W. Cohen

[1] *On topological completeness*, Bull. Amer. Math. Soc. **46** (1940) 706–710

L. W. Cohen and C. Goffman

[1] *On the metrization of uniform space*, Proc. Amer. Math. Soc. **1** (1950) 750–753

J. Colmez

[1] *Espaces à écart généralisé régulier*, C. R. Acad. Sci. Paris **224** (1947) 372–373

M. M. Day

[1] *Convergence, closure and neighborhoods*, Duke Math. J. **11** (1944) 181–199

J. Dieudonné

[1] *Une généralization des espaces compacts*, J. Math. Pures Appl. **23** (1944) 65–76

[2] *Sur un espace localement compact non metrisable*, Anais da Acad. Bras. Ci. **19** (1947) 67–69

[3] *Sur la complétion des groupes topologiques*, C. R. Acad. Sci. Paris **218** (1944) 774–776

[4] *On topological groups of homeomorphisms*, Amer. J. Math. **70** (1948) 659–680

[5] *Un exemple d'espace normal non susceptible d'une structure uniforme d'espace complet*, C. R. Acad. Sci. Paris **209** (1939) 145–147

[6] *Sur les espaces uniformes complets*, Ann. Sci. École Norm. Sup. **56** (1939) 227–291

J. Dixmier

[1] *Sur certains espaces considérés par M. H. Stone*, Summa Brasil. Math. **2** (1951) 151–182

R. Doss

[1] *On uniform spaces with a unique structure*, Amer. J. Math. **71** (1949) 19–23

C. H. Dowker

[1] *An embedding theorem for paracompact metric spaces*, Duke Math. J. **14** (1947) 639–645

[2] *On countably paracompact spaces*, Canadian J. Math. **3** (1951) 219–244

[3] *On a theorem of Hanner*, Ark. Mat. **2** (1952) 307–313

J. Dugundji

[1] *An extension of Tietze's theorem*, Pacific J. Math. **1** (1951) 353–367

S. Eilenberg

[1] *Sur le théorème de décomposition de la théorie de la dimension*, Fund. Math. **26** (1936) 146–149

S. Eilenberg and N. Steenrod

[1] *Foundations of algebraic topology*, Princeton (1952)

W. T. van Est and H. Freudenthal

[1] *Trennung durch stetige functionen in topologischen Räumen*, Indagationes Math. **13** (1951) 359–368

M. K. Fort, Jr.

[1] *A note on pointwise convergence*, Proc. Amer. Math. Soc. **2** (1951) 34–35

R. H. Fox

[1] *On topologies for function spaces*, Bull. Amer. Math. Soc. **51** (1945) 429–432

A. Fraenkel

[1] *Einleitung in die Mengenlehre* (Amer. Ed.) New York (1946)

M. Fréchet

[1] *Sur quelques points du Calcul Fonctionnel* (*These*) Rendiconti di Palermo **22** (1906) 1–74

[2] *Les espaces abstractes*, Paris (1926)

H. Freudenthal

[1] *Neuaufbau der Endentheorie*, Ann. of Math. (2) **43** (1942) 261–279

A. H. Frink

[1] *Distance functions and the metrization problem*, Bull. Amer. Math. Soc. **43** (1937) 133–142

D. Gale

[1] *Compact sets of functions and function rings*, Proc. Amer. Math. Soc. **1** (1950) 303–308

K. Gödel

[1] *The consistency of the continuum hypothesis*, Ann. of Math. Studies **3** (1940)

A. P. Gomes

[1] *Topologie induite par un pseudo-diamètre*, C. R. Acad. Sci. Paris **227** (1948) 107–109

L. M. Graves

[1] *The theory of functions of real variables*, New York (1946)

A. Grothendieck

[1] *Critères de compacité dans les espaces fonctionnels généraux*, Amer. J. Math. **74** (1952) 168–186

W. Gustin

[1] *Countable connected spaces*, Bull. Amer. Math. Soc. **52** (1946) 101–106

P. R. Halmos

[1] *Measure theory*, New York (1950)

O. Hanner
[1] *Retraction and extension of mappings of metric and non-metric spaces*, Ark. Math. **2** (1952) 315–360
E. Hewitt
[1] *On two problems of Urysohn*, Ann. of Math. (2) **47** (1946) 503–509
[2] *Rings of real-valued continuous functions I*, Trans. Amer. Math. Soc. **64** (1948) 45–99
F. Hausdorff
[1] *Grundzüge der Mengenlehre*, Leipzig (1914)
[2] *Die Mengen G_δ in vollständigen Räumen*, Fund. Math. **6** (1924) 146–148
E. Hille
[1] *Functional analysis and semi-groups*, A.M.S. Colloquium Publ. XXI, New York (1948)
S. T. Hu
[1] *Archimedean uniform spaces and their natural boundedness*, Portugaliae Math. **6** (1947) 49–56
W. Hurewicz and H. Wallman
[1] *Dimension theory*, Princeton (1941)
K. Iseki
[1] *On definitions of topological space*, J. Osaka Inst. Sci. Tech. **1** (1949) 97–98
S. Kakutani
[1] *Über die Metrization der topologischen Gruppen*, Proc. Imp. Acad. Japan **12** (1936) 82–84
G. K. Kalisch
[1] *On uniform spaces and topological algebra*, Bull. Amer. Math. Soc. **52** (1946) 936–939
M. Katětov
[1] *On H-closed extensions of topological spaces*, Časopis Pěst. Mat. Fys. **72** (1947) 17–32
J. L. Kelley
[1] *Convergence in topology*, Duke Math. J. **17** (1950) 277–283
[2] *The Tychonoff product theorem implies the axiom of choice*, Fund. Math. **37** (1950) 75–76
V. L. Klee
[1] *Invariant metrics in groups (solution of a problem of Banach)*, Proc. Amer. Math. Soc. **3** (1953) 484–487
B. Knaster and C. Kuratowski
[1] *Sur les ensembles connexes*, Fund. Math. **2** (1921) 206–255

A. Kolmogoroff
[1] *Zur Normierbarkeit eines allgemeinen topologischen linearen Räumes*, Studia Math. **5** (1934) 29–33

G. Köthe
[1] *Die Quotientenräume eines linearen vollkommenen Räumes*, Math. Z. **51** (1947) 17–35

S. B. Krishna Murti
[1] *A set of axioms for topological algebra*, J. Indian Math. Soc. (N.S.) **4** (1940) 116–119

C. Kuratowski
[1] *Topologie I* (2nd Ed.) Warsaw (1948)
[2] *Topologie II*, Warsaw (1950)
[3] *Une méthode d'élimination des nombres transfinis des raisonnements mathématiques*, Fund. Math. **3** (1922) 76–108

E. Landau
[1] *Grundlagen der Analysis*, (Amer. Ed.) New York (1946)

J. P. Lasalle
[1] *Topology based upon the concept of pseudo-norm*, Proc. Nat. Acad. Sci. U.S.A. **27** (1941) 448–451

S. Lefschetz
[1] *Algebraic topology*, A.M.S. Colloquium Publ. XXVII, New York (1942)

L. H. Loomis
[1] *On the representation of σ-complete Boolean algebras*, Bull. Amer. Math. Soc. **53** (1947) 757–760
[2] *Abstract harmonic analysis*, New York (1953)

E. J. McShane
[1] *Partial orderings and Moore-Smith limits*, Amer. Math. Monthly **59** (1952) 1–11
[2] *Integration*, Princeton (1944)
[3] *Order-preserving maps and integration processes*, Ann. of Math. Studies **31**, Princeton (1953)

E. Michael
[1] *A note on paracompact spaces*, Proc. Amer. Math. Soc. **4** (1953) 831–838
[2] *Topologies on spaces of subsets*, Trans. Amer. Math. Soc. **71** (1951) 151–182

A. Monteiro
[1] *Caractérisation de l'opération de fermeture par une seul axiome*, Portugaliae Math. **4** (1945) 158–160

[2] *Caractérisation des espaces de Hausdorff au moyen de l'opération de dérivation*, Portugaliae Math. **1** (1940) 333–339

E. H. Moore

[1] *Definition of limit in general integral analysis*, Proc. Nat. Acad. Sci. U.S.A. **1** (1915) 628

[2] *General analysis I, Pt. II*, Philadelphia (1939)

E. H. Moore and H. L. Smith

[1] *A general theory of limits*, Amer. J. Math. **44** (1922) 102–121

R. L. Moore

[1] *Foundations of point set theory*, A.M.S. Colloquium Publ. XIII, New York (1932)

K. Morita

[1] *Star-finite coverings and the star-finite property*, Math. Japonicae **1** (1948) 60–68

S. B. Myers

[1] *Normed linear spaces of continuous functions*, Bull. Amer. Math. Soc. **56** (1950) 233–241

[2] *Equicontinuous sets of mappings*, Ann. of Math. (2) **47** (1946) 496–502

[3] *Functional uniformities*, Proc. Amer. Math. Soc. **2** (1951) 153–158

A. D. Myškis

[1] *On the concept of boundary*, Mat. Sbornik N.S. **25** (1949) 387–414

[2] *The definition of boundary by means of continuous mappings*, Mat. Sbornik N.S. **26** (1950) 225–227

[3] *On the equivalence of certain methods of definition of boundary*, Mat. Sbornik N.S. **26** (1950) 228–236

L. Nachbin

[1] *Topological vector spaces*, Rio de Janeiro (1948)

J. Nagata

[1] *On a necessary and sufficient condition of metrizability*, J. Inst. Polytech. Osaka City Univ. **1** (1950) 93–100

[2] *On the uniform topology of bicompactifications*, J. Inst. Polytech. Osaka City Univ. **1** (1950) 28–39

H. Nakano

[1] *Topology and linear topological spaces*, Tokyo (1951)

J. von Neumann

[1] *On complete topological spaces*, Trans. Amer. Math. Soc. **37** (1935) 1–20

M. H. A. Newman

[1] *Elements of the topology of plane sets of points*, Cambridge (1939)

J. Novak

[1] *Regular space on which every continuous function is constant*, Časopis Pěst. Mat. Fys. **73** (1948) 58–68

B. J. Pettis

[1] *On continuity and openness of homomorphisms in topological groups*, Ann. of Math. (2) **51** (1950) 293–308

[2] *A note on everywhere dense subgroups*, Proc. Amer. Math. Soc. **3** (1952) 322–326

L. Pontrjagin

[1] *Topological groups*, Princeton (1939)

W. V. O. Quine

[1] *Mathematical logic*, Cambridge (U.S.A.) (1947)

A. Ramanathan

[1] *Maximal Hausdorff spaces*, Proc. Indian Acad. Sci. Sect. A, **26** (1947) 31–42

H. Ribeiro

[1] *Une extension de la notion de convergence*, Portugaliae Math. **2** (1941) 153–161

[2] *Sur les espace à métrique faible*, Portugaliae Math. **4** (1943) 21–40, also 65–68

[3] *Caractérisations des espaces réguliers normaux et complètement normaux au moyen de l'opération de dérivation*, Portugaliae Math. **2** (1940) 1–7

P. Samuel

[1] *Ultrafilters and compactification of uniform spaces*, Trans. Amer. Math. Soc. **64** (1948) 100–132

T. Shirota

[1] *On systems of structures of a completely regular space*, Osaka Math. J. **2** (1950) 131–143

[2] *A class of topological spaces*, Osaka Math. J. **4** (1952) 23–40

W. Sierpinski

[1] *General topology* (2nd Ed.) Toronto (1952)

[2] *Sur les ensembles complets d'un espace* (D), Fund. Math. **11** (1928) 203–205

Yu. M. Smirnov

[1] *A necessary and sufficient condition for metrizability of a topological space*, Doklady Akad. Nauk S.S.S.R. N.S. **77** (1951) 197–200

[2] *On metrization of topological spaces*, Uspehi Matem. Nauk **6** (1951) 100–111

[3] *On normally disposed sets of normal spaces*, Mat. Sbornik N.S. **29** (1951) 173–176

R. H. Sorgenfrey

[1] *On the topological product of paracompact spaces*, Bull. Amer. Math. Soc. **53** (1947) 631–632

A. H. Stone

[1] *Paracompactness and product spaces*, Bull. Amer. Math. Soc. **54** (1948) 977–982

M. H. Stone

[1] *Notes on integration I, II, III, IV*, Proc. Nat. Acad. Sci. U.S.A. **34** (1948) 336–342, 447–455, 483–490; **35** (1949) 50–58

[2] *Topological representations of distributive lattices and Brouwerian logics*, Časopis Pěst. Mat. Fys. **67** (1937) 1–27

[3] *The theory of representations for Boolean algebras*, Trans. Amer. Math. Soc. **40** (1936) 37–111

[4] *Boundedness properties in function lattices*, Canadian J. Math. **1** (1946) 176–186

[5] *The generalized Weierstrass approximation theorem*, Math. Mag. **21** (1948) 167–184

[6] *Applications of the theory of Boolean rings to general topology*, Trans. Amer. Math. Soc. **41** (1937) 375–481

E. C. Stopher, Jr.

[1] *Point set operators and their interrelations*, Bull. Amer. Math. Soc. **45** (1939) 758–762

E. Szpilrajn

[1] *Remarque sur les produits cartésiens d'espaces topologiques*, C. R. (Doklady) Acad. Sci. U.R.S.S. N.S. **31** (1941) 525–527

P. Szymanski

[1] *La notion des ensembles separé comme terme primitif de la topologie*, Mathematica Timişoara **17** (1941) 65–84

A. Tarski

[1] *Introduction to modern logic* (2nd Amer. Ed.), New York (1946)

H. Tong

[1] *On some problems of Čech*, Ann. of Math. (2) **50** (1949) 154–157

J. W. Tukey

[1] *Convergence and uniformity in topology*, Ann. of Math. Studies **2** (1940)

A. Tychonoff

[1] *Über einen Funktionenräum*, Math. Ann. **111** (1935) 762–766

[2] *Über die topologische Erweiterung von Räumen*, Math. Ann. **102** (1929) 544–561

H. Umegaki
[1] *On the uniform space*, Tohoku Math. J. (2) **2** (1950) 57–63
H. D. Ursell and L. C. Young
[1] *Remarks on the theory of prime ends*, Memoirs Amer. Math. Soc. **3** (1951)
P. Urysohn
[1] *Über die Machtigkeit der zusammenhängen Mengen*, Math. Ann. **94** (1925) 262–295
R. Vaidyanathaswamy
[1] *Treatise on set topology I*, Madras (1947)
A. D. Wallace
[1] *Separation spaces*, Ann. of Math. (2) **42** (1941) 687–697
[2] *Extensional invariance*, Trans. Amer. Math. Soc. **70** (1951) 97–102
H. Wallman
[1] *Lattices and topological spaces*, Ann. of Math. (2) **42** (1941) 687–697
A. Weil
[1] *Sur les espaces a structure uniforme et sur la topologie générale*, Actualités Sci. Ind. **551**, Paris (1937)
[2] *L'integration dans les groupes topologiques et ses applications*, Actualités Sci. Ind. **869**, Paris (1940)
G. T. Whyburn
[1] *Analytic topology*, A.M.S. Colloquium Publ. XXVIII, New York (1942)
[2] *Open and closed mappings*, Duke Math. J. **17** (1950) 69–74
R. L. Wilder
[1] *Topology of manifolds*, A.M.S. Colloquium Publ. XXXII, New York (1949)
E. Zermelo
[1] *Neuer Beweis für die Wohlordnung*, Math. Ann. **65** (1908) 107–128

INDEX

A CATALOG OF SELECTED

DOVER BOOKS

IN SCIENCE AND MATHEMATICS

Mathematics–Bestsellers

HANDBOOK OF MATHEMATICAL FUNCTIONS: with Formulas, Graphs, and Mathematical Tables, Edited by Milton Abramowitz and Irene A. Stegun. A classic resource for working with special functions, standard trig, and exponential logarithmic definitions and extensions, it features 29 sets of tables, some to as high as 20 places. 1046pp. 8 x 10 1/2. 0-486-61272-4

ABSTRACT AND CONCRETE CATEGORIES: The Joy of Cats, Jiri Adamek, Horst Herrlich, and George E. Strecker. This up-to-date introductory treatment employs category theory to explore the theory of structures. Its unique approach stresses concrete categories and presents a systematic view of factorization structures. Numerous examples. 1990 edition, updated 2004. 528pp. 6 1/8 x 9 1/4. 0-486-46934-4

MATHEMATICS: Its Content, Methods and Meaning, A. D. Aleksandrov, A. N. Kolmogorov, and M. A. Lavrent'ev. Major survey offers comprehensive, coherent discussions of analytic geometry, algebra, differential equations, calculus of variations, functions of a complex variable, prime numbers, linear and non-Euclidean geometry, topology, functional analysis, more. 1963 edition. 1120pp. 5 3/8 x 8 1/2. 0-486-40916-3

INTRODUCTION TO VECTORS AND TENSORS: Second Edition--Two Volumes Bound as One, Ray M. Bowen and C.-C. Wang. Convenient single-volume compilation of two texts offers both introduction and in-depth survey. Geared toward engineering and science students rather than mathematicians, it focuses on physics and engineering applications. 1976 edition. 560pp. 6 1/2 x 9 1/4. 0-486-46914-X

AN INTRODUCTION TO ORTHOGONAL POLYNOMIALS, Theodore S. Chihara. Concise introduction covers general elementary theory, including the representation theorem and distribution functions, continued fractions and chain sequences, the recurrence formula, special functions, and some specific systems. 1978 edition. 272pp. 5 3/8 x 8 1/2. 0-486-47929-3

ADVANCED MATHEMATICS FOR ENGINEERS AND SCIENTISTS, Paul DuChateau. This primary text and supplemental reference focuses on linear algebra, calculus, and ordinary differential equations. Additional topics include partial differential equations and approximation methods. Includes solved problems. 1992 edition. 400pp. 7 1/2 x 9 1/4. 0-486-47930-7

PARTIAL DIFFERENTIAL EQUATIONS FOR SCIENTISTS AND ENGINEERS, Stanley J. Farlow. Practical text shows how to formulate and solve partial differential equations. Coverage of diffusion-type problems, hyperbolic-type problems, elliptic-type problems, numerical and approximate methods. Solution guide available upon request. 1982 edition. 414pp. 6 1/8 x 9 1/4. 0-486-67620-X

VARIATIONAL PRINCIPLES AND FREE-BOUNDARY PROBLEMS, Avner Friedman. Advanced graduate-level text examines variational methods in partial differential equations and illustrates their applications to free-boundary problems. Features detailed statements of standard theory of elliptic and parabolic operators. 1982 edition. 720pp. 6 1/8 x 9 1/4. 0-486-47853-X

LINEAR ANALYSIS AND REPRESENTATION THEORY, Steven A. Gaal. Unified treatment covers topics from the theory of operators and operator algebras on Hilbert spaces; integration and representation theory for topological groups; and the theory of Lie algebras, Lie groups, and transform groups. 1973 edition. 704pp. 6 1/8 x 9 1/4. 0-486-47851-3

Mathematics–Logic and Problem Solving

PERPLEXING PUZZLES AND TANTALIZING TEASERS, Martin Gardner. Ninety-three riddles, mazes, illusions, tricky questions, word and picture puzzles, and other challenges offer hours of entertainment for youngsters. Filled with rib-tickling drawings. Solutions. 224pp. 5 3/8 x 8 1/2. 0-486-25637-5

MY BEST MATHEMATICAL AND LOGIC PUZZLES, Martin Gardner. The noted expert selects 70 of his favorite "short" puzzles. Includes The Returning Explorer, The Mutilated Chessboard, Scrambled Box Tops, and dozens more. Complete solutions included. 96pp. 5 3/8 x 8 1/2. 0-486-28152-3

THE LADY OR THE TIGER?: and Other Logic Puzzles, Raymond M. Smullyan. Created by a renowned puzzle master, these whimsically themed challenges involve paradoxes about probability, time, and change; metapuzzles; and self-referentiality. Nineteen chapters advance in difficulty from relatively simple to highly complex. 1982 edition. 240pp. 5 3/8 x 8 1/2. 0-486-47027-X

SATAN, CANTOR AND INFINITY: Mind-Boggling Puzzles, Raymond M. Smullyan. A renowned mathematician tells stories of knights and knaves in an entertaining look at the logical precepts behind infinity, probability, time, and change. Requires a strong background in mathematics. Complete solutions. 288pp. 5 3/8 x 8 1/2. 0-486-47036-9

THE RED BOOK OF MATHEMATICAL PROBLEMS, Kenneth S. Williams and Kenneth Hardy. Handy compilation of 100 practice problems, hints and solutions indispensable for students preparing for the William Lowell Putnam and other mathematical competitions. Preface to the First Edition. Sources. 1988 edition. 192pp. 5 3/8 x 8 1/2. 0-486-69415-1

KING ARTHUR IN SEARCH OF HIS DOG AND OTHER CURIOUS PUZZLES, Raymond M. Smullyan. This fanciful, original collection for readers of all ages features arithmetic puzzles, logic problems related to crime detection, and logic and arithmetic puzzles involving King Arthur and his Dogs of the Round Table. 160pp. 5 3/8 x 8 1/2. 0-486-47435-6

UNDECIDABLE THEORIES: Studies in Logic and the Foundation of Mathematics, Alfred Tarski in collaboration with Andrzej Mostowski and Raphael M. Robinson. This well-known book by the famed logician consists of three treatises: "A General Method in Proofs of Undecidability," "Undecidability and Essential Undecidability in Mathematics," and "Undecidability of the Elementary Theory of Groups." 1953 edition. 112pp. 5 3/8 x 8 1/2. 0-486-47703-7

LOGIC FOR MATHEMATICIANS, J. Barkley Rosser. Examination of essential topics and theorems assumes no background in logic. "Undoubtedly a major addition to the literature of mathematical logic." – *Bulletin of the American Mathematical Society.* 1978 edition. 592pp. 6 1/8 x 9 1/4. 0-486-46898-4

INTRODUCTION TO PROOF IN ABSTRACT MATHEMATICS, Andrew Wohlgemuth. This undergraduate text teaches students what constitutes an acceptable proof, and it develops their ability to do proofs of routine problems as well as those requiring creative insights. 1990 edition. 384pp. 6 1/2 x 9 1/4. 0-486-47854-8

FIRST COURSE IN MATHEMATICAL LOGIC, Patrick Suppes and Shirley Hill. Rigorous introduction is simple enough in presentation and context for wide range of students. Symbolizing sentences; logical inference; truth and validity; truth tables; terms, predicates, universal quantifiers; universal specification and laws of identity; more. 288pp. 5 3/8 x 8 1/2. 0-486-42259-3

Mathematics–Algebra and Calculus

VECTOR CALCULUS, Peter Baxandall and Hans Liebeck. This introductory text offers a rigorous, comprehensive treatment. Classical theorems of vector calculus are amply illustrated with figures, worked examples, physical applications, and exercises with hints and answers. 1986 edition. 560pp. 5 3/8 x 8 1/2. 0-486-46620-5

ADVANCED CALCULUS: An Introduction to Classical Analysis, Louis Brand. A course in analysis that focuses on the functions of a real variable, this text introduces the basic concepts in their simplest setting and illustrates its teachings with numerous examples, theorems, and proofs. 1955 edition. 592pp. 5 3/8 x 8 1/2. 0-486-44548-8

ADVANCED CALCULUS, Avner Friedman. Intended for students who have already completed a one-year course in elementary calculus, this two-part treatment advances from functions of one variable to those of several variables. Solutions. 1971 edition. 432pp. 5 3/8 x 8 1/2. 0-486-45795-8

METHODS OF MATHEMATICS APPLIED TO CALCULUS, PROBABILITY, AND STATISTICS, Richard W. Hamming. This 4-part treatment begins with algebra and analytic geometry and proceeds to an exploration of the calculus of algebraic functions and transcendental functions and applications. 1985 edition. Includes 310 figures and 18 tables. 880pp. 6 1/2 x 9 1/4. 0-486-43945-3

BASIC ALGEBRA I: Second Edition, Nathan Jacobson. A classic text and standard reference for a generation, this volume covers all undergraduate algebra topics, including groups, rings, modules, Galois theory, polynomials, linear algebra, and associative algebra. 1985 edition. 528pp. 6 1/8 x 9 1/4. 0-486-47189-6

BASIC ALGEBRA II: Second Edition, Nathan Jacobson. This classic text and standard reference comprises all subjects of a first-year graduate-level course, including in-depth coverage of groups and polynomials and extensive use of categories and functors. 1989 edition. 704pp. 6 1/8 x 9 1/4. 0-486-47187-X

CALCULUS: An Intuitive and Physical Approach (Second Edition), Morris Kline. Application-oriented introduction relates the subject as closely as possible to science with explorations of the derivative; differentiation and integration of the powers of x; theorems on differentiation, antidifferentiation; the chain rule; trigonometric functions; more. Examples. 1967 edition. 960pp. 6 1/2 x 9 1/4. 0-486-40453-6

ABSTRACT ALGEBRA AND SOLUTION BY RADICALS, John E. Maxfield and Margaret W. Maxfield. Accessible advanced undergraduate-level text starts with groups, rings, fields, and polynomials and advances to Galois theory, radicals and roots of unity, and solution by radicals. Numerous examples, illustrations, exercises, appendixes. 1971 edition. 224pp. 6 1/8 x 9 1/4. 0-486-47723-1

AN INTRODUCTION TO THE THEORY OF LINEAR SPACES, Georgi E. Shilov. Translated by Richard A. Silverman. Introductory treatment offers a clear exposition of algebra, geometry, and analysis as parts of an integrated whole rather than separate subjects. Numerous examples illustrate many different fields, and problems include hints or answers. 1961 edition. 320pp. 5 3/8 x 8 1/2. 0-486-63070-6

LINEAR ALGEBRA, Georgi E. Shilov. Covers determinants, linear spaces, systems of linear equations, linear functions of a vector argument, coordinate transformations, the canonical form of the matrix of a linear operator, bilinear and quadratic forms, and more. 387pp. 5 3/8 x 8 1/2. 0-486-63518-X

Mathematics-Probability and Statistics

BASIC PROBABILITY THEORY, Robert B. Ash. This text emphasizes the probabilistic way of thinking, rather than measure-theoretic concepts. Geared toward advanced undergraduates and graduate students, it features solutions to some of the problems. 1970 edition. 352pp. 5 3/8 x 8 1/2. 0-486-46628-0

PRINCIPLES OF STATISTICS, M. G. Bulmer. Concise description of classical statistics, from basic dice probabilities to modern regression analysis. Equal stress on theory and applications. Moderate difficulty; only basic calculus required. Includes problems with answers. 252pp. 5 5/8 x 8 1/4. 0-486-63760-3

OUTLINE OF BASIC STATISTICS: Dictionary and Formulas, John E. Freund and Frank J. Williams. Handy guide includes a 70-page outline of essential statistical formulas covering grouped and ungrouped data, finite populations, probability, and more, plus over 1,000 clear, concise definitions of statistical terms. 1966 edition. 208pp. 5 3/8 x 8 1/2. 0-486-47769-X

GOOD THINKING: The Foundations of Probability and Its Applications, Irving J. Good. This in-depth treatment of probability theory by a famous British statistician explores Keynesian principles and surveys such topics as Bayesian rationality, corroboration, hypothesis testing, and mathematical tools for induction and simplicity. 1983 edition. 352pp. 5 3/8 x 8 1/2. 0-486-47438-0

INTRODUCTION TO PROBABILITY THEORY WITH CONTEMPORARY APPLICATIONS, Lester L. Helms. Extensive discussions and clear examples, written in plain language, expose students to the rules and methods of probability. Exercises foster problem-solving skills, and all problems feature step-by-step solutions. 1997 edition. 368pp. 6 1/2 x 9 1/4. 0-486-47418-6

CHANCE, LUCK, AND STATISTICS, Horace C. Levinson. In simple, non-technical language, this volume explores the fundamentals governing chance and applies them to sports, government, and business. "Clear and lively ... remarkably accurate." – *Scientific Monthly*. 384pp. 5 3/8 x 8 1/2. 0-486-41997-5

FIFTY CHALLENGING PROBLEMS IN PROBABILITY WITH SOLUTIONS, Frederick Mosteller. Remarkable puzzlers, graded in difficulty, illustrate elementary and advanced aspects of probability. These problems were selected for originality, general interest, or because they demonstrate valuable techniques. Also includes detailed solutions. 88pp. 5 3/8 x 8 1/2. 0-486-65355-2

EXPERIMENTAL STATISTICS, Mary Gibbons Natrella. A handbook for those seeking engineering information and quantitative data for designing, developing, constructing, and testing equipment. Covers the planning of experiments, the analyzing of extreme-value data; and more. 1966 edition. Index. Includes 52 figures and 76 tables. 560pp. 8 3/8 x 11. 0-486-43937-2

STOCHASTIC MODELING: Analysis and Simulation, Barry L. Nelson. Coherent introduction to techniques also offers a guide to the mathematical, numerical, and simulation tools of systems analysis. Includes formulation of models, analysis, and interpretation of results. 1995 edition. 336pp. 6 1/8 x 9 1/4. 0-486-47770-3

INTRODUCTION TO BIOSTATISTICS: Second Edition, Robert R. Sokal and F. James Rohlf. Suitable for undergraduates with a minimal background in mathematics, this introduction ranges from descriptive statistics to fundamental distributions and the testing of hypotheses. Includes numerous worked-out problems and examples. 1987 edition. 384pp. 6 1/8 x 9 1/4. 0-486-46961-1

Mathematics–Geometry and Topology

PROBLEMS AND SOLUTIONS IN EUCLIDEAN GEOMETRY, M. N. Aref and William Wernick. Based on classical principles, this book is intended for a second course in Euclidean geometry and can be used as a refresher. More than 200 problems include hints and solutions. 1968 edition. 272pp. 5 3/8 x 8 1/2. 0-486-47720-7

TOPOLOGY OF 3-MANIFOLDS AND RELATED TOPICS, Edited by M. K. Fort, Jr. With a New Introduction by Daniel Silver. Summaries and full reports from a 1961 conference discuss decompositions and subsets of 3-space; n-manifolds; knot theory; the Poincaré conjecture; and periodic maps and isotopies. Familiarity with algebraic topology required. 1962 edition. 272pp. 6 1/8 x 9 1/4. 0-486-47753-3

POINT SET TOPOLOGY, Steven A. Gaal. Suitable for a complete course in topology, this text also functions as a self-contained treatment for independent study. Additional enrichment materials make it equally valuable as a reference. 1964 edition. 336pp. 5 3/8 x 8 1/2. 0-486-47222-1

INVITATION TO GEOMETRY, Z. A. Melzak. Intended for students of many different backgrounds with only a modest knowledge of mathematics, this text features self-contained chapters that can be adapted to several types of geometry courses. 1983 edition. 240pp. 5 3/8 x 8 1/2. 0-486-46626-4

TOPOLOGY AND GEOMETRY FOR PHYSICISTS, Charles Nash and Siddhartha Sen. Written by physicists for physics students, this text assumes no detailed background in topology or geometry. Topics include differential forms, homotopy, homology, cohomology, fiber bundles, connection and covariant derivatives, and Morse theory. 1983 edition. 320pp. 5 3/8 x 8 1/2. 0-486-47852-1

BEYOND GEOMETRY: Classic Papers from Riemann to Einstein, Edited with an Introduction and Notes by Peter Pesic. This is the only English-language collection of these 8 accessible essays. They trace seminal ideas about the foundations of geometry that led to Einstein's general theory of relativity. 224pp. 6 1/8 x 9 1/4. 0-486-45350-2

GEOMETRY FROM EUCLID TO KNOTS, Saul Stahl. This text provides a historical perspective on plane geometry and covers non-neutral Euclidean geometry, circles and regular polygons, projective geometry, symmetries, inversions, informal topology, and more. Includes 1,000 practice problems. Solutions available. 2003 edition. 480pp. 6 1/8 x 9 1/4. 0-486-47459-3

TOPOLOGICAL VECTOR SPACES, DISTRIBUTIONS AND KERNELS, François Trèves. Extending beyond the boundaries of Hilbert and Banach space theory, this text focuses on key aspects of functional analysis, particularly in regard to solving partial differential equations. 1967 edition. 592pp. 5 3/8 x 8 1/2. 0-486-45352-9

INTRODUCTION TO PROJECTIVE GEOMETRY, C. R. Wylie, Jr. This introductory volume offers strong reinforcement for its teachings, with detailed examples and numerous theorems, proofs, and exercises, plus complete answers to all odd-numbered end-of-chapter problems. 1970 edition. 576pp. 6 1/8 x 9 1/4. 0-486-46895-X

FOUNDATIONS OF GEOMETRY, C. R. Wylie, Jr. Geared toward students preparing to teach high school mathematics, this text explores the principles of Euclidean and non-Euclidean geometry and covers both generalities and specifics of the axiomatic method. 1964 edition. 352pp. 6 x 9. 0-486-47214-0